Singular Integrals
and Differentiability Properties
of Functions

PRINCETON MATHEMATICAL SERIES

Editors: Phillip A. Griffiths, Marston Morse, and Elias M. Stein

Singular Integrals
and Differentiability Properties
of Functions

ELIAS M. STEIN

PRINCETON UNIVERSITY PRESS
PRINCETON, NEW JERSEY
1970

Printed in the United States of America
by Princeton University Press, Princeton, New Jersey

To Elly

Preface

This book is an outgrowth of a course which I gave at Orsay during the academic year 1966-67.* My purpose in those lectures was to present some of the required background and at the same time clarify the essential unity that exists between several related areas of analysis. These areas are: the existence and boundedness of singular integral operators; the boundary behavior of harmonic functions; and differentiability properties of functions of several variables. As such the common core of these topics may be said to represent one of the central developments in n-dimensional Fourier analysis during the last twenty years, and it can be expected to have equal influence in the future. These possibilities are further highlighted by listing some of the fields (which are not treated here) where either the particular results of this book, or ideas closely related to the techniques presented here are continuing to find significant application. These include partial differential equations, holomorphic functions of several complex variables, and analysis on other groups, either non-commutative or commutative.

In this connection we should point out that the book *Introduction to Fourier Analysis on Euclidean Spaces*** details some of these applications, as well as much background and related material. It may therefore not be inappropriate to view the present volume as a companion to *Fourier Analysis*. Both books, however, may be read independently. In fact an effort has been made to make the present volume essentially self-contained, requiring only elementary facts from integration theory and Fourier transforms as prerequisites.

A brief description of the organization of the book is as follows. The first three chapters deal primarily with material which, for the most part, is beginning to find its way into several advanced texts and monographs, namely covering lemmas and maximal functions, the Marcinkiewicz interpolation theorem, singular integrals generalizing the Hilbert transform, and harmonic functions represented as Poisson integrals. In the last five chapters the topics are of a more advanced nature, including the Littlewood-Paley theory, multipliers, Sobolov spaces and their variants, extension theorems, further results about harmonic functions, and almost-everywhere differentiability theorems. Here part of the material is systematically organized for the first time, and, for example, the last two chapters contain several results whose details were hitherto unpublished.

---

* For the published lecture notes of this course see Stein [10].

** This work of G. Weiss and the author is referred to as *Fourier Analysis* in the rest of the text.

In any enterprise of this kind the author is faced with the task of balancing two aims which unfortunately are not always compatible. On the one hand there is the desire to facilitate the task of the serious student by providing all the background material and by presenting proofs in a way so that all details, no matter how unenlightening, are fully given. On the other hand there is the need to get on with the essential job of developing the basic ideas of the subject. In doing the latter it is sometimes best to be brief about certain technical details, and also at times to forego the urge to pursue various possible generalizations which could be formulated. Others may surely find fault with how I have weighed these alternatives. My justification would be based either on the ground of personal predilection (which allows no argument) or, in a more serious vein, in terms of my view of the present subject: that it has advanced to a high degree of sophistication and is still rapidly developing, but has not yet reached the level of maturity that would require it to be enshrined in a edifice of great perfection.

It is my pleasant task to acknowledge with gratitude those who have helped me in writing this book: Norman Weiss, who prepared the lecture notes (unpublished) of a course given at Princeton University in 1964-65, where an earlier version of some of this material was presented; Messrs. Bachvan and A. Somen who wrote the published lecture notes already alluded to; Misses Elizabeth Epstein and Florence Armstrong who typed the bulk of the manuscript; and Messrs. W. Beckner, C. Fefferman, and S. Gelbart who helped both mathematically and in proofreading. To all those and others unnamed, I express my thanks.

September 1970   E. M. STEIN

Notation

Principal Symbols

dx—Lebesgue measure on $\mathbf{R}^n$; also $m(E)$—measure of the set E

$L^p(\mathbf{R}^n)$—the L^p space with respect to the measure dx

C^k—the class of functions which have continuous derivatives up to and including total order k

$\mathscr{D}$—the space of indefinitely differentiable functions with compact support

$\mathscr{S}$—the space of indefinitely differential functions all of whose derivatives remain bounded when multiplied by any polynomial

^cE—complement of the set E

The symbols that follow are listed according to their first and other principal occurrence.

Chapter I, §1

$B(x, r)$—ball of radius r centered at x

$M(f)$—maximal function

§2

$\delta(x) = \delta(x, F)$—distance of x from the set F

$I(x)$, $I_*(x)$—integrals of Marcinkiewicz involving the distance function (See also Chapter VIII, §3.)

§3

$Q_1, \ldots, Q_k, \ldots$—cubes; also $\Omega = \bigcup_k Q_k$

Chapter II, §1

$C_0(\mathbf{R}^n)$, $\mathscr{B}(\mathbf{R}^n)$—continuous functions on $\mathbf{R}^n$ vanishing at infinity and the dual space of finite Borel measures

$\mu = \mu_1 * \mu_2$—the convolution of measures μ_1 and μ_2

$\hat{f}(y) = \mathscr{F}(f)(y)$—Fourier transform of f

§3
$H(f)$—Hilbert transform. (See also Chapter III, §1.)

§4
S^{n-1}, $d\sigma$—the unit sphere in $\mathbf{R}^n$ and its induced element of volume

§5
$L^p(\mathbf{R}^n, \mathcal{H})$—$L^p$ space of functions which take their values in $\mathcal{H}$

Chapter III, §1
R_j—Riesz transforms
$$c_n = \Gamma\left(\frac{n+1}{2}\right)\bigg/ \pi^{(n+1)/2}$$

§2
$\mathbf{R}^{n+1}_+$—the upper half-space $\{(x, y) \in \mathbf{R}^{n+1} : x \in \mathbf{R}^n, y > 0\}$
$$P_y(x) = \frac{c_n y}{(|x|^2 + y^2)^{(n+1)/2}}\text{—Poisson kernel}$$
Δ—Laplacean (see also Chapter VII), where it is used in the context of $\mathbf{R}^{n+1}_+$, or in Chapter III, §3, where it is used in the context of $\mathbf{R}^n$

§3
$\mathcal{H}_k$—solid spherical harmonics of degree k
H_k—surface spherical harmonics of degree k

Chapter IV, §1
g, g_1, g_x, g_k—variants of the Littlewood-Paley g functions

§2
g^*_λ—another variant. (See also Chapter VII, §3.)
Γ—cone $\{(x, y) : x \in \mathbf{R}^n, |x| < y\}$ (See also Chapter VII, §1.)
S—area integral of Lusin. (See also Chapter VII, §§2 and 3.)
$M_\mu(f) = (M(f^\mu))^{1/\mu}, \mu \geq 1$

§3
T_m—multiplier transformation with multiplier m
$\mathcal{M}_p$—algebra of L^p multipliers

§4
$S_\rho(f)$—"partial sum" operator
$S_\mathfrak{R}(f)$—analogue of above corresponding to a family of rectangles

§5
r_m—Rademacher function. (See also Appendix D.)

Chapter V, §1
I_α—Riesz potential

§2
$L^p_k(\mathbf{R}^n)$—Sobolov space

§3
$\mathscr{T}_\alpha(f) = G_\alpha(f)$—Bessel potential
$\mathscr{L}^p_\alpha(\mathbf{R}^n)$—space of Bessel potentials
$\omega_p(t)$—L^p modulus of continuity
$\tilde{\omega}_p(t)$—second-order L^p modulus of continuity

§4
$\Lambda_\alpha(\mathbf{R}^n)$—Lipschitz space

§5
$\Lambda^{p,q}_\alpha(\mathbf{R}^n)$—Besov space

Chapter VI, §1
$\Delta(x)$—regularized distance

§2
$\mathscr{E}_k$—Whitney extension operators

§3
$L^p_k(D)$—Sobolov space for the domain D
$\mathfrak{E}$—extension operator for the domain D

Chapter VII, §1
$\Gamma_\alpha(x^0)$—cone $\{(x, y) \in \mathbf{R}^{n+1}_+, |x - x^0| < \alpha y\}$
Γ^h_α—truncated cone, $\Gamma^h_\alpha(x^0) = \Gamma_\alpha(x^0) \cap \{0 < y < h\}$
$\mathscr{R} = \bigcup\limits_{x^0 \in E} \Gamma^h_\alpha(x^0)$. (See also Chapter VIII, §2.)

§3
H^p space—space of conjugate harmonic functions satisfying an appropriate L^p inequality
$\mathfrak{S}_q$—variant of the area integral

Contents

Singular Integrals
and Differentiability Properties
of Functions

CHAPTER I

Some Fundamental Notions of
Real-Variable Theory

The basic ideas of the theory of real variables are connected with the concepts of sets and functions, together with the processes of integration and differentiation applied to them. While the essential aspects of these ideas were brought to light in the early part of our century, some of their further applications were developed only more recently. It is from this latter perspective that we shall approach that part of the theory that interests us. In doing so, we distinguish several main features:

(1) The theorem of Lebesgue about the *differentiation of the integral*. The study of properties related to this process is best done in terms of a "maximal function" to which it gives rise; the basic features of the latter are expressed in terms of a "weak-type" inequality which is characteristic of this situation.

(2) Certain *covering* lemmas. In general the idea is to cover an arbitrary (open) set in terms of a disjoint union of cubes or balls, chosen in a manner depending on the problem at hand. One such example is a lemma of Whitney, (Theorem 3). Sometimes, however, it suffices to cover only a portion of the set, as in the simple covering lemma, which is used to prove the weak-type inequality mentioned above.

(3) *Behavior near a "general" point* of an arbitrary set. The simplest notion here is that of point of density. More refined properties are best expressed in terms of certain integrals first studied systematically by Marcinkiewicz.

(4) The *splitting of functions* into their large and small parts. This feature which is more of a technique than an end in itself, recurs often. It is especially useful in proving L^p inequalities, as in the first theorem of this chapter. That part of the proof of the first theorem is systematized in the Marcinkiewicz interpolation theorem discussed in §4 of this chapter and also in Appendix B.

3

1. *The maximal function*

1.1 According to the fundamental theorem of Lebesgue, the relation

$$(1) \qquad \lim_{r \to 0} \frac{1}{m(B(x, r))} \int_{B(x,r)} f(y) \, dy = f(x)$$

holds for almost every x, whenever f is locally integrable function defined on $\mathbf{R}^n$. The notation here used is that $B(x, r)$ is the ball of radius r, centered at x, and $m(B(x, r))$ denotes its measure. In order to study the limit (1) we consider its quantitative analogue, where "lim" is replaced by "sup"; this is the *maximal function, Mf.* Since the properties of this function are expressed in terms of relative size and do not involve any cancellation of positive and negative values, we replace f by $|f|$. Thus we define

$$(2) \qquad M(f)(x) = \sup_{r > 0} \frac{1}{m(B(x, r))} \int_{B(x,r)} |f(y)| \, dy$$

It is to be noticed that nothing excludes the possibility that $(Mf)(x)$ is infinite for any given x.

The passage from a limiting expression to a corresponding maximal function is a situation that recurs often. Our first example here, (2), will turn out to be the most fundamental one.

1.2 We shall now be interested in giving a concise expression for the relative size of a function. Thus let $g(x)$ be defined on $\mathbf{R}^n$ and for each α consider the set where $|g|$ is greater than α,

$$\{x : |g(x)| > \alpha\}.$$

The function $\lambda(\alpha)$, defined to be the measure of this set, is the sought-for expression. It is the *distribution function* of $|g|$.

In particular, the decrease of $\lambda(\alpha)$ as α grows describes the relative largeness of the function; this is the main concern locally. The increase of $\lambda(\alpha)$ as α tends to zero describes the relative smallness of the function "at infinity"; this is its importance globally, and is of no interest if, for example, the function is supported on a bounded set.

Any quantity dealing solely with the size of g can be expressed in terms of the distribution function $\lambda(\alpha)$. For example, if $g \in L^p$, then

$$\int_{\mathbf{R}^n} |g(y)|^p \, dy = - \int_0^\infty \alpha^p \, d\lambda(\alpha)$$

and if $g \in L^\infty$, then

$$\|g\|_\infty = \inf \{\alpha, \lambda(\alpha) = 0\}.$$

A related fact concerning the distribution function will have immediate application. It is this: If g is integrable, then

$$\lambda(\alpha) \leq A/\alpha \quad \text{where} \quad A = \int_{\mathbf{R}^n} |g(y)| \, dy.$$

In fact

$$\int_{\mathbf{R}^n} |g(y)| \, dy \geq \int_{|g| > \alpha} |g(y)| \, dy \geq \alpha\lambda(\alpha),$$

which proves the assertion.

1.3 With these definitions we can state our first theorem. It gives the main results for the maximal function, and has as a corollary the differentiability almost everywhere of the integral, expressed in (1).

THEOREM 1. *Let f be a given function defined on* $\mathbf{R}^n$
(a) *If* $f \in L^p(\mathbf{R}^n)$, $1 \leq p \leq \infty$, *then the function Mf is finite almost everywhere.*
(b) *If* $f \in L^1(\mathbf{R}^n)$, *then for every* $\alpha > 0$

$$m\{x : (Mf)(x) > \alpha\} \leq \frac{A}{\alpha} \int_{\mathbf{R}^n} |f| \, dx,$$

where A is a constant which depends only on the dimension n ($A = 5^n$ *will do*)
(c) *If* $f \in L^p(\mathbf{R}^n)$, *with* $1 < p \leq \infty$, *then* $Mf \in L^p(\mathbf{R}^n)$ *and*

$$\|Mf\|_p \leq A_p\|f\|_p,$$

where A_p *depends only p and the dimension n.*

COROLLARY 1. *If* $f \in L^p(\mathbf{R}^n)$, $1 \leq p \leq \infty$, *or more generally if f is locally integrable, then*

$$\lim_{r \to 0} \frac{1}{B(x, r)} \int_{B(x,r)} f(y) \, dy = f(x),$$

for almost every x.

1.4 Before we come to the proof of the theorem we make some clarifying comments.
(i) In contrast with the case $p > 1$, when $p = 1$ the mapping $f \to M(f)$ is not bounded on $L^1(\mathbf{R}^n)$. Thus if f is not identically zero Mf is *never* integrable on all of $\mathbf{R}^n$. This can be seen by making the simple observation

that $Mf(x) \geq c|x|^{-n}$, for $|x| \geq 1$. Moreover even if we limit our considerations to any bounded subset of $\mathbf{R}^n$, then the integrability of Mf holds only if stronger conditions than the integrability of f are required. (See §5.2 below.)

(ii) The result that is obtained, namely estimate (b), is weaker than the statement that $f \to M(f)$ is bounded on $L^1(\mathbf{R}^n)$, as the remarks in §1.2 show; for this reason (b) is referred to as a *weak-type* estimate. This estimate is the best possible (as far as order of magnitude) for the distribution function of $M(f)$, where f is an arbitrary function in $L^1(\mathbf{R}^n)$. That this is so can be seen by replacing the measure $|f(y)| \, dy$ in definition (2) by the measure $d\mu$, whose total measure of one is concentrated at the origin; ($d\mu$ is the "Dirac measure"). Then $M(d\mu)(x) = c|x|^{-n}$, where c^{-1} = volume of the unit ball. In this case the distribution function $\lambda(\alpha)$ is exactly $1/\alpha$. But we can always find a sequence $\{f_m(x)\}$ of positive integrable functions, whose L^1 norm is each one, and which converge weakly to the measure $d\mu$. So we cannot expect an estimate essentially stronger than (b), since in the limit a similar stronger version would have to hold for $M(d\mu)(x)$.

1.5 Proof of Theorem 1 and its corollary. Here we shall prove the theorem and its corollary, taking for granted the covering lemma of "Vitali-type" stated in §1.6 and proved in §1.7 below. With the definition of Mf, and with

$$E_\alpha = \{x : Mf(x) > \alpha\}$$

then for each $x \in E_\alpha$ there exists a ball of center x, which we call B_x, so that

$$(3) \qquad \int_{B_x} |f(y)| \, dy > \alpha m(B_x).$$

But on the one hand (3) gives $m(B_x) < (1/\alpha) \|f\|_1$, for all such x; on the other hand when x runs through the set E_α the union of the corresponding B_x covers E_α. Thus using the covering lemma (1.6) below from this family of balls we can extract a sequence of balls, which we designate by $\{B_k\}$; these balls are mutually disjoint and have the property that

$$(4) \qquad \sum_{k=0}^{\infty} m(B_k) \geq Cm(E_\alpha),$$

(e.g. the bound $C = 5^{-n}$ will work). Applying (3) and then (4) to each of the mutually disjoint balls we get

$$\int_{\cup B_k} |f(y)| \, dy > \alpha \sum_k m(B_k) \geq \alpha Cm(E_\alpha).$$

But since the first member of this inequality is majorized by $\|f\|_1$, on taking $A = 1/C$ we obtain the assertion (b) of the theorem; (and thus also part (a), when $p = 1$). We shall now prove simultaneously assertion (a) (the finiteness almost everywhere of $M(f)(x)$), and assertion (c) (the L^p inequality), for $1 < p \leq \infty$. The case $p = \infty$ is of course trivial, and here the bound is $A_\infty = 1$. Let us therefore suppose that $1 < p < \infty$. We shall use a simple example of the technique of splitting a function into its large and small parts, alluded to at the beginning of this chapter. Let us define $f_1(x)$ by $f_1(x) = f(x)$, if $|f(x)| \geq \alpha/2$, and $f_1(x) = 0$ otherwise. Then we have successively $|f(x)| \leq |f_1(x)| + \alpha/2$; $M(f)(x) \leq M(f_1)(x) + \alpha/2$, therefore

$$\{x : M(f)(x) > \alpha\} \subset \{x : M(f_1)(x) > \alpha/2\},$$

and finally

$$m(E_\alpha) = m\{x : Mf(x) > \alpha\} \leq \frac{2A}{\alpha} \|f_1\|_1,$$

which is

(5) $$m(E_\alpha) = m\{x : Mf(x) > \alpha\} \leq \frac{2A}{\alpha} \int_{|f| > \alpha/2} |f| \, dx.$$

The last inequality is obtained by applying conclusion (b) of the theorem which we may since $f_1 \in L^1$ whenever $f \in L^p$. We now set $g = M(f)$, and λ the distribution function of g. Then using the observations in (1.2) together with an integration by parts we have

$$\int_{\mathbf{R}^n} (Mf)^p \, dx = -\int_0^\infty \alpha^p \, d\lambda(\alpha) = p \int_0^\infty \alpha^{p-1} \lambda(\alpha) \, d\alpha$$

In particular, because of (5),

$$\|Mf\|_p^p = p \int_0^\infty \alpha^{p-1} m(E_\alpha) \, d\alpha \leq p \int_0^\infty \alpha^{p-1} \left(\frac{2A}{\alpha} \int_{|f| > \alpha/2} |f(x)| \, dx \right) d\alpha.$$

The double integral is evaluated by interchanging the orders of integration and integrating first with respect to α. The inner integral is then

$$\int_0^{2|f(x)|} \alpha^{p-2} \, d\alpha = \left(\frac{1}{p-1} \right) |2f(x)|^{p-1},$$

since $p > 1$. So the double integral has the value

$$\frac{2Ap}{p-1} \int_{\mathbf{R}^n} |f| \, |2f|^{p-1} \, dx = (A_p)^p \int_{\mathbf{R}^n} |f|^p \, dx,$$

which proves conclusion (c). Calculating the constants we get

$$A_p = 2 \left(\frac{5^n p}{p-1} \right)^{1/p}, \qquad 1 < p < \infty.$$

It is useful, for certain applications, to observe that

$$A_p = O\left(\frac{1}{p-1}\right), \quad \text{as} \quad p \to 1.$$

We now come to the proof of the corollary. We easily reduce the consideration to the case $p = 1$, by multiplying our original function by the characteristic function of a ball, and then exhausting $\mathbf{R}^n$ by a denumerable union of such balls. Let us denote by f_r the function

$$(6) \qquad f_r(x) = \frac{1}{m(B(x, r))} \int_{B(x,r)} f(y)\, dy, \quad r > 0.$$

We know* that if $r \to 0$, $\|f_r - f\|_1 \to 0$, whenever $f \in L^1(\mathbf{R}^n)$. Therefore $f_{r_k} \to f$, almost everywhere for a suitable sequence $\{r_k\} \to 0$. What remains to be seen, therefore, is that $\lim_{r \to 0} f_r(x)$ exists almost everywhere. For this purpose let us denote for each $g \in L^1$, and $x \in \mathbf{R}^n$

$$(7) \qquad \Omega g(x) = |\limsup_{r \to 0} g_r(x) - \liminf_{r \to 0} g_r(x)|$$

where g_r is defined like f_r. Ωg represents the oscillation of the family $\{g_r\}$, as $r \to 0$.

If g is continuous with compact support, then $g_r \to g$ uniformly, and thus Ωg is identically zero in this case.

Next if g is in $L^1(\mathbf{R}^n)$, then by conclusion (b) of the theorem

$$m\{x : 2M(g) > \varepsilon\} \leq \frac{2A}{\varepsilon} \|g\|_1.$$

However clearly

$$\Omega g(x) \leq 2Mg(x),$$

thus

$$(8) \qquad m\{x : \Omega g(x) > \varepsilon\} \leq \frac{2A}{\varepsilon} \|g\|_1, \quad g \in L^1(\mathbf{R}^n).$$

Finally any $f \in L^1(\mathbf{R}^n)$ can be written as $f = h + g$, where h is continuous with compact support and where the L^1 norm of g is at our disposal. But $\Omega f \leq \Omega h + \Omega g$, and $\Omega h \equiv 0$ since h is continuous. Therefore (8) shows that

$$m\{x : \Omega f(x) > \varepsilon\} \leq \frac{2A}{\varepsilon} \|g\|_1.$$

---

* This is a particular property of approximations of the identity. See Chapter III, §2.2 for a detailed discussion; the relevant part of that section can be used without fear of circularity.

Since the norm of g can be chosen to be arbitrarily small we get $\Omega f = 0$ almost everywhere, which means that $\lim_{r \to 0} f_r(x)$ exists almost everywhere.

The following summarizing comment about the proof of the corollary is worth making. The argument used was of a very general nature and occurs often. That is, the almost everywhere convergence is proved as a combination of two parts, one which is deep and already contains the essence of the result; it is expressed in terms of a maximal inequality like part (b) (or (c)) of the theorem. The second fact is usually much simpler to establish but it is just as essential. It is the convergence almost everywhere for a dense subset of the function space, in this case the continuous function on $\mathbf{R}^n$ with compact support.

1.6 A covering lemma. We have therefore completed the proof of Theorem 1 and its corollary, save for the crucial step of the covering lemma, which we postponed until now. Not only the simplicity of its statement, or the application we use it for, but also the many variants of it that can be found in the mathematical literature attest to the fundamental character of this lemma. The reader should note that its statement and proof are closely related to a more refined but probably better known lemma of Vitali.*

LEMMA. *Let E be a measurable subset of $\mathbf{R}^n$ which is covered by the union of a family of balls $\{B_j\}$, of bounded diameter. Then from this family we can select a disjoint subsequence, $B_1, B_2, \ldots B_k, \ldots$, (finite or infinite) so that*

$$\sum_k m(B_k) \geq C m(E)$$

Here C is a positive constant that depends only on the dimension n; $C = 5^{-n}$ will do.

1.7 We begin the proof of the lemma by describing the choice of $B_1, B_2, \ldots B_k, \ldots$. We choose B_1 so that it is essentially as large as possible; that is so that the diameter of $B_1 \geq \frac{1}{2} \sup \{$diameter $B_j\}$. Of course the choice of a B_1 satisfying these conditions, as well as the later choices of the other B_k, is not unique; but this non-uniqueness is of no consequence to us. Let us now suppose that $B_1, B_2, \ldots B_k$ have already been chosen. We are now forced to select B_{k+1} from those balls in the family $\{B_j\}$ which are disjoint with $B_1, B_2, \ldots B_k$. We choose one that again is essentially as large as possible. That is we take B_{k+1} to be disjoint from $B_1, \ldots B_k$, and the diameter of $B_{k+1} \geq \frac{1}{2} \{\sup_j$ diameter of B_j, with B_j disjoint from $B_1, B_2, \ldots B_k\}$.

---

* The lemma of Vitali may be found in §5.4 below.

In this way we get the sequence $B_1, B_2, \ldots B_k, \ldots$ of balls. In principle this sequence could be finite, and terminate at B_k; this would be the case if there were no balls in $\{B_j\}$ disjoint with $B_1, B_2, \ldots B_k$.

Now two cases present themselves, depending on whether $\sum m(B_k) = \infty$ or $\sum m(B_k) < \infty$. In the first case we have attained our conclusion whether $m(E)$ is infinite or finite. Let us therefore consider the case when $\sum m(B_k) < \infty$.

For this purpose we denote by B_k^* the ball having the same center as B_k, but whose diameter is *five* times as large. We claim that

$$(9) \qquad\qquad U_k B_k^* \supset E.$$

To prove (9) we have to show that $U_k B_k^* \supset B_j$, for any fixed B_j in our given family which covers E. We may certainly assume that our fixed B_j is not one of the sequence $B_1, B_2, \ldots B_k, \ldots$, for otherwise there is nothing to prove. Since $\sum m(B_k) < \infty$, then diam $(B_k) \to 0$, as $k \to \infty$, and so we take the *first* k, with the property that diam $(B_{k+1}) < \frac{1}{2}(\text{diam } B_j)$. Now the ball B_j must intersect one of the k previous balls $B_1, B_2, \ldots B_k$, or it should have been picked as the $k + 1^{\text{th}}$ ball instead of B_{k+1}, since its diameter is more than twice that of B_{k+1}. That is B_j intersects B_{j_0}, for some $1 \leq j_0 \leq k$, and $\frac{1}{2}(\text{diameter of } B_j) \leq \text{diameter of } B_{j_0}$. From an obvious geometric consideration it is then evident that B_j is contained in the ball that has the same center as B_{j_0}, but five times the diameter of B_{j_0}; i.e. $B_j \subset B_{j_0}^*$.

Thus we have proved (9), and so

$$m(E) \leq \sum m(B_k^*) = 5^n \sum m(B_k),$$

which proves the lemma.

1.8 Lebesgue set. The differentiation theorem just proved refers to the limits of averages taken with respect to balls. But this theorem has, as a rather simple consequence of itself, a generalization where the averages are taken over more general families of sets.

Let $\mathscr{F}$ be a family of measurable subsets of $\mathbf{R}^n$. We shall say that this family is *regular*, if there exists a constant $c > 0$, so that if $S \in \mathscr{F}$, then $S \subset B$, with $m(S) \geq cm(B)$, where B is an appropriate open ball centered at the origin. Examples of such regular families are: (1) the family $\mathscr{F}$ of all sets of the δU, $0 < \delta < \infty$, (which are the dilations of a fixed set U), where U is bounded and $m(U) > 0$. (2) the family of all cubes with the property that their distance from the origin is bounded by a constant multiple of their diameter. (3) any subfamily of such a family $\mathscr{F}$. In analogy with the special case of the family of all balls centered at the origin, we defined the appropriate maximal function

$$M_{\mathscr{F}}(f)(x) = \sup_{S \in \mathscr{F}} \frac{1}{m(S)} \int_S |f(x - y)| \, dy$$

Then clearly $M_{\mathscr{F}}(f)(x) \le c^{-1} Mf(x)$, and therefore $M_{\mathscr{F}}$ satisfies the same conclusion as those in theorem 1 for M. So a repetition of the argument of the corollary leads to the fact that whenever f is locally integrable

$$(10) \qquad \lim_{\substack{S \in \mathscr{F} \\ m(S) \to 0}} \frac{1}{m(S)} \int_S f(x - y) \, dy = f(x),$$

for almost every x.

All of this is very simple, but is not completely satisfying for the following reason. Given a fixed locally integrable function f, we have proved that the relation (10) holds almost everywhere, but the exceptional set (of measure zero) depends on the given family $\mathscr{F}$. It would be better if we could find one exceptional set of measure zero (depending on f), so that outside of it the relation (10) would hold for *every* regular family. This is the role of the complement of the *Lebesgue set* of f, where the latter set is defined as those x for which

$$(11) \qquad \lim_{r \to 0} \frac{1}{m(B(x, r))} \int_{B(x,r)} |f(y) - f(x)| \, dy = 0.$$

(Recall that $B(x, r)$ is the ball of radius r centered at x)

To see that the limit (11) is realized almost everywhere, we consider the relation

$$(11') \qquad \lim_{r \to 0} \frac{1}{m(B(x, r))} \int_{B(x,r)} |f(y) - c| \, dy = |f(x) - c|$$

which holds for each constant c, for almost every x. That is, there is an exceptional set E_c, with $m(E_c) = 0$, so that (11') is valid whenever $x \notin E_c$. Let $c_1, c_2, \ldots c_n, \ldots$ be an enumeration of the rationals. If $x \notin E = U_n E_{c_n}$, then (11') holds for any rational c, and so by continuity for every real c. In particular, x in the complement of the set E are in the Lebesgue set of f; that is, for those x, (11) is valid.

But

$$\left| \frac{1}{m(S)} \int_S f(x - y) \, dy - f(x) \right| = \left| \frac{1}{m(S)} \int_S [f(x - y) - f(x)] \, dy \right|$$

$$\le \frac{1}{m(S)} \int_S |f(x - y) - f(x)| \, dy$$

$$\le c^{-1} \frac{1}{m(B(x, r))} \int_{B(x,r)} |f(y) - f(x)| \, dy,$$

so that differentiability with respect to any regular family is established at every point of the Lebesgue set of f.

For a discussion of the case of non-regular families, see §5.3 below.

2. Behavior near general points of measurable sets

2.1 In this section we wish to treat various properties of measurable sets of positive measure which confirm the observation that a "general" point of such a set is almost completely surrounded by other points of the set. The simplest concrete example of this heuristic principle is contained in the notion of a *point of density*.

Suppose E is a given measurable set, and $x \in \mathbf{R}^n$. Then we say that x is a point of density of E, if

$$(12) \qquad \lim_{r \to 0} \frac{m\{E \cap B(x, r)\}}{m\{B(x, r)\}} = 1.$$

Of course for general x the limit need not have value 1, or may not even exist; but if the limit in (12) has value 0, then according to our definition x is a point of density of the complement of E. Let us now apply the differentiation theorem (the corollary to theorem 1 stated in §1.3), to the case when f is χ_E, the characteristic function of the set E. This gives us immediately the following proposition.

PROPOSITION 1. *For almost every point $x \in E$, the limit* (12) *holds; that is almost every point $x \in E$ is a point of density of E, and almost every point of the complement of E is not a point of density of E.*

Notice that if we had restricted our attention to the points of the Lebesgue set of χ_E, we should have obtained instead of Proposition 1 a similar but stronger conclusion. The balls in (12) could then have been replaced by regular families, in the sense of §1.8.

2.2 In order to continue, we shall now limit ourselves to sets E which are closed, but are still otherwise arbitrary. The reason for this limitation is obvious: In what follows the results will be expressed in terms of the distance from E; if E is not closed the distance from E is in reality the distance from $\bar{E}$, the closure of E, and clearly E and $\bar{E}$ may be quite different measure-theoretically. However, the limitation to closed sets is not a serious obstacle in applications. Closed sets are sufficiently general; in particular, any measurable set may be approximated by the closed sets it contains, so that the respective difference sets have measure as small as we wish.

To reflect our newly imposed restriction we shall denote a general closed subset of $\mathbf{R}^n$ by F, and we let $\delta(x) = \delta(x, F)$ represent the distance of the point x from F. Of course $\delta(x) = 0$ if and only if $x \in F$. Now it is clear that if $x \in F$, $\delta(x + y) \leq |y|$, since x is a point in F whose distance from $x + y$ is equal to $|y|$. However in general, this estimate of the distance from F can be improved; that is $\delta(x + y) = o(|y|)$, for most x in F. The relation of "little o" means that given any $\varepsilon > 0$, there exists a $\eta = \eta_\varepsilon$, so that $\delta(x + y) \leq \varepsilon |y|$, if $|y| \leq \eta$.

PROPOSITION 2. *Let F be a closed set. Then for almost every $x \in F$, $\delta(x + y) = o(|y|)$. This holds in particular if x is a point of density of F.*

We have formulated this proposition mainly because it is a simple illustration of the notion of point of density. We shall, however, also find an application for this proposition, but this is not until much later.

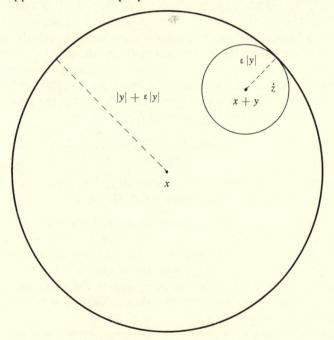

Figure 1. The point of density argument. The larger ball is $B(x, |y| + \varepsilon |y|)$, and the smaller ball is $B(x + y, \varepsilon |y|)$.

To prove the proposition, let x be a point of density of F and suppose ε is given, $\varepsilon > 0$. Consider the "small" ball of center $x + y$, and radius $\varepsilon |y|$; and the "large" ball of center x and radius $|y| + \varepsilon |y|$. Obviously

$B(x + y, \varepsilon |y|) \subset B(x, |y| + \varepsilon |y|)$. We claim that if $|y|$ is sufficiently small then there exists a $z \in F$, so that $z \in B(x + y, \varepsilon |y|)$. For otherwise $F \cap B(x + y, \varepsilon |y|) = \varnothing$, and

$$\frac{m(F \cap B(x, |y| + \varepsilon |y|))}{m(B(x, |y| + \varepsilon |y|))} \leq \frac{m(B(x, |y| + \varepsilon |y|)) - m(B(x + y, \varepsilon |y|))}{m(B(x, |y| + \varepsilon |y|))}$$

$$\leq 1 - \left(\frac{\varepsilon}{1 + \varepsilon}\right)^n,$$

which contradicts (12) if $|y|$ is small enough. Thus there exists a z in F which also is in $B(x + y, \varepsilon |y|)$; this means that within a distance of $\varepsilon |y|$ from the point $x + y$ we can find a point of F, i.e. $\delta(x + y) \leq \varepsilon |y|$.

2.3 Integral of Marcinkiewicz. We shall now present another expression of the principle that a general point of a measurable set is almost completely surrounded by other points of the set. This form will be independent of the theorem of differentiation, but for many problems it will have a significance which is equally important. In fact, the integrals considered below, first treated systematically by Marcinkiewicz, intervene in a decisive way in the theory of singular integrals, as discussed in the following chapter, as well as other problems treated in this book.

We consider as before a closed set, F; $\delta(x)$ denotes the distance of x from F, and we shall study the integral $I(x)$ given by

$$(13) \qquad I(x) = \int_{|y| \leq 1} \frac{\delta(x + y)}{|y|^{n+1}} \, dy.$$

THEOREM 2.   (a) *When* $x \in$ *complement of* F, *then* $I(x) = \infty$.
                (b) *For almost every* $x \in F$, $I(x) < \infty$.

The conclusion (a) is evident, since the complement of F is an open set. Then if x belongs to this complement $\delta(x + y) \geq c > 0$, for a neighborhood of the origin in y. The conclusion (b) is the interest of this theorem, and it states in effect that the estimate $\delta(x + y) = o(|y|)$ of Proposition 2 can be refined on the average, so as to lead to the convergence of the integral (13). But this is not to say that the convergence of (13) for a given x implies $\delta(x + y) = o(|y|)$.

The theorem will be a simple consequence of the following lemma, which is a more quantitative expression of the same fact.

LEMMA. *Let* F *be a closed set whose complement has finite measure. With* $\delta(x)$ *defined as above we let*

$$(14) \qquad I_*(x) = \int_{\mathbf{R}^n} \frac{\delta(x + y)}{|y|^{n+1}} \, dy.$$

Then $I_*(x) < \infty$ for almost every $x \in F$. Moreover

$$(15) \qquad \int_F I_*(x)\, dx \leq c \cdot m({}^cF).$$

2.4 In proving the lemma, we observe that it suffices to prove (15) since the integrand is positive. Also by the same positivity we can interchange the orders of integration in evaluating the left side of (15). This will accomplish the proof. In detail:

$$\int_F I_*(x)\, dx = \int_F \int_{\mathbf{R}^n} \frac{\delta(x+y)}{|y|^{n+1}}\, dy\, dx = \int_F \int_{\mathbf{R}^n} \frac{\delta(y)}{|x-y|^{n+1}}\, dy\, dx$$

$$= \int_F \int_{{}^cF} \frac{\delta(y)}{|x-y|^{n+1}}\, dy\, dx = \int_{{}^cF} \left(\int_F \frac{dx}{|x-y|^{n+1}} \right) \delta(y)\, dy.$$

Now consider

$$\int_F \frac{dx}{|x-y|^{n+1}} \quad \text{with} \quad y \in {}^cF.$$

The smallest value of $|x-y|$ (as x varies over F) is of course $\delta(y)$, which is the distance of y from F. Thus

$$\int_F \frac{dx}{|x-y|^{n+1}} \leq \int_{|x| \geq \delta(y)} \frac{dx}{|x|^{n+1}} \leq c(\delta(y))^{-1}.$$

This shows that

$$\int_F I_*(x)\, dx \leq \int_{{}^cF} c(\delta(y))^{-1} \delta(y)\, dy = cm({}^cF),$$

and the lemma is proved.

Theorem 2 is obtained from this lemma as follows. Let B_m denote the open ball of radius m center at the origin, and let $F_m = F \cup {}^cB_m$. Then F_m is closed but its complement has finite measure (since it is contained in B_m). Thus we can apply the lemma to F_m. So let δ_m denote the distance from F_m, and δ the distance from F. Observe that $\delta(x+y) = \delta_m(x+y)$, if $|y| \leq 1$ and $x \in B_{m-2}$. Hence the lemma implies that $I(x) < \infty$, for almost every $x \in F \cap B_{m-2}$. Letting $m \to \infty$ we get the desired result.

Among the several variants of the theorem and the lemma we present here one. (Another variant is discussed at the end of this chapter in §5.) We can replace $I(x)$ by

$$I^{(\lambda)}(x) = \int_{|y| \geq 1} \frac{\delta^\lambda(x+y)}{|y|^{n+\lambda}}\, dy,$$

where $\lambda > 0$. Similarly $I_*(x)$ can be replaced by

$$I_*^{(\lambda)}(x) = \int_{\mathbf{R}^n} \frac{\delta^\lambda(x+y)}{|y|^{n+\lambda}}\, dy, \qquad \lambda > 0.$$

In both cases similar conclusions are obtained with the above methods.

3. *Decomposition in cubes of open sets in* $\mathbf{R}^n$

3.1 The decomposition of a given set into a disjoint union of cubes (or balls) is a fundamental tool in the theory described in this chapter. We have already used this type of notion, in very rough form, in the covering lemma, §1.6.

3.1.1 We now pose ourselves the following related general problem which, however, does not involve measure theory, but deals with the geometric structure of general closed sets F in $\mathbf{R}^n$: Can the complement of F be realized as a disjoint union of cubes in a canonical way? For $n = 1$ the answer is of course *yes*, since every open set is in a unique way the union of disjoint open intervals. For $n \geq 2$, the situation is no longer that simple, since we can realize an arbitrary open set in an infinity of different ways as a disjoint union of cubes (by cubes we now mean *closed* cubes; by disjoint we mean that their *interiors* are disjoint). However there are decompositions, which while not canonical, are very satisfactory and useful substitutes. We have in mind the idea first introduced by Whitney and formulated as follows.

THEOREM 3.   *Let F be a non-empty closed set in* $\mathbf{R}^n$. *Then its complement* Ω *is the union of a sequence of cubes* Q_k, *whose sides are parallel to the axes, whose interiors are mutually disjoint, and whose diameters are approximately proportional to their distances from F. More explicitly:*

(i) $\Omega = {}^c F = \bigcup_{k=1}^{\infty} Q_k.$

(ii) $Q_j^0 \cap Q_k^0 = \varnothing$ *if* $j \neq k$.

(iii) *There exist two constants* $c_1, c_2 > 0$, (*we can take* $c_1 = 1$, *and* $c_2 = 4$), *so that*

$$c_1\,(diameter\ Q_k) \leq distance\ Q_k\ from\ F \leq c_2\,(diameter\ Q_k).$$

3.1.2 Our intention for stating the theorem at this stage is obviously pedagogical. We shall not, strictly speaking, need to apply it until later (Chapter VI), and since its proof is a little intricate we postpone it until that point.

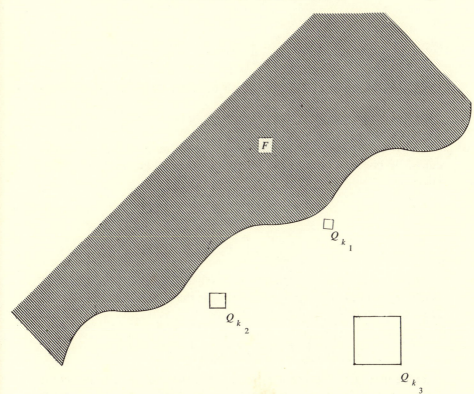

Figure 2. The decomposition of the complement of F into cubes, whose diameters equal approximately their respective distances from F.

A related reason we have presented the theorem here is because it will help to clarify the meaning of the theorem immediately following. We are referring to a fundamental lemma of Calderón and Zygmund which can be used to give another approach to the theory of the maximal function of §2, but whose main importance for us will be its application to singular integrals in the next chapter.

3.2 THEOREM 4. *Let f be a non-negative integrable function on* **R**ⁿ, *and let α be a positive constant. Then there exists a decomposition of* **R**ⁿ *so that*

(i) **R**ⁿ $= F \cup \Omega$, $F \cap \Omega = \emptyset$.

(ii) $f(x) \leq \alpha$ *almost everywhere on F.*

(iii) Ω *is the union of cubes,* $\Omega = \bigcup_{k} Q_k$, *whose interiors are disjoint, and* *so that for each Q_k*

(16) $$\alpha < \frac{1}{m(Q_k)} \int_{Q_k} f(x)\, dx \leq 2^n\alpha.$$

3.3 We decompose $\mathbf{R}^n$ into a mesh of equal cubes, whose interiors are disjoint, and whose common diameter is so large that $\dfrac{1}{m(Q')} \int_{Q'} f \leq \alpha$, for every cube Q' in this mesh.

Let Q' be a fixed cube in this mesh. We divide it into 2^n congruent cubes, by bisecting each of the sides of Q'. Let Q'' be one of these new cubes.

First case: $$\frac{1}{m(Q'')} \int_{Q''} f\, dx \leq \alpha.$$

Second case: $$\frac{1}{m(Q'')} \int_{Q''} f\, dx > \alpha.$$

In the second case one does not sub-divide Q'' any further, and Q'' is selected as one of the cubes Q_k appearing in the statement of the theorem. One has for it the inequality (16), because

$$\alpha < \frac{1}{m(Q'')} \int_{Q''} f\, dx \leq \frac{1}{2^{-n}m(Q')} \int_{Q'} f\, dx \leq 2^n\alpha.$$

In the first case we proceed with the sub-division of Q'', and repeat this process until (if ever) we are forced into the second case. We denote by $\Omega = \bigcup_k Q_k$, the union of cubes obtained from the second case, where we start the process with all possible cubes Q' of our initial mesh. We claim that $f(x) \leq \alpha$ almost everywhere in $F = {}^c\Omega$. In fact for almost every point $x \in F$ we have because of the theorem of differentiation (see the variant in §1.8), that

$$f(x) = \lim_{Q} \frac{1}{m(Q)} \int_Q f(y)\, dy,$$

where the limit is taken over all cubes Q which contain x, and the diameter of Q tends to zero. But each of the cubes that enter in our decomposition which contains an $x \in F$, is a cube for which the first alternative holds. This proves the theorem.

3.4 We now state an immediate corollary, whose interest is that it contains that part of the theorem that we shall apply in the next chapter.

COROLLARY. *Suppose f, α, F, Ω, and Q_k have the same meaning as in Theorem 4. Then there exists two constants A and B (depending only on* *the dimension n), so that (i) and (ii) hold and*

(a) $$m(\Omega) \leq \frac{A}{\alpha} \|f\|_1$$

(b) $$\frac{1}{m(Q_k)} \int_{Q_k} f \, dx \leq B\alpha.$$

In fact by (16) we can take $B = 2^n$, and also because of (16)

$$m(\Omega) = \sum m(Q_k) < \frac{1}{\alpha} \int_\Omega f(x) \, dx \leq \frac{1}{\alpha} \|f\|_1.$$

This proves the corollary with $A = 1$, and $B = 2^n$.

3.5 It is possible however to give another proof of the corollary without using Theorem 4 from which it was deduced, but by using Theorem 1, the maximal theorem, and also the theorem about the decomposition of an arbitrary open set as a union of disjoint cubes. This more indirect method of proof has the advantage of clarifying the roles of the sets F and Ω into which $\mathbf{R}^n$ was divided. We know that in F, $f(x) \leq \alpha$, but this fact does not determine F. The set F is however determined, in effect, by the fact that the maximal function satisfies $Mf(x) \leq \alpha$ on it. So we choose $F = \{x : Mf(x) \leq \alpha\}$, and $\Omega = E_\alpha = \{x : Mf(x) > \alpha\}$. Then by Theorem 1, part (b) we know that $m(\Omega) \leq \frac{A}{\alpha} \|f\|_1$, with in fact $A = 5^n$. Notice that since by definition F is closed, we can choose cubes Q_k according to Theorem 3, so that $\Omega = \bigcup Q_k$, and whose diameters are approximately proportional to their distances from F. Let Q_k then be one of these cubes, and p_k a point of F so that

$$\text{distance } (F, Q_k) = \text{distance } (p_k, Q_k).$$

Let B_k be the smallest ball whose center is p_k and which contains the interior of Q_k. Let us set

$$\gamma_k = \frac{m(B_k)}{m(Q_k)}.$$

We have because $p_k \in \{x : Mf(x) \leq \alpha\}$,

$$\alpha \geq (Mf)(p_k) \geq \frac{1}{m(B_k)} \int_{B_k} f \, dx \geq \frac{1}{\gamma_k m(Q_k)} \int_{Q_k} f \, dx.$$

But elementary geometry and the inequality (iii) of Theorem 3 then show that $\gamma_k = \frac{m(B_k)}{m(Q_k)} \leq$ constant, for all k. Thus we have another proof of the corollary.

Notice that this second proof of the lemma also rewarded us with an unexpected benefit: the cubes Q_k are at a distance from F comparable to their diameters.

3.6   A final remark about the affinity of the present theorem with Theorem 1. It may be seen that the former also implies the latter, without the use of the covering lemma in §1.6. For this see §5.1 at the end of this chapter.

4. *An interpolation theorem for* L^p

4.1   We wish here to formalize a part of the reasoning used in the proof of Theorem 1. What we have in mind is that part of the argument that took us from the inequality (b) to the L^p inequality. This idea will lead us to the Marcinkiewicz interpolation theorem—or more precisely, a basic special case of that theorem. The more extended form of that interpolation theorem will be presented later in Appendix B.

We shall require several definitions. Let T be a mapping from $L^p(\mathbf{R}^n)$ to $L^q(\mathbf{R}^n)$, $1 \leq p \leq \infty$, $1 \leq q \leq \infty$. Then T is *of type* (p, q) if

$$(17) \qquad \|T(f)\|_q \leq A \|f\|_p, \qquad f \in L^p(\mathbf{R}^n)$$

where A does not depend on f. Similarly T is of *weak-type* (p, q) if

$$(18) \qquad m\{x : |Tf(x)| > \alpha\} \leq \left(\frac{A \|f\|_p}{\alpha}\right)^q, \quad \text{when} \quad q < \infty$$

where A does not depend on f or α, $\alpha > 0$.

If $q = \infty$ we shall say that T is of weak-type (p, q) if it is of type (p, q). Notice that (17) implies (18) so that the notion of type (p, q) is stronger than the notion of weak-type (p, q). In fact, if $q < \infty$,

$$\alpha^q m\{x : |Tf(x)| > \alpha\} \leq \int_{\mathbf{R}^n} |Tf|^q \, dx = \|Tf\|_q^q \leq (A \|f\|_p)^q.$$

It will also be necessary to treat operators T defined on several L^p spaces simultaneously. Thus we define $L^{p_1}(\mathbf{R}^n) + L^{p_2}(\mathbf{R}^n)$ to be the space of all functions f, so that $f = f_1 + f_2$, with $f_1 \in L^{p_1}(\mathbf{R}^n)$ and $f_2 \in L^{p_2}(\mathbf{R}^n)$. Suppose now $p_1 < p_2$. Then we observe that $L^p(\mathbf{R}^n) \subset L^{p_1}(\mathbf{R}^n) + L^{p_2}(\mathbf{R}^n)$, for all p such that $p_1 \leq p \leq p_2$. In fact let $f \in L^p(\mathbf{R}^n)$ and let γ be a fixed positive constant. Set

$$f_1(x) = \begin{cases} f(x) & \text{if} \quad |f(x)| > \gamma \\ 0 & \text{if} \quad |f(x)| \leq \gamma \end{cases}$$

$$f_2(x) = \begin{cases} f(x) & \text{if} \quad |f(x)| \leq \gamma \\ 0 & \text{if} \quad |f(x)| > \gamma. \end{cases}$$

Then $\int |f_1(x)|^{p_1}\, dx = \int |f_1(x)|^p\, |f_1(x)|^{p_1-p}\, dx \leq \gamma^{p_1-p} \int |f(x)|^p\, dx$, since $p_1 - p \leq 0$. Similarly

$$\int |f_2(x)|^{p_2}\, dx = \int |f_2(x)|^p |f_2(x)|^{p_2-p}\, dx \leq \gamma^{p_2-p} \int |f(x)|^p dx,$$

so $f_1 \in L^{p_1}$ and $f_2 \in L^{p_2}$, with $f = f_1 + f_2$.

The idea we have just used of splitting f into two parts according to their respective size, is the main idea of the proof of the theorem that follows.

4.2 THEOREM 5. *Suppose that $1 < r \leq \infty$. If T is a sub-additive mapping from $L^1(\mathbf{R}^n) + L^r(\mathbf{R}^n)$ to the space of measurable functions on $\mathbf{R}^n$ which is simultaneously of weak type $(1, 1)$ and weak-type (r, r), then T is also of type (p, p), for all p such that $1 < p < r$. More explicitly: Suppose that for all $f, g \in L^1(\mathbf{R}^n) + L^r(\mathbf{R}^n)$*

(i) $|T(f + g)(x)| \leq |Tf(x)| + |Tg(x)|$

(ii) $m\{x : |Tf(x)| > \alpha\} \leq \dfrac{A_1}{\alpha} \|f\|_1, \qquad f \in L^1(\mathbf{R}^n)$

(iii) $m\{x : |Tf(x)| > \alpha\} \leq \left(\dfrac{A_r}{\alpha} \|f\|_r\right)^r, \qquad f \in L^r(\mathbf{R}^n)$

(if $r < \infty$; when $r = \infty$ we assume that the form (17) holds). Then

$$\|T(f)\|_p \leq A_p \|f\|_p, \qquad f \in L^p(\mathbf{R}^n)$$

for all $1 < p < r$, where A_p depends only on $A_1, A_2, p,$ and r.

4.3 We prove the theorem under the restriction that $r < \infty$. In the case $r = \infty$ the argument presented below needs a slight modification which we leave as an exercise to the interested reader; this case is anyway contained implicitly in the proof given in §1.5.

Let $f \in L^p(\mathbf{R}^n)$. We wish to estimate the distribution function $\lambda(\alpha) = m\{x : |Tf(x)| > \alpha\}$. Fix α for the moment. As we saw above we can split $f = f_1 + f_2$, so that $f_1 \in L^1(\mathbf{R}^n)$ and $f_2 \in L^r(\mathbf{R}^n)$ where the splitting was obtained by cutting $|f|$, in effect, at the altitude $\gamma, \gamma > 0$. At that stage γ was arbitrary but we now fix it to be equal to α. Since $|T(f)(x)| \leq |Tf_1(x)| + |Tf_2(x)|$, we have

$$\{x : |Tf(x)| > \alpha\} \subset \{x : |Tf_1(x)| > \alpha/2\} \cup \{x : |Tf_2(x)| > \alpha/2\}$$

so

$$\lambda(\alpha) = m\{x : |Tf(x)| > \alpha\} \leq m\{x : |Tf_1(x)| > \alpha/2\} + m\{x : |Tf_2(x)| > \alpha/2\},$$

and therefore by the assumptions (ii) and (iii)

$$\lambda(\alpha) \leq \frac{A_1}{\alpha/2} \int |f_1(x)| \, dx + \frac{A_r^r}{(\alpha/2)^r} \int |f_2(x)|^r \, dx.$$

Because of the definitions of f_1 and f_2 we get

(19) $$\lambda(\alpha) \leq \frac{2A_1}{\alpha} \int_{|f|>\alpha} |f| \, dx + \frac{(2A_r)^r}{\alpha^r} \int_{|f|\leq\alpha} |f(x)|^r \, dx.$$

Now we know that

$$\int_{\mathbf{R}^n} |Tf|^p \, dx = -\int_0^\infty \alpha^p \, d\lambda(\alpha) = p \int_0^\infty \alpha^{p-1}\lambda(\alpha) \, d\alpha,$$

so we need only multiply both sides of (19) by $p\alpha^{p-1}$ and integrate with respect to α. To do this observe that

$$\int_0^\infty \alpha^{p-1}\alpha^{-1} \left\{ \int_{|f|>\alpha} |f| \, dx \right\} d\alpha = \int_{\mathbf{R}^n} |f| \int_0^{|f|} \alpha^{p-2} \, d\alpha \, dx$$

$$= \frac{1}{p-1} \int_{\mathbf{R}^n} |f| \, |f|^{p-1} \, dx$$

since $p > 1$. Similarly

$$\int_0^\infty \alpha^{p-1}\alpha^{-r} \int_{|f|\leq\alpha} |f|^r \, dx \, d\alpha = \int_{\mathbf{R}^n} |f|^r \int_{|f|}^\infty \alpha^{p-1-r} \, d\alpha \, dx$$

$$= \frac{1}{r-p} \int_{\mathbf{R}^n} |f|^r \, |f|^{p-r} \, dx$$

since $p < r$. Putting the two together we get

$$\|T(f)\|_p \leq A_p \|f\|_p, \quad \text{with} \quad (A_p)^p = \frac{2A_1}{p-1} + \frac{(2A_r)^r}{r-p}.$$

One should remark that as in the case of the maximal function, the bound A_p satisfies the inequality $A_p \leq A/(p-1)$, as $p \to 1$.

One example of this theorem is of course Theorem 1, part (c). Part (b) of Theorem 1 tells us that the operator $f \to M(f)$ is of weak-type $(1, 1)$, while the fact that $f \to M(f)$ is of type (∞, ∞) is obvious. Another important application of the Marcinkiewicz interpolation theorem occurs in the theory of singular integrals, which are the subject of the next chapter.

5. *Further results*

5.1 Theorem 4 (in §3), may be used to give another proof of the fundamental inequality for the maximal function in part (b) of Theorem 1. In fact for $f \geq 0$,

$f \in L^1(\mathbf{R}^n)$ and $\alpha > 0$, let $\Omega = \bigcup_k Q_k$, be as given in Theorem 4. Then

$$m(\Omega) \leq \frac{1}{\alpha} \int f \, dx.$$

Let Q_k^* be the cube with the same center as Q_k but with twice the diameter. Then clearly $m\left(\bigcup_k Q_k^*\right) \leq \frac{2^n}{\alpha} \int f \, dx$, and it can be seen that $Mf(x) \leq c\alpha$, if $x \notin \bigcup Q_k^*$, for an appropriate constant c; that is $m\{x : Mf(x) > c\alpha\} \leq \frac{2^n}{\alpha} \int f \, dx$. For details see Calderón and Zygmund [1], pp. 114–115.

5.2 (a) Suppose that f is supported in a finite ball $B \subset \mathbf{R}^n$. Then $M(f) \in L^1(B)$ if $|f| \log (2 + |f|)$ is integrable over B. In fact

$$\int_B Mf \, dx \leq m(B) + \int_{Mf \geq 1} Mf \, dx$$

while

$$\int_{Mf \geq 1} Mf \, dx = \int_1^\infty \lambda(\alpha) \, d\alpha + \lambda(1),$$

where $\lambda(\alpha) = m\{x : Mf(x) > \alpha\} \leq \frac{2A}{\alpha} \int_{|f| > \alpha/2} |f| \, dx$, by (5) in §1. See Wiener [1].

(b) The above inequality for $\lambda(\alpha)$ can be reserved, essentially. In fact for an appropriate positive constant, c, $m\{x : Mf(x) > c\alpha\} \geq \frac{2^{-n}}{\alpha} \int_{|f| > \alpha} |f| \, dx$. To prove this apply Theorem 4 to $|f|$ and α. This leads to the cubes Q_k, where $2^n \alpha \geq \frac{1}{m(Q_k)} \int_{Q_k} |f| \, dx > \alpha$. Thus if $x \in Q_k$, $M(f)(x) > c\alpha$, and so

$$m\{x : M(f)(x) > c\alpha\} \geq \sum m(Q_k) \geq \frac{2^{-n}}{\alpha} \int_{\bigcup_k Q_k} |f| \, dx.$$

But $|f(x)| \leq \alpha$ if $x \notin \bigcup Q_k$; hence $\int_{\bigcup Q_k} |f| \, dx \geq \int_{|f| > \alpha} |f| \, dx$. This establishes the desired inequality.

(c) Part (a) has a converse. If f is supported in a ball B, then $M(f) \in L^1(B)$ implies that $|f| (\log (2 + |f|)$ is integrable over B. To prove this integrate the above inequality for $m(x : Mf(x) > c\alpha)$ in part (b) as in the direct part of the theorem. For (b) and (c) see Stein [12].

(d) More generally, if f is supported in B, then

$$M(f) \log (2 + Mf)^k \in L^1(B) \Longleftrightarrow |f| (\log 2 + |f|)^{k+1} \in L^1(B), \qquad k \geq 0.$$

5.3 We consider the question whether

(*) $\quad \lim_{\text{diam}(S) \to 0} \frac{1}{m(S)} \int_S f(x - y) \, dy = f(x)$, almost everywhere, with $S \in \mathscr{F}$,

where $\mathscr{F}$ is an appropriate family of *rectangles* containing the origin.

(a) When $\mathscr{F}$ is the family of *all* rectangles, (*) may be false even if f is bounded. See O. Nikodym [1], and Busemann and Feller [1].

(b) When $\mathscr{F}$ is the family of all rectangles with sides parallel to the axes, (*) is false for some integrable f. See Saks [1].

(c) However, for $\mathscr{F}$ the family of all rectangles with sides parallel to the axes, (*) holds if $f \in L^p(\mathbf{R}^n)$, $p > 1$. In fact if we define

$$\tilde{M}(f)(x) = \sup_{S \in \mathscr{F}} \frac{1}{m(S)} \int_S |f(x - y)| \, dy,$$

for such a family then $\|\tilde{M}f\|_p \leq A_p \|f\|_p$, $1 < p \leq \infty$. This inequality can be proved by an n-fold application of the one-dimensional L^p maximal inequalities of Theorem 1.

(d) Let $\mathscr{F}$ be a one parameter monotonic family of rectangles with sides parallel to the axes, i.e. $\mathscr{F} = \{S_t\}_{0 < t < \infty}$, with $S_{t_1} \subset S_{t_2}$, if $t_1 \leq t_2$.

Then (*) holds for $f \in L^p(\mathbf{R}^n)$, $1 \leq p$. This follows from the fact that for such a monotonic family of rectangles an analogue of the covering lemma (1.6) holds. For (c), (d), and further related results see Zygmund [8], Chapter XVII.

5.4 *Vitali covering theorem.* Suppose that a measurable set E is covered by a collection of balls $\{B_\alpha\}$, in the sense that for each $x \in E$, and each $\varepsilon > 0$, there exists a $B_{\alpha_0} \in \{B_\alpha\}$, so that $x \in B_{\alpha_0}$, and $m(B_{\alpha_0}) < \varepsilon$. Then there is a disjoint subsequence of these balls $B_1, B_2, \ldots, B_k, \ldots$, so that

$$m(E - \bigcup_k B_k) = 0.$$

For this and related generalizations see Saks [2], Chapter 4.

5.5 (F. Riesz,) Let $F(x)$ be a real-valued continuous and bounded function defined on the line $\mathbf{R}^1$, and suppose α is positive. Let Ω be the set of those x, for which $\sup_{h > 0} \dfrac{F(x + h) - F(x)}{h} > \alpha$. Then $\Omega = \bigcup_{k=1}^\infty I_k$, with $I_k = (a_k, b_k)$, and $\dfrac{F(b_k) - F(a_k)}{b_k - a_k} = \alpha$. This lemma gives another proof of Theorem 4, in one dimension, if we set $F(x) = \int_0^x f(t) \, dt$. The inequality (16) is then replaced by the identity $\dfrac{1}{m(I_k)} \int_{I_k} f \, dx = \alpha$. See Riesz and Nagy [1], Chapter 1.

5.6 (a) A strengthened form of the inequality (15) is as follows: Let $\psi \geq 0$, then $\int_F I_*(x)\psi(x) \, dx \leq \int_{c_F} (M\psi)(x) \, dx$, where $M\psi$ is the maximal function of ψ. This shows that $I_*(x) \in L^p(F)$ for all $1 \leq p < \infty$. If we use (§5.2) for $\psi \log(2 + \psi)$ integrable (but otherwise arbitrary), then it also shows that $\int_F e^{aI_*(x)} \, dx < \infty$, for an appropriate $a > 0$.

(b) A variant of $I_*(x)$ is $\mathscr{I}_*(x) = \displaystyle\int_{\mathbf{R}^n} \frac{\delta(x + y) \, dy}{[|y| + \delta(x+y)]^{n+1}}$. Then (i) $\mathscr{I}_*(x) \geq cI_*(x)$, $x \in F$; (ii) $\mathscr{I}_*(x) < \infty$ for almost every $x \in \mathbf{R}^n$; (iii) Further there exists a positive constant a, so that for every finite ball B, $\int_B e^{a\mathcal{f}_\bullet(x)} dx < \infty$. In this connection see Carleson [3].

5.7 Suppose that $f \in L^p(\mathbf{R}^n)$, $1 \leq p < \infty$. Then a slight modification of the argument of §1.8 shows that

$$\lim_{r \to 0} \frac{1}{m(B(x, r))} \int_{B(x,r)} |f(y) - f(x)|^p \, dy = 0 \quad \text{for almost every } x.$$

5.8 Let $f_1, f_2, \ldots, f_n, \ldots$ be a sequence of functions in $L^p(\mathbf{R}^n)$ with the property that $\left(\sum_j |f_j(x)|^2\right)^{\frac{1}{2}} \in L^p(\mathbf{R}^n)$. Denote by Mf_j the maximal function of f_j. Then $\left\|\left(\sum_j |Mf_j|^2\right)^{\frac{1}{2}}\right\|_p \leq A_p \left\|\left(\sum_j |f_j|^2\right)^{\frac{1}{2}}\right\|_p$, $1 < p < \infty$. The estimates for A_p are $A_p = O((p-1)^{-1})$, as $p \to 1$, and $A_p = O(p^{\frac{1}{2}})$ as $p \to \infty$. These are best possible. If the f_j are taken to be the characteristic functions of disjoint cubes, then the above result essentially contains §5.6. See Fefferman and Stein [1].

Notes*

Section 1. For the basic facts about integration and differentiation, see Saks [2]. The original maximal theorem, for $n = 1$, is due to Hardy-Littlewood [1], and its n-dimensional version is in Weiner [1]. For the covering lemma in §1.6 see Weiner [1] and Marcinkiewicz and Zygmund [3]. The reader may find it instructive to compare this lemma with a more refined version found in *Fourier Analysis*, Chapter II, section 3, which is based on further ideas of Besicovitch. See also Edwards and Hewitt [1] and Stein [11] for other generalizations.

Section 2. The integral of Marcinkiewicz arose first in Marcinkiewicz [1], [2], and [3]; see also Zygmund [8], Chapter IV. A systematic use for its n-dimensional form is found in Calderón and Zygmund [7].

Section 3. The decomposition theorem in §3.2 is in Calderón and Zygmund [1]. Its close connection with the Whitney decomposition seems to have been pointed out first in Stein [10].

Section 5. The interpolation theorem 5 is due to Marcinkiewicz [5]. The more general version presented in Appendix B is due to Zygmund, but the proof given there is that of R. Hunt [1]. See also the more extended treatment given in *Fourier Analysis*, Chapter V.

* *Fourier Analysis* refers to Stein and Weiss [4].

CHAPTER II

Singular Integrals

A basic example which lies at the source of the theory of singular integrals is given by the Hilbert transform. This transformation of f is defined by

$$(1) \qquad \frac{1}{\pi} \int_{-\infty}^{\infty} \frac{f(x-y)}{y} \, dy,$$

where the non-absolutely convergent integral is interpreted by a suitable limiting process. We shall here single out several features of the theory of the Hilbert transform so that in terms of their aspects we can describe the n-dimensional singular integrals treated in this chapter.

(a) *The L^2 theory.* We are dealing here, as in the general case, with an operator that commutes with translations. For this reason the tools of convolution, Fourier transforms, and Plancherel theorem—in brief, the basic implements of harmonic analysis in $\mathbf{R}^n$—are unavoidable; it is with a resumé of these that we begin the development in this chapter.

(b) *The L^p theory.* A fundamental property of the operator (1), as well as the generalizations we treat, is that each is a bounded operator on L^p, $1 < p < \infty$. In the case of Hilbert transforms this classical theorem was proved by M. Riesz, using complex function theory. This approach is inappropriate in the general context, and there the L^p theory will be obtained as a consequence of the L^1 theory.

(c) *The L^1 theory.* The Hilbert transform is not a bounded operator on L^1. There is for it, however, a substitute result, namely that it is of weak-type $(1, 1)$. There is a similar situation in the general case. The (real-variable) techniques for proving the weak-type result were initiated by Besicovitch and Titchmarsh in the case of the Hilbert transform, and were further developed by Calderón and Zygmund's treatment of the n-dimensional theory. It is the presentation of those methods that may be said to represent the core of the present chapter. Needless to say, we shall make decisive use of the general real-variable theory of Chapter **I**.

(d) *Special properties of the Hilbert transform.* Among these are:

(i) The operator (1) commutes not only with translations, but also with dilations $x \rightarrow \delta x$, $\delta > 0$. It is therefore not surprising that the theorems

26

describing the n-dimensional generalizations are essentially invariant under dilations. Further, those operators which, like (1), are left fixed by dilations represent an important sub-class for which the theory is more explicit and far-reaching. These are the subject matter of §4.

(ii) The connection with analytic functions. There is a special relation between the transform (1) (or certain of its n-dimensional variants), and analytic functions (or their generalizations). The meaning of this relation and its concomitant properties of invariance with respect to rotations will be described in the next chapter.

1. *Review of certain aspects of harmonic analysis in* **R**^{n}

We state here, without proof, certain elementary facts taken from the theory of harmonic analysis in **R**^{n}, which incidentally find their natural generality in the setting of locally compact abelian groups.

1.1 Together with the spaces $L^p(\mathbf{R}^n)$, $1 \leq p \leq \infty$ already used, we consider the space $C_0(\mathbf{R}^n)$ of continuous functions tending to zero at infinity, with the usual supremum norm; also its dual space $\mathscr{B}(\mathbf{R}^n)$, which as is well known can be identified with the Banach space of all finite measures $d\mu$, with norm $\|d\mu\| = \int_{\mathbf{R}^n} |d\mu|$. The space $L^1(\mathbf{R}^n)$ can be identified as a subspace of $\mathscr{B}(\mathbf{R}^n)$ by the isometry $f(x) \to f(x)\,dx$, where dx is Lebesgue measure.

A basic operation is that of that of convolution. Thus if $\mu_1, \mu_2 \in \mathscr{B}$, then $\mu = \mu_1 * \mu_2$ is defined by

$$\mu(f) = \int_{\mathbf{R}^n} \int_{\mathbf{R}^n} f(x + y)\, d\mu_1(x)\, d\mu_2(y).$$

We have $\mu_1 * \mu_2 = \mu_2 * \mu_1$ and $\|\mu\| \leq \|\mu_1\| \|\mu_2\|$. The operation of convolution when one of the factors is restricted to $L^1(\mathbf{R}^n)$ has its range also in $L^1(\mathbf{R}^n)$. Hence if $f \in L^1(\mathbf{R}^n)$, $g = f * \mu = \int_{\mathbf{R}^n} f(x - y)\, d\mu(y)$ converges absolutely for almost every x, and $g \in L^1(\mathbf{R}^n)$ with

$$\|g\|_1 \leq \|f\|_1 \|\mu\|.$$

Similarly if $f \in L^p(\mathbf{R}^n)$ then $\int_{\mathbf{R}^n} f(x - y)\, d\mu(y)$ is also in L^p and $\|g\|_p \leq \|f\|_p \|\mu\|$. It is to be noted that the transformation

$$f \to \int f(x - y)\, d\mu(y),$$

with $\mu \in \mathscr{B}$, which we have just asserted is bounded in $L^p(\mathbf{R}^n)$, also commutes with translations, $x \to x + h$. This class of transformations is characterized in the following theorem.

1.2 PROPOSITION 1. *Let T be a bounded linear transformation mapping $L^1(\mathbf{R}^n)$ to itself. Then a necessary and sufficient condition that T commutes with translations is that there exists a measure μ in $\mathscr{B}(\mathbf{R}^n)$ so that $T(f) = f * \mu$, for all $f \in L^1(\mathbf{R}^n)$. One has then $\|T\| = \|\mu\|$.*

1.3 For each measure $\mu \in \mathscr{B}(\mathbf{R}^n)$ we can define its Fourier transform $\hat{\mu}(y)$ by

$$\hat{\mu}(y) = \int_{\mathbf{R}^n} e^{2\pi i x \cdot y} \, d\mu(x).^*$$

In particular, the Fourier transform is defined for all $f \in L^1(\mathbf{R}^n)$, with $\hat{f}(y) = \int_{\mathbf{R}^n} e^{2\pi i x \cdot y} f(x) \, dx \in C_0(\mathbf{R}^n)$. The Fourier transform has the fundamental property that if $\mu = \mu_1 * \mu_2$, then $\hat{\mu}(y) = \hat{\mu}_1(y)\hat{\mu}_2(y)$.

If $f \in L^1(\mathbf{R}^n) \cap L^2(\mathbf{R}^n)$, then $\hat{f} \in L^2(\mathbf{R}^n)$ and $\|\hat{f}\|_2 = \|f\|_2$. The Fourier transform can then be extended to all of $L^2(\mathbf{R}^n)$ by continuity, so that it is unitary on $L^2(\mathbf{R}^n)$. By continuity we also have that if $g = f * \mu$, with $f \in L^2(\mathbf{R}^n)$ and $\mu \in \mathscr{B}(\mathbf{R}^n)$, then $\hat{g}(y) = \hat{f}(y)\hat{\mu}(y)$.

1.4 The L^2 analogue of Proposition 1 is the following theorem.

PROPOSITION 2. *Let T be a bounded linear transformation mapping $L^2(\mathbf{R}^n)$ to itself. Then a necessary and sufficient condition that T commutes with translation is that there exists a bounded measurable function $m(y)$ (a "multiplier") so that $(T\hat{f})(y) = m(y)\hat{f}(y)$, for all $f \in L^2(\mathbf{R}^n)$. One has then $\|T\| = \|m\|_\infty$.*

Notice that in the special case where T is also bounded on $L^1(\mathbf{R}^n)$, then $m(y) = \hat{\mu}(y)$, where $Tf = f * \mu$.

2. Singular integrals: the heart of the matter

2.1 The two propositions above show that the structure of translation invariant transformations that are bounded on L^1 or L^2 is both simple and well understood. The study of translation invariant operators that are bounded on some L^p, $p \neq 2$, but not all p, is both more arduous and still unfinished. However, for an important class of transformations much has been done. This class consists of convolution operators with a singular kernel, each having its only singularities at a finite point (the origin), and at infinity. The analogous study of kernels whose singularities are situated

---

* The conventional choice of the factor $+2\pi$ in the exponential should be taken into account when comparing formulae which occur below with those in other texts.

on varieties more general than isolated points is an important problem which it seems must be left to a future theory.

The theorem below represents the essence of the main result. It is stated, however, with somewhat less generality than the fuller development in §3. We have here set ourselves this less exacting task, since it will facilitate the understanding of the main ideas of the theory.

2.2 THEOREM 1. *Let $K \in L^2(\mathbf{R}^n)$. We suppose:*
(a) *The Fourier transform of K is essentially bounded*

$$(1) \qquad\qquad |\hat{K}(x)| \leq B.$$

(b) *K is of class C^1 outside the origin and*

$$(2) \qquad\qquad |\nabla K(x)| \leq B/|x|^{n+1}.$$
For $f \in L^1 \cap L^p$, let us set

$$(3) \qquad\qquad (Tf)(x) = \int_{\mathbf{R}^n} K(x - y)f(y)\, dy.$$

Then there exists a constant A_p, so that

$$(4) \qquad\qquad \|T(f)\|_p \leq A_p\|f\|_p, \qquad 1 < p < \infty.$$

One can thus extend T to all of L^p by continuity. The constant A_p depends only on p, B, and the dimension n. In particular, it does not depend on the L^2 norm of K.

It is important to make the following remark. The assumption that $K \in L^2$ is made for the purpose of having a direct definition of Tf on a dense subset of L^p (in this case $L^1 \cap L^p$), and it could be replaced by other assumptions (such as $K \in L^1 + L^2$).

In the applications the hypothesis $K \in L^2$ is of no consequence since it can be dispensed with by appropriate limiting process; and this is because the final bounds in Theorem 1 do not depend on the L^2 norm of K. See Theorem 2 in §3.2 below.

2.3 PROOF: FIRST STEP. *T is of weak type (2, 2).*

If we use the Fourier transform we see that $(\widehat{Tf})(y) = \hat{K}(y)\hat{f}(y)$, for $f \in L^1 \cap L^2$, and so by assumption (a) and the Plancherel theorem,

$$(5) \qquad\qquad \|T(f)\|_2 \leq B\, \|f\|_2.$$

Because of (5) T has unique extension to all of L^2, where (5) is still valid. So by the remarks in §4.1 of Chapter I we obtain

$$(6) \qquad m\{x : |Tf(x)| > \alpha\} \leq (B^2/\alpha^2)\int_{\mathbf{R}^n} |f|^2\, dx, \qquad f \in L^2(\mathbf{R}^n).$$

2.4 SECOND STEP: *T is of weak type* (1, 1).

We treat Tf, for $f \in L^1(\mathbf{R}^n)$, by decomposing f, as $f = g + b$, where g is "good" and is equal to f on the set where f is essentially small; b is the "bad" part, carried on the set where f is essentially large. The good part g turns out to be in $L^2(\mathbf{R}^n)$, and the L^2 result above, (6), then gives an appropriate estimate for $T(g)$. One can already perceive the idea that is used for dealing with the large part b, when one considers a portion of the Hilbert transform. Thus in the integral

$$(7) \qquad\qquad \int_{-L}^{L} \frac{b(y)\, dy}{x - y}$$

the principal obstacle to an elementary (but favorable) estimate is the appearance of the logarithm when one integrates $1/x$, that is the fact that $\int_{h}^{L} \frac{dx}{x} \sim \log 1/h$, when $h \to 0$. The idea is to replace (7) by

$$(8) \qquad\qquad \int_{-L}^{L} \left[\frac{1}{x - y} - \frac{1}{x} \right] b(y)\, dy,$$

which we may if $\int_{-L}^{L} b(y)\, dy = 0$. Observe that $\left| \dfrac{1}{x - y} - \dfrac{1}{x} \right| \sim \dfrac{L}{x^2}$, when x is clearly separated from the interval $[-L, L]$, (say, for example, if $|x| \geq 2L$), and on the other hand $L \displaystyle\int_{|x| \geq 2L} \frac{dx}{x^2} \leq 1$.

In this way one can avoid the difficulty of the logarithm if the integrals of b on suitable intervals (cubes in the case of $\mathbf{R}^n$) are zero. This is the property of b expressed in (11) below.

Once we have replaced (7) by (8), we must add the analogues of (8), taken for each cube that occurs. The resulting sum turns out to be majorized by the integral of Marcinkiewicz concerning the distance function ((14) of Chapter I), and an appeal to the lemma of §2.3 in that chapter will then supply the necessary information to complete our estimates.

2.4.1 We turn to the details. We need to find a constant C so that*

$$(9) \qquad m\{x : (Tf(x)) > \alpha\} \leq \frac{C}{\alpha} \int_{\mathbf{R}^n} |f(x)|\, dx; \qquad f \in L(\mathbf{R}^n).$$

---

* In order to facilitate the writing of inequalities we shall adopt the following convention that until the end of this chapter, C will denote a general constant (not necessarily the same at different occurrences) but which depends only on the constant B of the hypothesis of the theorem and the dimension n.

To do this fix α, and for this α and $|f(x)|$ apply the corollary of Theorem 4 in Chapter I, §3.4. Then we have $\mathbf{R}^n = F \cup \Omega$, $F \cap \Omega = \emptyset$; $|f(x)| \leq \alpha$, $x \in F$; $\Omega = \bigcup_{j=1}^{\infty} Q_j$, with the interiors of the Q_j mutually disjoint;

$$m(\Omega) \leq \frac{C}{\alpha} \int_{\mathbf{R}^n} |f|\, dx, \text{ and } \frac{1}{m(Q_j)} \int_{Q_j} |f(x)|\, dx \leq C\alpha.$$

We therefore set

(10)
$$g(x) = \begin{cases} f(x), & \text{for } x \in F \\ \dfrac{1}{m(Q_j)} \displaystyle\int_{Q_j} f(x)\, dx, & \text{for } x \in Q_j^0 \end{cases}$$

which defines $g(x)$ almost everywhere. This and the fact that $f(x) = g(x) + b(x)$ has as a consequence

(11)
$$b(x) = 0, \qquad \text{for } x \in F$$
$$\int_{Q_j} b(x)\, dx = 0, \quad \text{for each cube } Q_j.$$

Now since $Tf = Tg + Tb$, it follows then that

$$m\{x : |Tf(x)| > \alpha\} \leq m\{x : |Tg(x)| > \alpha/2\} + m\{x : |Tb(x)| > \alpha/2\}$$

and it suffices to establish separately for both terms of the right-side inequalities analogous to our desired inequality (9).

2.4.2 Estimate for Tg: One has $g \in L^2(\mathbf{R}^n)$, because by (10)

$$\|g\|_2^2 = \int_{\mathbf{R}^n} |g(x)|^2\, dx = \int_F |g(x)|^2\, dx + \int_\Omega |g(x)|^2\, dx$$
$$\leq \int_F \alpha\, |f(x)|\, dx + C^2\alpha^2 m(Q)$$
$$\leq (C^2 A + 1)\alpha \, \|f\|_1.$$

If we now apply inequality (6) of the L^2 theory to g, we obtain

(12)
$$m\{x : |Tg(x)| > \alpha/2\} \leq \frac{C}{\alpha} \|f\|_1.$$

2.4.3 Estimate for Tb: Let us write

$$b_j(x) = \begin{cases} b(x), & x \in Q_j \\ 0, & x \notin Q_j \end{cases}.$$

Then $b(x) = \sum_j b_j(x)$, and $(Tb)(x) = \sum_j (Tb_j)(x)$, with

(13) $$Tb_j(x) = \int_{Q_j} K(x - y)b_j(y)\,dy.$$

We shall be able to obtain a favorable estimate of (13) when $x \in F$ (= the complement of $\bigcup_j Q_j$). First,

$$Tb_j(x) = \int_{Q_j} [K(x - y) - K(x - y^j)]b_j(y)\,dy,$$

with y^j the center of the cube Q_j, since $\int_{Q_j} b_j(y)\,dy = 0$. Because $|\nabla K| \le B\,|x|^{-n-1}$, it follows that $|K(x - y) - K(x - y^j)| \le C\dfrac{\text{diameter }(Q_j)}{|x - \bar{y}^j|^{n+1}}$ where $\bar{y}^j$ is a (variable) point on the straight-line segment connecting y^j with $y\ (\in Q_j)$.

Next we use the remark made in §3.4 of Chapter I that the diameter of Q_j is comparable to its distance from F. This means that if x is a fixed point in F the set of distances $\{|x - y|\}$, as y varies over Q_j, are all comparable with each other. Hence

$$|Tb_j(x)| \le C \text{ diameter }(Q_j) \int_{Q_j} \frac{|b(y)|\,dy}{|x - y|^{n+1}}.$$

However $\int_{Q_j} |b(y)|\,dy \le \int_{Q_j} |f(y)|\,dy + C\alpha \int_{Q_j} dy$, so $\int_{Q_j} |b(y)|\,dy \le (1 + C)\alpha m(Q_j)$. This has the following consequence. If $\delta(y)$ denotes the distance of y from F, because $\text{diameter}(Q_j)m(Q_j) \le C\int_{Q_j} \delta(y)\,dy$, then

$$|Tb_j(x)| \le C\alpha \int_{Q_j} \frac{\delta(y)}{|x - y|^{n+1}}\,dy, \qquad x \in F.$$

Finally,

(14) $$|(Tb)(x)| \le C\alpha \int_{\mathbb{R}^n} \frac{\delta(y)}{|x - y|^{n+1}}\,dy, \qquad x \in F.$$

This majorization of T in terms of the Marcinkiewicz integral is what we promised earlier. The rest is now relatively easy.

Using the lemma in §2.3 we are led to the fact that

(15) $$\int_F |Tb(x)|\,dx \le C\alpha m(\Omega) \le C\,\|f\|_1.$$

From the inequality it follows immediately that

(16) $$m\{x \in F : |Tb(x)| > \alpha/2\} \le \frac{2C}{\alpha}\,\|f\|_1.$$

Since, however, $m(^cF) = m(\Omega) \leq \dfrac{C}{\alpha} \|f\|_1$, we have obtained the estimate for Tb, that is

$$m\{x : |Tb(x)| > \alpha/2\} \leq \frac{C}{\alpha} \|f\|_1.$$

If we combine it with (12), which is the similar result for Tg, we get (9), that is T is of weak-type $(1, 1)$.

2.5 FINAL STEP: THE L^p INEQUALITIES.

(a) For $p = 2$, see §2.3.

(b) For $1 < p < 2$, it suffices to verify the hypotheses of the interpolation theorem (§4.2 in Chapter I), for the case $r = 2$. T is well defined for $L^1(\mathbf{R}^n) + L^2(\mathbf{R}^n)$ and is also linear. It is of weak-type $(1, 1)$ by §2.4 and of weak-type $(2, 2)$ by §2.3, with bounds that depend only on B and the dimension n; (B appears in the hypothesis of the present theorem). Thus the interpolation theorem shows that

$$\|T(f)\|_p \leq A_p \|f\|_p, \qquad 1 < p < 2, \qquad f \in L^p,$$

where A_p depends only on B, p, and n.

(c) For $2 < p < \infty$, we will exploit the duality between L^p and L^q, $1/p + 1/q = 1$, and the fact that the theorem is proved for L^q. Observe the following: if a function ψ is locally integrable and if $\sup |\int \psi\varphi\, dx| = A < \infty$, where the sup is taken over all continuous φ with compact support which verify $\|\varphi\|_q \leq 1$, then $\psi \in L^p$ and $\|\psi\|_p = A$.

This being so, take $f \in L^1 \cap L^p$, $(2 < p < \infty)$, and φ of the type described above. Since $K \in L^2$, and because of our choice of f and φ, the double integral

$$\int_{\mathbf{R}^n}\int_{\mathbf{R}^n} K(x - y)f(y)\varphi(x)\, dx\, dy$$

converges absolutely; its value is therefore

$$I = \int_{\mathbf{R}^n} f(y)\left(\int_{\mathbf{R}^n} K(x - y)\varphi(x)\, dx\right) dy.$$

But the theorem is valid for $1 < q < 2$ (with the kernel $K(-x)$ instead of $K(x)$, but with the same constant A_q). Therefore $\int_{\mathbf{R}^n} K(x - y)\varphi(x)\, dx$ belongs to L^q, and its L^q norm is majorized by $A_q \|\varphi\|_q = A_q$. Hölder's inequality then shows that $|\int_{\mathbf{R}^n} (Tf)\varphi\, dx| = |I| \leq A_q \|f\|_p$, and taking the supremum of all the φ's indicated above gives the result that

$$\|Tf\|_p \leq A_q \|f\|_p, \qquad 2 < p < \infty.$$

We have, therefore, completed the proof of the theorem.

3. Singular integrals: some extensions and variants of the preceding

3.1   The hypotheses of Theorem 1 were of two different kinds. One dealt with the L^2 theory, that is hypothesis (a); it was already formulated in the most general, but not the most useful, way. The second hypothesis, (b), which is used to deal with the weak-type (1, 1) estimate may be improved somewhat. The interest in formulating this improvement is that it represents what seems to be essentially the weakest condition for which the type of argument used in Theorem 1 holds. The condition is the following

$$(2') \qquad \int_{|x| \geq 2|y|} |K(x - y) - K(x)| \, dx \leq B', \qquad |y| > 0$$

That condition (2) implies (2′) is a simple matter that the reader may verify without difficulty. We have the following corollary of the method of proof of Theorem 1.

COROLLARY.   *The results of Theorem 1 hold with condition (b) (equation (2)) replaced by (2′) above, and with the bound B′ replacing the bound B.*

The argument is the same as that of Theorem 1, except that in the proof of the weak-type (1, 1) inequality certain changes are made which we shall now indicate.

In this variant of the argument we shall not use the observation that diameters of the cubes Q_j are comparable with their distances from $F = \left(\bigcup_j Q_j \right)^c$, (if only because we want to show that this fact is not really necessary here!). We get around this point by the simple device of considering for each cube Q_j the cube Q_j^* which has the same center y^j, but which is expanded $2n^{\frac{1}{2}}$ times. We have:

(i) $Q_j \subset Q_j^*$; if $\Omega^* = \cup Q_j^*$, $\Omega \subset \Omega^*$, and $m(\Omega^*) \leq (2n^{\frac{1}{2}})^n m(\Omega)$; if $F^* = {}^c\Omega^*$, then $F^* \subset F$.

(ii) If $x \notin Q_j^*$, then $|x - y^j| \geq 2|y - y^j|$ for all $y \in Q_j$, as an obvious geometric consideration shows.

The other difference is that we do not majorize $|Tb(x)|$ by the distance integral, but we estimate it directly; as a consequence a favorable estimate is obtained on the set F^*, instead of F.

As in the theorem,

$$Tb_j(x) = \int_{Q_j} [K(x - y) - K(x - y^j)] b_j(y) \, dy$$

we get

$$\int_{F^*} |Tb(x)| \, dx \leq \sum_j \int_{x \notin Q_j^*} \int_{y \in Q_j} |K(x-y) - K(x-y^j)| \, |b(y)| \, dy \, dx.$$

However by (ii) for $y \in Q_j$,

$$\int_{x \in Q_j^*} |K(x-y) - K(x-y^j)| \, dx \leq \int_{|x'| \geq 2|v'|} |K(x'-y') - K(x')| \, dx' \leq B'$$

if we invoke the hypothesis. So

(17) $$\int_{F^*} |Tb(x)| \, dx \leq B' \sum_j \int_{Q_j} |b(y)| \, dy \leq C \, \|f\|_1.$$

This brings us back to (15) in the proof of Theorem 1, and the rest is then as before.

3.2 There is still an element which may be considered unsatisfactory in our formulation, and this is because of the following related points: The L^2 boundedness of T has been assumed and not obtained as a consequence of some condition on the kernel K; an extraneous condition such as $K \in L^2$ subsists in the hypothesis; and for this reason our results do not directly treat the "principal-value" singular integrals, those which exist because of the cancellation of positive and negative values. However, from what we have done it is now a relatively simple matter to obtain the following theorem which covers the cases of interest.

THEOREM 2. *Suppose the kernel $K(x)$ satisfies the conditions*

(18)
$$|K(x)| \leq B |x|^{-n}, \qquad\qquad 0 < |x|$$
$$\int_{|x| \geq 2|v|} |K(x-y) - K(x)| \, dx \leq B, \qquad 0 < |y|$$

and

(19) $$\int_{R_1 < |x| < R_2} K(x) \, dx = 0, \qquad 0 < R_1 < R_2 < \infty.$$

For $f \in L^p(\mathbf{R}^n)$, $1 < p < \infty$, let

(20) $$T_\varepsilon(f)(x) = \int_{|y| \geq \varepsilon} f(x-y) K(y) \, dy, \qquad \varepsilon > 0.$$

Then

(21) $$\|T_\varepsilon(f)\|_p \leq A_p \, \|f\|_p$$

with A_p independent of f and ε. Also for each $f \in L^p(\mathbf{R}^n)$, $\lim_{\varepsilon \to 0} T_\varepsilon(f) = T(f)$ exists in L^p norm. The operator T so defined also satisfies the inequality (21).

The cancellation property alluded to is contained in condition (19). This hypothesis, together with (18), allows us to prove the L^2 boundedness and from this the L^p convergence of the truncated integrals (21).

3.3   For the L^2 boundedness we have the following lemma.

LEMMA.   *Suppose K satisfies the conditions of the above theorem with bound B.*

Let $K_\varepsilon(x) = \begin{cases} K(x) & \text{if } |x| \geq \varepsilon \\ 0 & \text{if } |x| < \varepsilon \end{cases}$. *Then obviously* $K_\varepsilon \in L^2(\mathbf{R}^n)$; *for the Fourier transforms we have the estimates*

$$(22) \qquad\qquad \sup_y |\hat{K}_\varepsilon(y)| \leq CB, \qquad \varepsilon > 0$$

where C depends only on the dimension n.

We prove the inequality (22) first for the special case $\varepsilon = 1$.

Observe, and this requires only a semi-trivial estimate, that $K_1(x)$ satisfies the same conditions (18) and (19) as $K(x)$, except that the bound B must be replaced by CB, where C depends only on the dimension n.
Next,

$$\hat{K}_1(y) = \lim_{R \to \infty} \int_{|x| \leq R} e^{2\pi i x \cdot y} K_1(x)\, dx$$

$$= \int_{|x| \leq 1/|y|} e^{2 i \pi x \cdot y} K_1(x)\, dx + \lim_{R \to \infty} \int_{1/|y| \leq |x| \leq R} e^{2\pi i x \cdot y} K_1(x)\, dx$$

$$= I_1 + I_2.$$

However, $\int_{|x| \leq 1/|y|} e^{2\pi i x \cdot y} K_1(x)\, dx = \int_{|x| \leq 1/|y|} [e^{2\pi i x \cdot y} - 1] K_1(x)\, dx$, because K_1 satisfies condition (19). Hence $|I_1| \leq C |y| \int_{|x| \leq 1/|y|} |x|\, |K_1(x)|\, dx \leq C'B$, in view of (18).

To estimate I_2 choose $z = z(y)$ so that $e^{2\pi i y \cdot z} = -1$. This choice can be realized if $z = \dfrac{1}{2} \dfrac{y}{|y|^2}$, with $|z| = \dfrac{1}{2|y|}$. But $\int_{\mathbf{R}^n} K_1(x) e^{2\pi i y \cdot z} dx = \frac{1}{2} \int_{\mathbf{R}^n} [K_1(x) - K_1(x - z)] e^{2\pi i z \cdot y} dx$, so

$$\lim_{R \to \infty} \int_{1/|y| < |x| \leq R} K_1(x) e^{2\pi i x \cdot y}\, dx$$

$$= \frac{1}{2} \lim_{R \to \infty} \int_{1/|y| \leq |x| \leq R} [K_1(x) - K_1(x - z)] e^{2\pi i x \cdot y}\, dx$$

$$- \frac{1}{2} \int_{\left\{ \substack{1/|y| \leq |x + z| \\ |x| \leq 1/|y|} \right\}} K_1(x) e^{2\pi i x \cdot y}\, dx.$$

The last integral is taken over a region contained in the spherical shell, $\frac{1}{2}|y| \leq |x| \leq 1/|y|$, and is bounded since $|K_1(x)| \leq B|x|^{-n}$. The first integral on the right hand side is majorized by $\frac{1}{2}\int_{|x| \geq 1/|y|} |K_1(x - z) - K_1(x)|\, dx$. But since $|z| = (2|y|)^{-1}$, the condition analogous to (18) applied to K_1 shows this integral is also bounded by CB. If we add the bounds for I_1 and I_2 we obtain the proof of our lemma for K_1. To pass to the case of general K_ε we use a simple observation whose significance carries over to the whole theory presented in this chapter.

Let τ_ε be the dilation by the factor ε, $\varepsilon > 0$, that is $(\tau_\varepsilon f)(x) = f(\varepsilon x)$. Thus if T is a convolution operator $T(f) = \varphi * f = \int_{\mathbf{R}^n} \varphi(x - y)f(y)\, dy$, then $\tau_{\varepsilon^{-1}} T \tau_\varepsilon$ is the convolution operator with the kernel φ_ε, where $\varphi_\varepsilon(x) = \varepsilon^{-n}\varphi(\varepsilon^{-1}x)$. In our case if T corresponds to the kernel $K(x)$, $\tau_{\varepsilon^{-1}} T \tau_\varepsilon$ corresponds to the kernel $\varepsilon^{-n}K(\varepsilon^{-1}x)$. Notice that if K satisfies the assumptions of our theorem, then $\varepsilon^{-n}K(\varepsilon^{-1}x)$ also satisfies these assumptions, and with the *same bounds*. (A similar remark holds for the assumptions of all the theorems in this chapter.) Now with our K given, let $K' = \varepsilon^n K(\varepsilon x)$. Then K' satisfies the conditions of our lemma with the same bound B, and so if $K_1' = \begin{cases} K_1'(x) & \text{if } |x| > 1 \\ 0 & \text{if } |x| < 1 \end{cases}$, then we know that $|\widehat{K_1'(y)}| \leq CB$. The Fourier transform of $\varepsilon^{-n}K_1'(\varepsilon^{-1}x)$ is $\widehat{K_1'(\varepsilon y)}$ and this is again bounded by CB; however $\varepsilon^{-n}K_1'(\varepsilon^{-1}x) = K_\varepsilon(x)$, therefore the lemma is completely proved.

3.4 We can now prove Theorem 2. Since K satisfies the conditions (18) and (19), then $K_\varepsilon(x)$ satisfies the same conditions with bounds not greater than CB. We pointed this out for K_1 in the proof of the lemma, and the passage from K_1 to K_ε follows by the dilation argument also given there. Clearly however, each $K_\varepsilon \in L^2(\mathbf{R}^n)$, $\varepsilon > 0$. So an application of the corollary in §3.1 proves (21); that is, the L^p norms of the operators are uniformly bounded. Next suppose f_1 is a continuous function with compact support which has one continuous derivative. Then

$$T_\varepsilon(f_1)(x) = \int_{|y| \geq \varepsilon} K(y)f_1(x - y)\, dy$$

$$= \int_{|y| \geq 1} K(y)f_1(x - y)\, dy + \int_{1 \geq |y| \geq \varepsilon} K(y)[f_1(x - y) - f_1(x)]\, dy,$$

because of the cancellation condition (19). The first integral represents an L^p function since it is the convolution of an L^1 function, f_1, with an L^p function $K(y)$, since $|K(y)| \leq B|y|^{-n}$, if $|y| \geq 1$. The second integral is supported in a fixed compact set of x, and converges uniformly in x as $\varepsilon \to 0$ because $|f_1(x - y) - f_1(x)| \leq A|y|$, in view of the differentiability of f_1. To summarize, $T_\varepsilon(f_1)$ converges in L^p norm, if $\varepsilon \to 0$.

Finally an arbitrary $f \in L^p$ can be written as $f = f_1 + f_2$ where f_1 is of the type described above and $\|f_2\|_p$ is small. If we apply the basic inequality (21) for f_2 in place of f we see that $\lim_{\varepsilon \to 0} T_\varepsilon(f)$ exists in L^p norm; that the limiting operator T also satisfies the inequality (21) is then obvious, and this completes the proof of the theorem.

We should point out that the kernel $K(x) = \dfrac{1}{\pi x}$, $x \in \mathbf{R}^1$, clearly satisfies the hypotheses of Theorem 2. We have therefore proved the existence of the Hilbert transform in the sense that if $f \in L^p(\mathbf{R}^1)$, $1 < p < \infty$, then

$$\lim_{\varepsilon \to 0} \frac{1}{\pi} \int_{|y| \geq \varepsilon} \frac{f(x - y)}{y}\, dy$$

exists in the L^p norm and the resulting operator is bounded in L^p. A closer study of this example and certain important n-dimensional analogues will be taken up in the following section.

4. *Singular integral operators which commute with dilations*

4.1 A basic class of operators in any abelian group is the set of operators that commute with (group) translations. However $\mathbf{R}^n$ is not only an abelian group, and so besides translations possesses certain other distinguished groups of transformations that act on it and which are related to its Euclidean structure. The transformations we have in mind here are the dilations $\tau_\varepsilon : x \to \varepsilon x$, $\varepsilon > 0$ and their corresponding action on functions $\tau_\varepsilon f(x) = f(\varepsilon x)$, discussed before.

We shall be interested in those operators which not only commute with translations but also with dilations. Among these we shall study the class of singular integral operators, falling under the scope of Theorem 2.

If T corresponds to the kernel $K(x)$, then as we have already pointed out, $\tau_{\varepsilon^{-1}} T \tau_\varepsilon$ corresponds to the kernel $\varepsilon^{-n} K(\varepsilon^{-1} x)$. So if $\tau_{\varepsilon^{-1}} T \tau_\varepsilon = T$ we are back to the requirement $K(\varepsilon x) = \varepsilon^{-n} K(x)$, $\varepsilon > 0$; that is, K is homogeneous of degree $-n$. Put another way

$$(23) \qquad\qquad K(x) = \frac{\Omega(x)}{|x|^n},$$

with Ω homogeneous of degree 0, i.e. $\Omega(\varepsilon x) = \Omega(x)$, $\varepsilon > 0$. This condition on Ω is equivalent with the fact that it is constant on rays emanating from the origin; Ω is, in particular, completely determined by its restriction to the unit sphere S^{n-1}. Let us try to reinterpret the conditions of Theorem 2 in terms of Ω. First, by (18), $\Omega(x)$ must be bounded, and consequently integrable on S^{n-1}. The cancellation condition (19) is then the same as the condition

(24) $$\int_{S^{n-1}} \Omega(x)\, d\sigma = 0$$

where $d\sigma$ is the induced Euclidean measure on S^{n-1}. The precise condition (18) is not easily restated in terms of Ω; what is evident, however, is that it requires a certain continuity of Ω. Here we shall content ourselves in treating the case where Ω satisfies the following "Dini-type" condition suggested by (18):

(25)   if $\displaystyle\sup_{\substack{|x-x'|\leq\delta \\ |x|=|x'|=1}} |\Omega(x) - \Omega(x')| = \omega(\delta)$,   then $\displaystyle\int_0^1 \frac{\omega(\delta)\, d\delta}{\delta} < \infty$.

Of course any Ω which is of class C^1, or even merely Lipschitz continuous, satisfies the condition (25)*.

4.2 THEOREM 3.   *Let Ω be homogeneous of degree 0, and suppose that Ω satisfies the cancellation property (24), and the smoothness property (25) above. For $1 < p < \infty$, and $f \in L^p(\mathbf{R}^n)$ let*

$$T_\varepsilon(f)(x) = \int_{|y|\geq\varepsilon} \frac{\Omega(y)}{|y|^n} f(x - y)\, dy.$$

(a) *Then there exists a bound A_p (independent of f or ε) so that*

$$\|T_\varepsilon(f)\|_p \leq A_p \|f\|_p.$$

(b) $\displaystyle\lim_{\varepsilon\to 0} T_\varepsilon(f) = T(f)$ *exists in L^p norm, and*

$$\|T(f)\|_p \leq A_p \|f\|_p.$$

(c) *If $f \in L^2(\mathbf{R}^n)$, then the Fourier transforms of f and $T(f)$ are related by*

$(Tf)\hat{\ }(x) = m(x)\hat{f}(x)$, *where m is a homogeneous function of degree 0. Explicitly,*

(26)
$$m(x) = \int_{S^{n-1}} \left[\frac{\pi i}{2} \operatorname{sign}(x \cdot y) + \log(1/|x \cdot y|)\right]\Omega(y)\, d\sigma(y), \qquad |x| = 1.$$

The conclusions (a) and (b) of the theorem are immediate consequences of Theorem 2, once we have shown that any $K(x)$ of the form $\dfrac{\Omega(x)}{|x|^n}$ satisfies $\int_{|x|\geq 2|y|} |K(x - y) - K(x)|\, dx \leq B$, if Ω is as in condition (25). However

$$K(x - y) - K(x) = \left(\frac{\Omega(x - y) - \Omega(x)}{|x - y|^n}\right) + \Omega(x)\left[\frac{1}{|x - y|^n} - \frac{1}{|x|^n}\right].$$

* For a generalization of this condition see §6.10, at the end of this chapter.

The second group of terms satisfies the correct estimate, since

$$\int_{|x|\geq 2|y|}\left|\frac{1}{|x-y|^n}-\frac{1}{|x|^n}\right|dx \leq C$$

and Ω is bounded. To estimate the first group of terms, we notice that the distance between the projections of $x-y$ and x on the unit sphere,

$\left|\dfrac{x-y}{|x-y|}-\dfrac{x}{|x|}\right|$, is bounded by $C\left|\dfrac{y}{x}\right|$, if $|x|\geq 2\,|y|$. So the integral

corresponding to the first group of terms is dominated by

$$C'\int_{|x|\geq 2|y|}\omega\left(C\frac{|y|}{|x|}\right)\frac{dx}{|x|^n}=C''\int_0^{c/2}\frac{\omega(\delta)}{\delta}\,d\delta < \infty.$$

4.3   Since T is a bounded operator on L^2 which commutes with translations, we know by the proposition in §1.4 that T can be realized in terms of a multiplier m, so that $(Tf)^{\wedge}$ is obtained by multiplying $\hat{f}$ by m. For such operators, the fact that they commute with dilations is equivalent with the property that the multiplier is homogeneous of degree 0. For our particular operators we have not only the existence of m but an explicit expression of the multiplier in terms of the kernel. This formula is deduced as follows. Let

$$0<\varepsilon<\eta<\infty, \quad\text{and}\quad K_{\varepsilon,\eta}(x)=\begin{cases}\dfrac{\Omega(x)}{|x|^n}, & \text{if } \varepsilon\leq|x|\leq\eta \\ \\ 0 & \text{otherwise}\end{cases}$$

Clearly $K_{\varepsilon,\eta}\in L^1(\mathbf{R}^n)$. If $f\in L^2(\mathbf{R}^n)$ then $(K_{\varepsilon,\eta}*f)^{\wedge}=K_{\varepsilon,\eta}^{\wedge}(y)\hat{f}(y)$.

We shall prove two facts about $K_{\varepsilon,\eta}^{\wedge}(y)$.
(i) $\sup|K_{\varepsilon,\eta}^{\wedge}(x)|\leq A$, with A independent of ε and η,
(ii) if $x\neq 0$, $\lim\limits_{\substack{\varepsilon\to 0 \\ \eta\to 0}}K_{\varepsilon,\eta}^{\wedge}(x)=m(x)$, (see (26)).

For this purpose it is convenient to introduce polar coordinates. Let $x=Rx', R=|x|, x'=x/|x|\in S^{n-1}$, and $y=ry', r=|y|, y'=y/|y|\in S^{n-1}$. We shall also need the auxiliary integral

$$I_{\varepsilon,\eta}(x,y')=\int_\varepsilon^\eta[\exp(2\pi i\,Rrx'\cdot y')-\cos(2\pi\,Rr)]\frac{dr}{r}, \quad R>0.$$

Its imaginary part, $\displaystyle\int_\varepsilon^\eta\frac{\sin 2\pi\,Rr(x'\cdot y')}{r}\,dr$, is as an integration by parts

shows, uniformly bounded, and converges to

$$\left(\int_0^\infty\frac{\sin t}{t}\,dt\right)\text{sign}(x'\cdot y')=\frac{\pi}{2}\cdot\text{sign}(x'\cdot y').$$

Its real part is bounded in absolute value by $C \log 1/|x' \cdot y'| + C$, as again an integration by parts shows. Also $\lim_{\substack{\varepsilon \to 0 \\ \eta \to \infty}} Re(I_{\varepsilon,\eta}(x \cdot y') = \log 1/|x' \cdot y'|$, since $\lim_{\substack{\varepsilon \to 0 \\ \eta \to \infty}} \int_{\varepsilon}^{\eta} \frac{h(\lambda r) - h(\mu r)}{r} \, dr = h(0) \log (\mu/\lambda)$, if $\mu, \lambda > 0$, and h is an appropriate function. In this case $h(r) = \cos 2\pi r$, $\lambda = R |x' \cdot y'|$, and $\mu = R$.

Now $K_{\varepsilon,\eta}^{\wedge}(x) = \int_{S^{n-1}} \left(\int_{\varepsilon}^{\eta} e^{2\pi i R r x' \cdot y'} \Omega(y') \frac{dr}{r} \right) d\sigma(y')$. Since

$$\int_{S^{n-1}} \Omega(y) \, d\sigma(y') = 0$$

we can introduce the factor $\cos 2\pi r R$ (which does not depend on y') in the integral defining $K_{\varepsilon,\eta}^{\wedge}(x)$. This gives

$$K_{\varepsilon,\eta}^{\wedge}(x) = \int_{S^{n-1}} I_{\varepsilon,\eta}(x, y') \Omega(y') \, d\sigma(y').$$

Because of the properties of $I_{\varepsilon,\eta}$ just proved

$$|K_{\varepsilon,\eta}^{\wedge}(x)| \leq A \int_{S^{n-1}} [1 + \log 1/|x', y'|] \, |\Omega(y')| \, d\sigma(y')$$

which proves (i) (the uniform boundedness of the $K_{\varepsilon,\eta}^{\wedge}(x)$), since Ω is itself bounded. In view of the limit of $I_{\varepsilon,\eta}(x)$, as $\varepsilon \to 0$, $\eta \to \infty$ just ascertained, and the dominated convergence theorem, we get

$$\lim_{\substack{\varepsilon \to 0 \\ \eta \to \infty}} K_{\varepsilon,\eta}^{\wedge}(x) = m(x),$$

if $x \neq 0$, that is (ii).

By the Plancherel theorem then, if $f \in L^2(\mathbf{R}^n)$, $K_{\varepsilon,\eta} * f$ converges in L^2 norm as $\eta \to \infty$, and $\varepsilon \to 0$, and the Fourier transform of this limit is $m(x)\hat{f}(x)$. However if we keep ε fixed and let $\eta \to \infty$, then clearly $\int K_{\varepsilon,\eta}(y)f(x - y) \, dy$ converges everywhere to $\int_{|y| \leq \varepsilon} K(y)f(x - y) \, dy$, which is $T_\varepsilon(f)$.

Letting now $\varepsilon \to 0$, we obtain the conclusion (c) and our theorem is completely proved.

4.4 It is to be noted that the proof of part (c) holds under very general conditions on Ω. Write $\Omega = \Omega_e + \Omega_o$ where Ω_e is the even part of Ω, $\Omega_e(x) = \Omega_e(-x)$, and $\Omega_o(x)$ is the odd part, $\Omega_o(-x) = -\Omega_o(x)$. Then because of the uniform boundedness of the sine integral we required only $\int_{S^{n-1}} |\Omega_o(y')| \, d\sigma(y') < \infty$, i.e. the integrability of the odd part. For the even part, the proof requires the uniform boundedness of

$$\int_{S^{n-1}} |\Omega_e(y')| \log 1/|x', y'| \, d\sigma(y').$$

This observation is suggestive of certain generalizations of Theorem 2, (see §6.5).

It goes without saying that the transformations described in Theorem 3 are not bounded on either $L^1(\mathbf{R}^n)$ or $L^\infty(\mathbf{R}^n)$. In the case of the Hilbert transform this can be seen immediately by the explicit example of the transform of the characteristic function of the interval (a, b), which has the value $\dfrac{1}{\pi} \log \left| \dfrac{x - b}{x - a} \right|$. Other examples are described in §6.1.

4.5 Theorem 3 guaranteed the existence of the singular integral transformation

$$(27) \qquad\qquad \lim_{\varepsilon \to 0} \int_{|y| \geq \varepsilon} \frac{\Omega(y)}{|y|^n} f(x - y) \, dy$$

in the sense of convergence in the L^p norm. The natural counterpart of this result is that of convergence almost everywhere. In the classical case corresponding to the Hilbert transform the results of almost everywhere convergence predated the L^p results, and the former were obtained as a consequence of Fatou's theorem guaranteeing the boundary values almost everywhere of bounded harmonic functions. In our present situation the almost everywhere results will be, in effect, a consequence of the existence of the limit (27) in the L^p norm, already proved. As in other questions involving almost everywhere convergence, it is best to consider also the corresponding maximal function.

THEOREM 4. *Suppose that Ω satisfies the conditions of the previous theorem. For $f \in L^p(\mathbf{R}^n)$, $1 \leq p < \infty$, consider*

$$T_\varepsilon(f)(x) = \int_{|y| \geq \varepsilon} \frac{\Omega(y)}{|y|^n} f(x - y) \, dy, \qquad \varepsilon > 0.$$

(The integral converges absolutely for every x.)

(a) $\lim\limits_{\varepsilon \to 0} T_\varepsilon(f)(x)$ *exists for almost every x.*

(b) *Let* $T^*(f)(x) = \sup\limits_{\varepsilon > 0} |T_\varepsilon(f)(x)|$. *If* $f \in L^1(\mathbf{R}^n)$, *then the mapping* $f \to T^*f$ *is of weak type* (1, 1).

(c) *If* $1 < p < \infty$, *then* $\|T^*(f)\|_p \leq A_p \|f\|_p$.

4.6 The argument for Theorem 4 presents itself in three stages. At first there is the proof of inequality (c) which can be obtained as a relatively easy consequence of the L^p norm existence of $\lim_{\varepsilon \to 0} T_\varepsilon$, already proved, and certain general properties of "approximations to the identity." We shall therefore postpone the proof of (c) to the next chapter where we deal with these matters more systematically.

4.6.1 This brings us to the second, and most difficult stage of the proof, leading to conclusion (b). Here the argument proceeds in the main as in the proof of the weak-type (1, 1) result for singular integrals, in particular the variant given in §3.1. We review it with deliberate brevity so as to avoid a repetition of details already examined. For a given $\alpha > 0$, we split $f = g + b$ as in §2.4. We also consider for each cube Q_j its mate Q_j^*, which has the same center y^j but which is expanded $2n^{\frac{1}{2}}$ times. The following additional geometric remarks concerning these cubes are nearly obvious.

(iii) Suppose $x \in {}^cQ_j^*$ (in particular this is so if $x \in F^*$) and assume that for some $y \in Q_j$, $|x - y| = \varepsilon$. Then the closed ball centered at x, of radius $\gamma_n \varepsilon$, contains Q_j; i.e. $\overline{B(x, r)} \supset Q_j$, if $r = \gamma_n \varepsilon$.

(iv) Under the same hypotheses as (iii), we have that $|x - y| \geq \gamma_n' \varepsilon$, for every $y \in Q_j$.

γ_n and γ_n' depend only on the dimension n, and not the particular cube Q_j.

4.6.2 With these observations, and following the development in §2.4 we shall prove that if $x \in F^*$,

$$(28) \quad \sup_{\varepsilon > 0} |T_\varepsilon(b(x))| \leq \sum_j \int_{Q_j} |K(x - y) - K(x - y^j)| \, |b(y)| \, dy$$

$$+ C \sup_{r > 0} \frac{1}{m(B(x, r))} \int_{B(x,r)} |b(y)| \, dy,$$

with $K(x) = \dfrac{\Omega(x)}{|x|^n}$.

Thus the addition of the maximal function to the right side of (28) is the main new element of the proof. To prove (28), fix $x \in F^*$, and $\varepsilon > 0$. Now the cubes Q_j fall into three classes:

(a) for all $y \in Q_j$, $|x - y| < \varepsilon$

(b) for all $y \in Q_j$, $|x - y| > \varepsilon$

(c) there is a $y \in Q_j$, so that $|x - y| = \varepsilon$.

We now examine $T_\varepsilon b(x)$.

$$(29) \quad T_\varepsilon b(x) = \sum_j \int_{Q_j} K_\varepsilon(x - y) b(y) \, dy.$$

Case (a). $K_\varepsilon(x - y) = 0$ if $|x - y| < \varepsilon$, and so the integral over the cube Q_j in (29) is zero.

Case (b). $K_\varepsilon(x - y) = K(x - y)$, if $|x - y| > \varepsilon$, and therefore this integral over Q_j equals

$$\int_{Q_j} K(x - y)b(y) \, dy = \int_{Q_j} [K(x - y) - K(x - y^j)]b(y) \, dy.$$

This term is majorized in absolute value by

$$\int_{Q_j} |K(x - y) - K(x - y^j)| \, |b(y)| \, dy,$$

which expression appears in the right side of (28).

Case (c). We write simply

$$\left| \int_{Q_j} K_\varepsilon(x - y)b(y) \, dy \right| \leq \int_{Q_j} |K_\varepsilon(x - y)| \, |b(y)| \, dy$$

$$= \int_{B(x,r) \cap Q_j} |K_\varepsilon(x - y)| \, |b(y)| \, dy,$$

by (iii), with $r = \gamma_n \varepsilon$. However $|K_\varepsilon(x - y)| \leq \left| \dfrac{\Omega(x - y)}{|x - y|^n} \right| \leq \dfrac{B}{(\gamma_n')^n \varepsilon^n}$, by (iv) and the fact that Ω is bounded. So

$$\left| \int_{Q_j} K_\varepsilon(x - y)b(y) \, dy \right| \leq C \frac{1}{m[B(x, r)]} \int_{B(x,r) \cap Q_j} |b(y)| \, dy$$

in this case. If we add over all cubes Q_j we finally obtain

$$|T_\varepsilon b(x)| \leq \sum_j \int_{Q_j} |K(x - y) - K(x - y^j)| \, |b(y)| \, dy$$

$$+ C \frac{1}{B(x, r)} \int_{B(x,r)} |b(y)| \, dy, \qquad r = \gamma_n \varepsilon.$$

The taking of the supremum over ε gives (28).

This inequality can be written in the form

$$|T^* b(x)| \leq \Sigma + CMb(x), \qquad x \in F^*,$$

and so

$$m\{x \in F^* : |T^* b(x)| > \alpha/2\} \leq m\{x \in F^* : \Sigma > \alpha/4\}$$
$$+ m\{x \in F^* : CMb(x) > \alpha/4\}.$$

The measures of both sets appearing in the right-hand side of the just-written inequality are bounded by $\dfrac{c}{\alpha} \|b\|_1$. In the first case this is because an inequality similar to (17) of §3.1 holds for Σ; for the second set it is because of the weak-type estimate for the maximal function M (theorem in §1.3, Chapter I). The weak type $(1, 1)$ property of T^* then follows as in the proof of the same property for T, in §3.1 (or in greater detail in §2.4 following equation (15)).

4.6.3 The final stage of the proof of the theorem, the passage from the inequalities of T^* to the existence of the limits almost everywhere, follows the familiar pattern described in §1.5 of Chapter I. More precisely, for any $f \in L^p(\mathbf{R}^n)$, $1 \leq p < \infty$, let

$$\Lambda(f)(x) = |\limsup_{\varepsilon \to 0} T_\varepsilon(f)(x) - \limsup_{\varepsilon \to 0} T_\varepsilon(f)(x)|.$$

Clearly $\Lambda(f)(x) \leq 2(T^*f)(x)$. Now write $f = f_1 + f_2$ where f_1 has compact support and is of class C^1, and $\|f_2\|_p \leq \delta$. We have already remarked in §3.4 that $T_\varepsilon f_1$ converges uniformly as $\varepsilon \to 0$, so $\Lambda f_1(x) \equiv 0$. But $\Lambda(f)(x) \leq \Lambda(f_1)(x) + \Lambda(f_2)(x)$, so

$$\|\Lambda(f_2)\|_p \leq 2A_p \|f_2\|_p \leq 2A_p \delta, \quad \text{if} \quad 1 < p < \infty.$$

This shows $\Lambda f_2 = 0$, almost everywhere, thus $\Lambda f = 0$ almost everywhere, and so $\lim_{\varepsilon \to 0} T_\varepsilon f$ exists almost everywhere if $1 < p < \infty$. In the case $p = 1$, we get similarly

$$m\{x : \Lambda f(x) > \alpha\} \leq \frac{A}{\alpha} \|f_2\|_1 \leq \frac{A\delta}{\alpha},$$

and so again $\Lambda f(x) = 0$ almost everywhere, which implies that $\lim_{\varepsilon \to 0} T_\varepsilon(f)(x)$ exists almost everywhere.

5. *Vector-valued analogues*

5.1 It is interesting to point out that the results of this chapter, where our functions were assumed to take real or complex values, can be extended to the case of functions taking their values in a Hilbert space. We present this generalization because it can be put to good use in several problems. An indication of this usefulness is given in the Littlewood-Paley theory in Chapter IV.

We begin by reviewing quickly certain aspects of integration theory in this context.

Let $\mathcal{H}$ be a separable Hilbert space. Then a function $f(x)$, from $\mathbf{R}^n$ to $\mathcal{H}$ is *measurable* if the scalar valued functions $(f(x), \varphi)$ are measurable, where $(\cdot, \cdot)$ denotes the inner product of $\mathcal{H}$, and φ denotes an arbitrary vector of $\mathcal{H}$. If $f(x)$ is such a measurable function, then $|f(x)|$ is also measurable (as a function with non-negative values), where $|\cdot|$ denotes the norm of $\mathcal{H}$. Thus $L^p(\mathbf{R}^n, \mathcal{H})$ is defined as the equivalence classes of measurable functions $f(x)$ from $\mathbf{R}^n$ to $\mathcal{H}$, with the property that the norm $\|f\|_p = (\int_{\mathbf{R}^n} |f(x)|^p \, dx)^{1/p}$ is finite, when $p < \infty$; when $p = \infty$ there is a similar definition, except $\|f\|_\infty = \text{ess sup} |f(x)|$.

Next, let $\mathcal{H}_1$ and $\mathcal{H}_2$ be two separable Hilbert spaces, and let $B(\mathcal{H}_1, \mathcal{H}_2)$ denote the Banach space of bounded linear operators from $\mathcal{H}_1$ to $\mathcal{H}_2$, with the usual operator norm. We say that a function $f(x)$, from $\mathbf{R}^n$ to $B(\mathcal{H}_1, \mathcal{H}_2)$ is measurable if $f(x)\varphi$ is an $\mathcal{H}_2$-valued measurable function for every $\varphi \in \mathcal{H}_1$. In this case also $|f(x)|$ is measurable and we can define the space $L^p(\mathbf{R}^n, B(\mathcal{H}_1, \mathcal{H}_2))$, as before; (here again $|\cdot|$ denotes the norm, this time in $B(\mathcal{H}_1, \mathcal{H}_2)$). The usual facts about convolution hold in this setting. For example, suppose $K(x) \in L^q(\mathbf{R}^n, B(\mathcal{H}_1, \mathcal{H}_2))$ and $f(x) \in L^p(\mathbf{R}^n, \mathcal{H}_1)$. Then $g(x) = \int_{\mathbf{R}^n} K(x - y)f(y) \, dy$ converges in the norm of $\mathcal{H}_2$ for almost every x, and

$$|g(x)| \leq \int_{\mathbf{R}^n} |K(x - y)f(y)| \, dy \leq \int_{\mathbf{R}^n} |K(x - y)| \, |f(y)| \, dy.$$

Also $\|g\|_r \leq \|K\|_q \|f\|_p$, if $1/r = 1/p + 1/q - 1$, with $1 \leq r \leq \infty$.

5.2 Suppose that $f(x) \in L^1(\mathbf{R}^n, \mathcal{H})$. Then we can define its Fourier transform $\hat{f}(y) = \int_{\mathbf{R}^n} e^{2\pi i x \cdot y} f(x) \, dx$ which is an element of $L^\infty(\mathbf{R}^n, \mathcal{H})$. If $f \in L^1(\mathbf{R}^n, \mathcal{H}) \cap L^2(\mathbf{R}^n, \mathcal{H})$, then $\hat{f}(y) \in L^2(\mathbf{R}^n, \mathcal{H})$ with $\|\hat{f}\|_2 = \|f\|_2$. The Fourier transform can then be extended by continuity to a unitary mapping of the Hilbert space $L^2(\mathbf{R}^n, \mathcal{H})$ to itself.

These facts can be obtained easily from the scalar-valued case by introducing an arbitrary orthonormal basis in $\mathcal{H}$.

5.3 Now suppose that $\mathcal{H}_1$ and $\mathcal{H}_2$ are two given Hilbert spaces. Assume $f(x)$ takes values in $\mathcal{H}_1$, and $K(x)$ takes values in $B(\mathcal{H}_1, \mathcal{H}_2)$. Then

$$Tf(x) = \int_{\mathbf{R}^n} K(y)f(x - y) \, dy,$$

whenever defined, takes values in $\mathcal{H}_2$.

THEOREM 5. *The results in this chapter, in particular Theorem 1, its corollary, and Theorems 2 to 4 are valid in the more general context where f takes its value in $\mathcal{H}_1$, K takes its values in $B(\mathcal{H}_1, \mathcal{H}_2)$ and (Tf) and $T_\varepsilon(f)$ take their value in $\mathcal{H}_2$, and where throughout the absolute value $|\cdot|$ is replaced by the appropriate norm in $\mathcal{H}_1$, $B(\mathcal{H}_1, \mathcal{H}_2)$ or $\mathcal{H}_2$ respectively.*

This theorem is not in any *obvious* way a corollary of the scalar-valued case treated. Its proof, however, consists in nothing but an identical repetition of the arguments given for the scalar-valued case, if we take into account the remarks made in the above paragraphs. This seemingly bold assertion may, in fact, be verified without difficulty by a patient review of the proofs; but if the reader is not so disposed, he may find the necessary details in the literature cited at the end of the chapter.

Several clarifying observations emerge from such a verification:

(a) The final bounds obtained do not depend on the Hilbert spaces $\mathscr{H}_1$ or $\mathscr{H}_2$, but only on B, p, and n, as in the scalar-valued case.

(b) Most of the argument goes through in the even greater generality of Banach space-valued functions, appropriately defined. The Hilbert space structure is used only in the L^2 theory when applying the variant of Plancherel's formula described in §5.2.

The Hilbert space structure also enters in the following corollary:

COROLLARY. *With the same assumptions as in Theorem 5, if in addition*

$$\|T(f)\|_2 = c\,\|f\|_2, \qquad c > 0, \qquad f \in L^2(\mathbf{R}^n, \mathscr{H}_1)$$

then $\|f\|_p \le A'_p\,\|T(f)\|_p$, *if* $f \in L^p(\mathbf{R}^n, \mathscr{H}_1)$, *if* $1 < p < \infty$.

Proof. We remark that the $L^2(\mathbf{R}^n, \mathscr{H}_j)$ are Hilbert spaces. In fact, let $(,)_j$ denote the inner product of $\mathscr{H}_j$, $j = 1, 2$, and let $\langle,\rangle_j$ denote the corresponding inner product in $L^2(\mathbf{R}^n, \mathscr{H}_j)$; that is

$$\langle f, g \rangle_j = \int_{\mathbf{R}^n} (f(x), g(x))_j\, dx.$$

Now T is a bounded linear transformation from the Hilbert space $L^2(\mathbf{R}^n, \mathscr{H}_1)$ to the Hilbert space $L^2(\mathbf{R}^n, \mathscr{H}_2)$, and so by the general theory of inner products there exists a unique adjoint transformation $\tilde{T}$, from $L^2(\mathbf{R}^n, \mathscr{H}_2)$ to $L^2(\mathbf{R}^n, \mathscr{H}_1)$, which satisfies the characterizing property

$$\langle Tf_1, f_2 \rangle_2 = \langle f_1, \tilde{T}f_2 \rangle_1, \quad \text{with} \quad f_j \in L^2(\mathbf{R}^n, \mathscr{H}_j).$$

But our assumption is equivalent with the identity

$$\langle Tf, Tg \rangle_2 = c^2 \langle f, g \rangle_1, \quad \text{for all} \quad f, g \in L^2(\mathbf{R}^n, \mathscr{H}_1).$$

Thus using the definition of the adjoint, $\langle \tilde{T}Tf, g \rangle_1 = c^2 \langle f, g \rangle_1$, and so the assumption can be restated as

$$(30) \qquad\qquad \tilde{T}Tf = c^2 f, \qquad f \in L^2(\mathbf{R}^n, \mathscr{H}_1).$$

$\tilde{T}$ is again an operator of the same kind as T but it takes function with values in $\mathscr{H}_2$ to functions with values in $\mathscr{H}_1$, and its kernel $\tilde{K}(x)$, is

$\tilde{K}(x) = K^*(-x)$, where here * denotes the adjoint of an element in $B(\mathcal{H}_1, \mathcal{H}_2)$.

This is obvious on the formal level since

$$\langle Tf_1, f_2 \rangle_2 = \int_{\mathbf{R}^n} \int_{\mathbf{R}^n} (K(x - y)f_1(y), f_2(x))_2 \, dy \, dx$$

$$= \int_{\mathbf{R}^n} \int_{\mathbf{R}^n} (f_1(y), K^*(-(y - x))f_2(x))_1 \, dx \, dy$$

$$= \langle f_1, \tilde{T}f_2 \rangle_1.$$

The rigorous justification of this identity is achieved by a simple limiting argument. We will not tire the reader with the routine details.

This being said we have only to add the remark that $K^*(-x)$ satisfies the same conditions as $K(x)$, and so we have for it similar conclusions as for K (with the same bounds). Thus by (30),

$$c^2 \|f\|_p = \|\tilde{T}Tf\|_p \leq A_p \|Tf\|_p.$$

This proves the corollary with $A'_p = A_p/c^2$.

This corollary applies in particular to the singular integrals of §4; then the condition required is that the multiplier $m(x)$ have constant absolute value. This is the case, for example, when T is the Hilbert transform, $K(x) = \dfrac{1}{\pi x}$, and $m(x) = i \operatorname{sign} x$. For a generalization of this remark see §6.6 below.

6. *Further results*

6.1 Let $K(x) = \dfrac{\Omega(x)}{|x|^n}$ be as in Theorem 3, with $\Omega \neq 0$.

(a) If $f \in L^1(\mathbf{R}^n)$, $f \geq 0$, then $Tf \notin L^1(\mathbf{R}^n)$, if $f \not\equiv 0$. *Hint:* $m(x)\hat{f}(x)$ cannot be continuous at 0, since $m(x)$ is homogeneous of degree 0 and non-constant, and $\hat{f}(0) > 0$.

(b) There exists a continuous f, which vanishes outside the unit ball B, such that $T(f)$ is unbounded near every point of B.

6.2 (a) If A_p is the L^p bound for T in Theorem 1, 2, or 3, then $A_p \leq \dfrac{A}{p-1}$ for $1 < p \leq 2$ and $A_p \leq Ap$, for $2 \leq p < \infty$. (See the remark at the end of §4, Chapter I.)

(b) If f is supported in a ball B, and $|f| \log (2 + |f|)$ is integrable over B, then Tf is integrable over B.

(c) If f is bounded and supported on B, then $e^{a|Tf|}$ is integrable over B, for suitable $a > 0$. *Hint:* Write $\displaystyle \int e^{a|Tf|} \, dx = \sum \frac{a^n}{n!} \|Tf\|_n^n$ and use part (a).

(d) The same result holds for the maximal operator T^* of Theorem 4. For these results see Calderón and Zygmund [1], and Zygmund [8], Chapter XII.

6.3 Let $(Tf)(x) = \lim\limits_{\varepsilon \to 0} \int_{|x-y|>\varepsilon} K(x,y)f(y)\,dy$, with $|K(x,y)| \leq \dfrac{A}{|x-y|^n}$.
Suppose T is bounded on $L^p(\mathbf{R}^n)$. Then T is bounded on the L^p space taken with respect to the measure $|x|^\alpha\,dx$, (instead of dx) where $-n < \alpha < n(p-1)$. See Stein [2]; this easily implies the same result for $(1 + |x|)^\alpha\,dx$.

6.4 The following approach unifies the maximal function and differentiation theorems of Chapter I, and many of the singular integrals of the present chapter.

Let $L(x)$ be integrable on $\mathbf{R}^n$, and suppose $L(x) = 0$, if $|x| \geq 2$,

$$\int_{\mathbf{R}^n} L(x)\,dx = 0,$$

and

$$\int_{\mathbf{R}^n} |L(x-y) - L(x)|\,dx \leq B\,|y|.$$

For any pair of integers i, j, define $L_{i,j}$ by $L_{i,j}(x) = \sum\limits_{k=-i}^{k=j} 2^{nk}L(2^k x)$. Write $T_{i,j}f = L_{i,j} * f$. Then if $T_* f = \sup\limits_{i,j} |T_{i,j}f(x)|$ we have

(a) $f \to T_* f$ is of weak-type $(1, 1)$
(b) $f \to T_* f$ is bounded on L^p, $1 < p < \infty$
(c) if $f \in L^p$, $1 \leq p < \infty$, $\lim\limits_{\substack{i \to \infty \\ j \to \infty}} T_{i,j}f$ exists almost everywhere, and also in L^p

   norm, if $1 < p$.

Two interesting examples are:

(i) $L(x) = 1 - 2^n$, for $|x| \leq 1$ and $L(x) = 1$, for $1 \leq |x| \leq 2$, 0 otherwise. Then

$$(T_{i,0}f)(x) = 2^{-ni} \int_{|y|\leq 2^{i+1}} f(x-y)\,dy - 2^n \int_{|y|\leq 1} f(x-y)\,dy$$

(ii) $L(x) = \dfrac{\Omega(x)}{|x|^n}$, for $1 \leq |x| \leq 2$, 0 otherwise. Then

$$(T_{i,j}f)(x) = \int_{2^{-i}\leq|y|\leq 2^{j+1}} \frac{\Omega(y)}{|y|^n} f(x-y)\,dy.$$

(For details see Cotlar [2].)

6.5 (a) Let y' be a unit vector in $\mathbf{R}^n$. The Hilbert transform in the direction y' can be defined as $\lim\limits_{\varepsilon \to 0} H_{y'}^{(\varepsilon)}(f)(x)$, where

$$H_{y'}^{(\varepsilon)}(f)(x) = \int_{|t|\geq\varepsilon} \frac{f(x-y't)\,dt}{t} = \int_\varepsilon^\infty \frac{[f(x-y't) - f(x+y't)]\,dt}{t}.$$

Then $\|H_{y'}^{(\varepsilon)} f\|_p \leq A_p \|f\|_p$ for $f \in L^p(\mathbf{R}^n), 1 < p < \infty$, with A_p independent of y' and ε.

(b) Suppose $\Omega(y')$ is homogeneous of degree 0, is integrable over the unit sphere S^{n-1} and is odd, i.e. $\Omega(y') = -\Omega(-y')$. Let

$$T_\varepsilon(f)(x) = \int_{|y| \geq \varepsilon} \frac{\Omega(y)}{|y|^n} f(x - y)\, dy.$$

Then

$$T_\varepsilon = \tfrac{1}{2} \int_{S^{n-1}} \Omega(y') H_{y'}^{(\varepsilon)}\, d\sigma(y'),$$

and so

$$\|T_\varepsilon(f)\|_p \leq \left(\tfrac{1}{2} A_p \int_{S^{n-1}} |\Omega(y')|\, d\sigma(y') \right) \|f\|_p.$$

(c) A similar but more difficult result holds if Ω is even. It is then required that $|\Omega(y')| \log (2 + |\Omega(y')|)$ be integrable over S^{n-1}.

(d) The behavior of $T_\varepsilon(f)$ for $f \in L^1(\mathbf{R}^n)$, for these general Ω considered here, remains open.

For details on (a), (b), and (c) above see Calderón and Zygmund [3]. This part of the theory is also presented in *Fourier Analysis*, Chapter VI, in somewhat lesser generality.

6.6 Suppose $m(x)$ is homogeneous of degree 0, and continuous on S^{n-1}. For $f \in L^2(\mathbf{R}^n)$ define Tf by $(Tf)^\wedge(x) = m(x)\hat{f}(x)$. Suppose $\|T(f)\|_p \leq A_p \|f\|_p$, for $f \in L^2 \cap L^p$ for some p, $1 < p < \infty$. If $|m(x)| \geq c > 0$, then also $\|f\|_p \leq B_p \|Tf\|_p$. (See Calderón and Zygmund [4], Hörmander [1], Benedek, Calderón and Panzone [1].

6.7 Let T be the Hilbert transform (1), and χ_E denote the characteristic function of a subset E of $\mathbf{R}^1$, of finite measure. Then the distribution function of $T\chi_E$ depends only on the measure of E; more precisely if $\lambda(\alpha)$ is this distribution function, then $\lambda(\alpha) = \dfrac{2m(E)}{\sin h\, \pi\alpha}$. See Stein and Weiss [1].

6.8 As already pointed out, the dilations $x \to \varepsilon x = (\varepsilon x_1, \varepsilon x_2, \ldots, \varepsilon x_n)$ play an important role in this chapter. There are variants of many of the results of this chapter where this type of homogeneity is replaced by a non-isotropic one, i.e. $x \to \tilde{\varepsilon}x = (\varepsilon^{a_1} x_1, \varepsilon^{a_2} x_2, \ldots, \varepsilon^{a_n} x_n)$, where $a_1, a_2, \ldots, a_n$ are fixed positive exponents, and $\varepsilon > 0$. Then the action on the kernels $K(x) \to \varepsilon^n K(\varepsilon x)$ is replaced by $K(x) \to \varepsilon^a K(\tilde{\varepsilon}x)$, where $a = a_1 + a_2 \cdots + a_n$. For details see Jones [1], Fabes and Rivière [1], Kree [1], and Besov, Il'in, and Lizorkin [1].

6.9 Let T and $K(x) = \dfrac{\Omega(x)}{|x|^n}$ be as in Theorem 3. Suppose that $0 < \alpha < 1$, and f is a continuous function of compact support which satisfies

$$|f(x + t) - f(x)| \leq A|t|^\alpha.$$

Then if $g(x) = T(f)$, we also have $|g(x + t) - g(x)| \leq B|t|^\alpha$. (Hint: If Ω is sufficiently smooth the proof is "elementary"; see Privalov [1], Calderón and Zygmund [2], and in general, Taibleson [1].

6.10 Let Ω be homogeneous of degree 0, integrable on the unit sphere and $\int_{S^{n-1}} \Omega \, d\sigma = 0$. Suppose that $\sup_{|r| \leq \delta} \int_{S^{n-1}} |\Omega(r(x')) - \Omega(x')| \, d\sigma \leq \omega(\delta)$, with $\int_0^\varepsilon \dfrac{\omega(\delta) \, d\delta}{\delta} < \infty$. Here r designates rotations about the origin, and $|r|$ denotes the distance of r from the identity rotation, measured by any smooth Riemannian metric on the group of rotations. Then

(a) $\int_{S^{n-1}} |\Omega| \log^+ |\Omega| \, d\sigma < \infty$. Therefore the L^p theory of §6.5 applies.

(b) If $K(x) = \dfrac{\Omega(x)}{|x|^n}$, then $\int_{|x| \geq 2|y|} |K(x - y) - K(x)| \, dx \leq B$, and so the L^1

theory of §3.1 also applies. See Calderón, M. Weiss, and Zygmund [1].

6.11 A slight modification of the argument of §3.3 proves the following. Suppose $K(x)$ is a given function which satisfies the assumptions

(i) $\int_{|x| \leq R} |x| \, |K(x)| \, dx \leq BR, \qquad 0 < R < \infty$

(ii) $\int_{|x| \geq 2|y|} |K(x - y) - K(x)| \, dx \leq B$

(iii) $\left| \displaystyle\int_{R_1 < |x| < R_2} K(x) \, dx \right| \leq B, \qquad 0 < R_1 < R_2 < \infty.$

Let $K_{\varepsilon,\eta}(x) = K(x)$ if $\varepsilon < |x| < \eta$,

$\qquad = 0$ otherwise.

Then $|\hat{K}_{\varepsilon,\eta}(x)| \leq CB$, with C independent of ε and η. See Benedek, Calderón, and Panzone [1].

6.12 Suppose f is bounded and has bounded support. Let $I^s(f)(x) = \int_{\mathbf{R}^n} f(x - y) \, |y|^{-n+\sigma+it} \, dy$, $\sigma > 0$, $s = \sigma + it$. Then for each $\varepsilon > 0$

$$I^s(f)(x) = \int_{|y| \leq \varepsilon} \{f(x - y) - f(x)\} \, |y|^{-n+s} \, dy + \int_{|y| \geq \varepsilon} f(x - y) \, |y|^{-n+s} \, dy$$
$$+ \omega_{n-1}(\varepsilon^s/s) f(x).$$

This shows that $I^{it}f(x) = \lim_{\sigma \to 0} I^{\sigma+it}(f)(x)$, exists if $t \neq 0$ and if in addition f is of class C^1. Finally we may apply §6.11 above with $K(x) = |x|^{-n+it}$, and Theorem 1 of the present chapter. This shows that the operator

$$f \to I_\varepsilon^{it}(f) = \int_{|y| \geq \varepsilon} f(x - y) \, |y|^{-n+it} \, dy,$$

is bounded on $L^p(\mathbf{R}^n)$, $1 < p < \infty$, uniformly in ε. By choosing an appropriate sequence of ε tending to zero (i.e. such that $\varepsilon^{it} \to 0$, for fixed $t \neq 0$) we see that I^{it} can be extended to be a bounded operator on $L^p(\mathbf{R}^n)$, $1 < p < \infty$. It can also be seen that by the use of §3.3 of Chapter III we have

$$(I^{it}(f))\hat{\ }(x) = \gamma_{0,it} \, |x|^{-it} \hat{f}(x),$$

with $\gamma_{0,it} = \pi^{n/2-it} \dfrac{\Gamma(it/2)}{\Gamma(n/2 - it/2)}$.

For related results see Muckenhoupt [1].

6.13   Let $K_1(x) = \dfrac{\Omega_1(x)}{|x|^n}$, and $K_2(x) = \dfrac{\Omega_2(x)}{|x|^m}$ be two kernels of the type considered in §4, defined respectively for $\mathbf{R}^n$ and $\mathbf{R}^m$. On $L^p(\mathbf{R}^{n+m})$ define the transformation $f \to T_{\varepsilon,\delta}(f)$, by $T_{\varepsilon,\delta} = T_\varepsilon^1 \otimes T_\delta^2$ where

$$T_\varepsilon^1(f)(x^1) = \int_{|y|^1 \ge \varepsilon} \frac{\Omega_1(y^1)}{|y^1|^n} f(x^1 - y^1)\, dy^1,$$

and $T_\delta^2(f)(x^2) = \displaystyle\int_{|y^2| \ge \delta} \dfrac{\Omega_2(y^2)f(x^2 - y^2)}{|y^2|^m}\, dy^2$. ($T_{\varepsilon,\delta}$ can be taken to be the composition of T_ε^1 acting on the first n variables, and T_δ^2 acting on the last m variables of functions on $\mathbf{R}^{n+m}$.)

(a) If $f \in L^p(\mathbf{R}^{n+m})$, then $\lim_{\substack{\varepsilon \to 0 \\ \delta \to 0}} T_{\varepsilon,\delta}(f)(x) = T(f)(x)$ exists almost everywhere and in L^p norm, when $1 < p < \infty$.

(See Sokol-Sokolowski [1], and Cotlar [1]. The latter paper however contains an inaccurate deduction in the case of $L \log L$).

(b) There is also a similar but more refined result which holds if f is in the class $L \log L$. For the case when T^1 and T^2 are one-dimensional Hilbert transforms, see Zygmund [3]; his method, however, uses complex function-theory. For the general case see Fefferman [2].

6.14   Let K be a distribution of compact support, which equals a locally integrable function away from the origin, and suppose that its Fourier transform $\hat{K}$ is a function. Assume that for a fixed θ, $0 \le \theta < 1$, we have

(i) $|\hat{K}(x)| \le A(1 + |x|)^{-n\theta/2}$

(ii) $\displaystyle\int_{|x| \ge 2|y|^{1-\theta}} |K(x - y) - K(x)|\, dx \le A.$

Then the operator $f \to K * f$, initially defined on C^∞ functions with compact support, extends to a transformation which is of weak-type $(1, 1)$ and is bounded on L^p, $1 < p < \infty$. Fefferman [1].

An example arises for the operator $\displaystyle\lim_{\varepsilon \to 0} \int_{\varepsilon \le |y| \le 1} K(y)f(x - y)\, dy$, where $K(x) = |x|^{-n} \exp \{i|x|^{-\gamma}\}$, with $\gamma > 0$. This is closely related to the multiplier transformations described in §7.4 of Chapter IV.

Notes

Section 1.   A detailed exposition of the material reviewed in this section may be found in *Fourier Analysis*, Chapter I. For a treatment from the point of view of abstract locally compact abelian groups, see Rudin [1], and Hewitt and Ross [1].

Sections 2, 3, 4, and 5.   The first approach (in one-dimension) was by complex function-theory methods. For details see Zygmund [7, Chapter VII] and [8, Chapter VII], where further historical references may be found. The real-variable theory for the Hilbert transform goes back to Besicovitch [1] and [2], Titchmarsh [1], and Marcinkiewicz [1]. The present n-dimensional theory originates in Calderón-Zygmund [1]. It was elaborated by Cotlar [2], Stein [3], Hörmander [1], Schwartz [1], and Benedeck, Calderón, and Panzone [1], among others. The reader is also referred to the survey papers of Zygmund [5] and Calderón [7].

CHAPTER III

Riesz Transforms, Poisson Integrals, and Spherical Harmonics

The reader who has followed us to this point has already had to deal with some of the more technical aspects of the theory. He has had to climb, step by step, in a direction that might well have seemed dry and unrewarding. It is understandable if at several places he has possibly felt reluctant to continue.

The purpose of the present chapter is in part to reassure the reader by scanning with him the landscape he has already mastered. At the same time we will take the opportunity to introduce him to some of the tools we will need in our further efforts.

Thus our presentation here will naturally differ in manner from that of the first two chapters. In fact, major stress will be laid on the significant formal aspects of the theory, and certain important examples will be studied in detail. In back of these formal aspects and special examples lie two considerations, which we briefly indicate. It is in the nature of things that the group of rotations that acts on $\mathbf{R}^n$ should play a decisive role in its harmonic analysis, as do the groups of translations and dilations. If from this point of view we consider the simplest, non-trivial, "invariant" operators we are lead to the Riesz transforms. Related to this is the intimate connection of classical harmonic analysis (that of $\mathbf{R}^1$) with complex function theory. The attempts to extend this as far as possible to $\mathbf{R}^n$ via the theory of harmonic functions leads us back to the Riesz transforms.

1. *The Riesz transforms*

1.1 We begin by some observations about the Hilbert transform

$$Hf(x) = \lim_{\varepsilon \to 0} \frac{1}{\pi} \int_{|y| \geq \varepsilon} \frac{f(x-y)}{y} \, dy.$$

54

Here we are in the case of $\mathbf{R}^1$, with $K(x) = 1/\pi x$, $\Omega(x) = \dfrac{1}{\pi}\,\text{sign}\, x = \dfrac{1}{\pi}\dfrac{x}{|x|}$. Then according to formula (26) of §4.2 of the previous chapter, we see immediately that in terms of the Fourier transform, $(Hf)^{\wedge}(x) = m(x)\hat{f}(x)$, where the multiplier $m(x)$ is given by $m(x) = i\,\text{sign}\, x$. From this, it is clear that H is unitary on $L^2(\mathbf{R}^1)$, and $H^2 = -I$.

Recall now the operation of dilation $(\tau_\delta f)(x) = f(\delta x)$, which in the case of one variable we find convenient to define for all non-zero δ, positive and negative. Then as is obvious, if $\delta > 0$,

$$(H\tau_\delta)f(x) = \lim_{\varepsilon \to 0} \frac{1}{\pi} \int_{|y| \geq \varepsilon} \frac{f(\delta x - \delta y)}{y}\, dy$$

$$= \lim_{\varepsilon' \to 0} \frac{1}{\pi} \int_{|y| \geq \varepsilon'} \frac{f(\delta x - y)}{y}\, dy = (\tau_\delta H)f(x),$$

so $H\tau_\delta = \tau_\delta H$; and it is equally obvious that $\tau_\delta H = -H\tau_\delta$, if $\delta < 0$.

These simple considerations of dilation "invariance" and the obvious translation invariance in fact characterize the Hilbert transform.

PROPOSITION 1.   *Suppose T is a bounded operator on $L^2(\mathbf{R}^1)$ which satisfies the following properties:*

   (a) *T commutes with translations*
   (b) *T commutes with positive dilations*
   (c) *T anticommutes with the reflection $f(x) \to f(-x)$.*

Then T is a constant multiple of the Hilbert transform.

The proof involves no difficulties. In fact since T commutes with translations, then according to the proposition in §1.4 of the previous chapter there is a bounded function $m(x)$, so that $(Tf)^{\wedge}(x) = m(x)\hat{f}(x)$. Let us also denote by $\mathscr{F}$ the Fourier transform, $\mathscr{F}f = \hat{f}$. Then

$$(\mathscr{F}\tau_\delta f)(y) = \int_{-\infty}^{\infty} e^{2\pi ixy} f(\delta x)\, dx$$

$$= |\delta|^{-1} \int_{-\infty}^{\infty} e^{2\pi ixy/\delta} f(x)\, dx = |\delta|^{-1}(\tau_{\delta^{-1}}\mathscr{F}f)(y),$$

so $\mathscr{F}\tau_\delta = |\delta|^{-1}\tau_{\delta^{-1}}\mathscr{F}$.

The definition of the multiplier may be written symbolically as $\mathscr{F}T = m\mathscr{F}$, (where by m we mean the operator of multiplication by m!). However the assumptions (b) and (c) may be rewritten as $T\tau_\delta = \text{sign}\,(\delta)\tau_\delta T$, which when inserted in the above gives

$$\tau_\delta m = \tau_\delta \mathscr{F}T\mathscr{F}^{-1} = |\delta|^{-1}\mathscr{F}\tau_{\delta^{-1}}T\mathscr{F}^{-1} = \delta^{-1}\mathscr{F}T\tau_{\delta^{-1}}\mathscr{F}^{-1}$$

$$= \text{sign}\,(\delta)\mathscr{F}T\mathscr{F}^{-1}\tau_\delta = \text{sign}\,(\delta)m\tau_\delta.$$

So $\tau_\delta m = \text{sign}(\delta) m \tau_\delta$, which means $m(\delta x) = \text{sign}(\delta) m(x)$, if $\delta \neq 0$.

This shows that $m(x) = \text{constant} \times \text{sign}(x)$, and the proposition is proved.

The proof of the proposition shows incidentally that the only bounded linear transformations on $L^2(\mathbf{R}^1)$ which *commute* with all the operations described, namely translations and positive and negative dilations, are constant multiples of the identity operator. This remark together with the proposition attest graphically to the special role of the Hilbert transform in harmonic analysis of $\mathbf{R}^1$. We now look for the operators in $\mathbf{R}^n$ which have the analogous structural characterization.

1.2   We begin by making a few remarks about the interaction of dilations and rotations with the n-dimensional Fourier transform. With $\mathscr{F}f = \hat{f}$, and $\delta > 0$

$$(\mathscr{F}\tau_\delta)f(x) = \int_{\mathbf{R}^n} e^{2\pi i x \cdot y} f(\delta y)\, dy$$

$$= \delta^{-n} \int_{\mathbf{R}^n} e^{2\pi i x \cdot y/\delta} f(y)\, dy = \delta^{-n}(\tau_{\delta^{-1}}\mathscr{F})f(x).$$

So, symbolically,

(1) $$\mathscr{F}\tau_\delta = \delta^{-n}\tau_{\delta^{-1}}\mathscr{F}.$$

Next let ρ denote any rotation (proper or improper) about the origin in $\mathbf{R}^n$. Denote also by ρ its induced action on functions, $\rho(f)(x) = f(\rho^{-1}x)$ Then

$$(\mathscr{F}\rho)f(x) = \int_{\mathbf{R}^n} e^{2\pi i x \cdot y} f(\rho^{-1}y)\, dy$$

$$= \int_{\mathbf{R}^n} e^{2\pi i x \cdot \rho y} f(y)\, dy = \int_{\mathbf{R}^n} e^{2\pi i \rho^{-1} x \cdot y} f(y)\, dy = (\rho\mathscr{F})f(x),$$

and

(2) $$\mathscr{F}\rho = \rho\mathscr{F}.$$

We shall also need the following elementary observation. Let $m(x) = (m_1(x), m_2(x), \ldots, m_n(x))$ be an n-tuple of functions defined on $\mathbf{R}^n$. For any rotation ρ, write $\rho = (\rho_{jk})$ for its matrix realization. Suppose that m transforms like a vector. Symbolically this can be written as

$$m(\rho^{-1}x) = \rho(m(x)),$$

or more explicitly

(3) $$m_j(\rho^{-1}x) = \sum_k \rho_{jk} m_k(x), \qquad \text{for every rotation } \rho.$$

LEMMA. *Suppose m is homogeneous of degree 0, i.e. $m(\delta x) = m(x)$, for $\delta > 0$. If m transforms according to (3) then $m(x) = c\dfrac{x}{|x|}$ for some constant c; that is*

$$(4) \qquad\qquad m_j(x) = c\frac{x_j}{|x|}.$$

To prove the assertion we notice that it suffices to consider x on the unit sphere. Now let $e_1, e_2, \ldots, e_n$ denote the usual unit vectors along the axes. Set $c = m_1(e_1)$. We can see that $m_j(e_1) = 0$, if $j \neq 1$. In fact, for any rotation ρ having e_1 fixed (3) gives us that $m_j(e_1) = \sum\limits_{k=2}^{n} \rho_{jk} m_k(e_1)$, $j = 2, \ldots, n$. That is, the $n - 1$ dimensional vector $(m_2(e_1), m_3(e_1), \ldots, m_n(e_1))$ is left fixed by all the rotations on this $n - 1$ dimensional vector space. So $m_2(e_1) = m_3(e_1) \cdots = m_n(e_1) = 0$. Inserting this again in (3) gives $m_j(\rho^{-1} e_1) = \rho_{j1} m_1(e_1) = c\rho_{j1}$. But if $\rho^{-1} e_1 = x$, then $\rho_{j1} = x_j$, so $m_j(x) = cx_j$, $(|x| = 1)$, which proves the lemma.

It is curious to observe that the full group of rotations is needed only in the case $n = 1$, and $n = 2$. In the case $n \geq 3$ the proper rotations would have sufficed, since then the subgroup of proper rotations in one less dimension is still transitive on its unit sphere (S^{n-2}).

We are now in a position to define the n Riesz transforms. We set for $f \in L^p(\mathbf{R}^n)$, $1 \leq p < \infty$,

$$(5) \qquad R_j(f)(x) = \lim_{\varepsilon \to 0} c_n \int_{|y| \geq \varepsilon} \frac{y_j}{|y|^{n+1}} f(x - y)\, dy, \qquad j = 1, \ldots, n,$$

with $c_n = \dfrac{\Gamma\left(\dfrac{n+1}{2}\right)}{\pi^{(n+1)/2}}$.

Thus R_j is defined by the kernel $K_j(x) = \dfrac{\Omega_j(x)}{|x|^n}$, and $\Omega_j(x) = c_n \dfrac{x_j}{|x|}$.

We shall next derive the multipliers which correspond to the Riesz transforms, and which in fact justify their definition. Let us recall the formula (26) of §4.2 of the previous chapter. It is

$$(6) \qquad\qquad m(x) = \int_{S^{n-1}} \Gamma(x \cdot y)\Omega(y)\, d\sigma(y), \qquad |x| = 1,$$

with $\Gamma(t) = \dfrac{\pi i}{2} \operatorname{sign} t + \log |1/t|$. Notice that the mapping (6) from Ω to m commutes with rotations, and this is nothing but an immediate consequence of the fact that the kernel $\Lambda(x \cdot y)$ depends only on the inner product of x and y. It is clear however that the kernels

$$(K_1(x), \ldots, K_n(x)) = c_n\left(\frac{x_1}{|x|^{n+1}}, \frac{x_2}{|x|^{n+1}}, \ldots, \frac{x_n}{|x|^{n+1}}\right)$$

satisfy the transformation law (3) (with K_j in place of m_j). Then, in view of the commutability of the mapping $K_j \to m_j$ with rotations just alluded to, it follows that the multipliers also satisfy (3). However the m_j are each homogeneous of degree 0, so the lemma shows that $m_j(x) = c\,\dfrac{x_j}{|x|}$. In this particular case (and because our choice of the constant c_n) we have $c = i$. By evaluating the m_j at a fixed point in (6) this last assertion is equivalent with

$$(7) \qquad \frac{2\Gamma\left(\dfrac{n+1}{2}\right)}{\pi^{(n+3)/2}} = \int_{S^{n-1}} |\cos\theta|\, d\sigma(y)$$

where θ denotes the angle made by the variable unit vector y with a fixed direction. One may either evaluate this integral directly, or this calculation can be avoided by appealing to the general result, Theorem 5, proved below. In either case we get

$$(8) \qquad (R_j f)^{\wedge}(x) = i\,\frac{x_j}{|x|}\,\hat{f}(x), \qquad j = 1, \ldots, n.$$

We can express the transformation law (3) acting on the Riesz transforms in a more intrinsic manner. More precisely

$$(9) \qquad \rho R_j \rho^{-1} f = \sum_k \rho_{jk} R_k f,$$

which is the statement that under rotations in $\mathbf{R}^n$, the Riesz operators transform in the same manner as the components of a vector. The verification of (9) is immediate. It may be done using the direct definition (5) of the Riesz transforms or, because of (8), in terms of their Fourier transforms. Thus if we denote symbolically $R_j^{\wedge} = m_j$, then (9) becomes $\rho(m_j \rho^{-1}(f)) = \sum_k \rho_{jk} m_k f$ which is $m_j(\rho^{-1}x) = \sum_k \rho_{jk} m_k(x)$. These observations have a converse.

PROPOSITION 2. *Let* $T = (T_1, T_2, \ldots, T_n)$ *be an n-tuple of bounded transformations on* $L^2(\mathbf{R}^n)$. *Suppose*
    (a) *Each* T_j *commutes with the translation of* $\mathbf{R}^n$
    (b) *Each* T_j *commutes with the dilations of* $\mathbf{R}^n$
    (c) *For every rotation* $\rho = (\rho_{jk})$ *of* $\mathbf{R}^n$, $\rho T_j \rho^{-1} f = \sum_k \rho_{jk} T_k f$.
Then the T_j *are a constant multiple of the Riesz transforms, i.e. there exists a constant* c, *so that* $T_j = cR_j, j = 1, \ldots, n$.

All the elements of the proof have already been discussed. We bring them together: (i) Since the T_j are bounded on $L^2(\mathbf{R}^n)$ and commute with translations they can be each realized by bounded multipliers; symbolically $T_j^{\wedge} = m_j$. (ii) Since the T_j commute with dilations and because of the relation (1) between dilation and the Fourier transform, we see that $m_j(\delta x) = m_j(x)$, $\delta > 0$; that is, each m_j is homogeneous of degree 0. (iii) Finally, assumption (c) has as a consequence the relation (3), and so by the lemma we can obtain the desired conclusion.

1.3 An application. One of the important applications of the Riesz transforms is that they can be used to mediate between various combinations of partial derivatives of a function. This service of the Riesz transforms will be particularly striking in Chapter V. We shall here content ourselves with two very simple illustrations, which examples have an interest on their own and have already the characteristic features of a general type of estimate which can be made in the theory of elliptic differential operators.

PROPOSITION 3. *Suppose f is of class C^2 and has compact support. Let $\Delta f = \sum\limits_{j=1}^{n} \dfrac{\partial^2 f}{\partial x_j^2}$. Then we have the a priori bound*

$$(10) \qquad \left\| \frac{\partial^2 f}{\partial x_j\, \partial x_k} \right\|_p \le A_p \|\Delta f\|_p, \qquad 1 < p < \infty.$$

This proposition is an immediate consequence of the L^p boundedness of the Riesz transforms and the identity

$$(11) \qquad \frac{\partial^2 f}{\partial x_j\, \partial x_k} = -R_j R_k \, \Delta f.$$

To prove (11) we use the Fourier transform. Thus if $\hat{f}(x)$ is the Fourier transform of f, $\hat{f}(x) = \int_{\mathbf{R}^n} e^{2\pi i x \cdot y} f(y)\, dy$, then the Fourier transform of $\dfrac{\partial f}{\partial x_j}$ is $-2\pi i x_j\, \hat{f}(x)$, and so

$$\left(\frac{\partial^2 f}{\partial x_j\, \partial x_k} \right)^{\wedge}(x) = -4\pi^2 x_j\, x_k \hat{f}(x)$$

$$= -\left(\frac{i x_j}{|x|} \right)\left(\frac{i x_k}{|x|} \right)(-4\pi |x|^2)\hat{f}(x) = -(R_j R_k \, \Delta f)^{\wedge},$$

which gives (11).

Another application of interest, this time for potential theory in two-dimensions, is the following.

PROPOSITION 4.   *Suppose f is of class C^1 in $\mathbf{R}^2$ and has compact support. Then we have the a priori bound*

$$\left\| \frac{\partial f}{\partial x_1} \right\|_p + \left\| \frac{\partial f}{\partial x_2} \right\|_p \leq A_p \left\| \frac{\partial f}{\partial x_1} + i \frac{\partial f}{\partial x_2} \right\|_p, \qquad 1 < p < \infty.$$

Needless to say, this proposition is significant only if f is *complex valued*.

The proof of Proposition 4 is very much like that of the previous one, except that here the identity used is

$$\frac{\partial f}{\partial x_j} = -R_j(R_1 - iR_2)\left(\frac{\partial f}{\partial x_1} + i \frac{\partial f}{\partial x_2} \right), \qquad j = 1, 2.$$

A more systematic presentation of these particular facts will be given in §3.5, below.

2. Poisson integrals and approximations to the identity

2.1   We shall now introduce a notion that will be indispensable in much of our further work. We have in mind the theory of harmonic functions. The setting for the application of this theory will be as follows. We shall think of $\mathbf{R}^n$ as the boundary hyperplane of the $(n + 1)$ dimensional upper-half space $\mathbf{R}^{n+1}_+$. In coordinate notation,

$$\mathbf{R}^{n+1}_+ = \{(x, y) : x \in \mathbf{R}^n, y > 0\}.$$

We shall consider the *Poisson integral* of a function f given on $\mathbf{R}^n$. This Poisson integral is effectively the solution to the Dirichlet problem for $\mathbf{R}^{n+1}_+$: find a harmonic function $u(x, y)$ on $\mathbf{R}^{n+1}_+$, whose boundary values on $\mathbf{R}^n$ (in the appropriate sense) are $f(x)$.

The formal solution of this problem can be given neatly in the context of the L^2 theory.

In fact, let $f \in L^2(\mathbf{R}^n)$, and let $\hat{f}$ be its Fourier transform. Consider

$$(12) \qquad u(x, y) = \int_{t \in \mathbf{R}^n} \hat{f}(t) e^{-2\pi i t \cdot x} e^{-2\pi |t| y} \, dt, \qquad y > 0.$$

This integral converges absolutely, because $\hat{f} \in L^2(\mathbf{R}^n)$, and $e^{-2\pi |t| y}$ is rapidly decreasing (in $|t|$), for $y > 0$. For the same reason the integral above may be differentiated with respect to x and y any number of times by carrying out the operation under the sign of integration. This gives

$$\Delta u = \frac{\partial^2 u}{\partial y^2} + \sum_{k=1}^{n} \frac{\partial^2 u}{\partial x_k^2} = 0,$$

because the factor $e^{-2\pi i t \cdot x} e^{-2\pi |t| y}$ satisfies this property for each fixed t.

Also by Plancherel's theorem $u(x, y) \to f(x)$ in $L^2(\mathbf{R}^n)$ norm, as $y \to 0$.

This solution of the problem can also be written without explicit use of the Fourier transform. For this purpose define the *Poisson kernel* $P_y(x)$ by

(13) $$P_y(x) = \int_{\mathbf{R}^n} e^{-2\pi i t \cdot x} e^{-2\pi |t| y} \, dt, \qquad y > 0.$$

Then the function $u(x, y)$ obtained above can be written as a convolution

(14) $$u(x, y) = \int_{\mathbf{R}^n} P_y(t) f(x - t) \, dt.$$

We shall say that u is the *Poisson integral* of f.

The Poisson kernel has an explicit expression.

PROPOSITION 5.

(15) $$P_y(x) = \frac{c_n y}{(|x|^2 + y^2)^{\frac{n+1}{2}}}, \qquad c_n = \frac{\Gamma\left(\dfrac{n+1}{2}\right)}{\pi^{\frac{n+1}{2}}},$$

c_n is the same constant that appears in the definition of the Riesz transforms ((5) above). The well-known formula (15) may be proved as follows. We use two identities:

(α) $$\int_{\mathbf{R}^n} e^{-\pi \delta |t|^2} e^{-2\pi i t \cdot x} \, dt = \delta^{-n/2} e^{-\pi |x|^2 / \delta}, \qquad \delta > 0$$

(β) $$e^{-\gamma} = \frac{1}{\sqrt{\pi}} \int_0^\infty \frac{e^{-u}}{\sqrt{u}} e^{-\gamma^2 / 4u} \, du, \qquad \gamma > 0.$$

The first, (α), is immediately reducible by a change of variables to the very well-known special case $\delta = 1$. The second, (β), expresses the exponential $e^{-\gamma}$ as a weighted average of the family of exponential $e^{-\gamma^2/4u}$, $0 < u < \infty$, and is an important instance of the principle of *subordination.** To prove (β), write $e^{-\gamma} = \dfrac{1}{\pi} \displaystyle\int_{-\infty}^\infty \dfrac{e^{i\gamma x}}{1 + x^2} \, dx$, and express the factor $\dfrac{1}{1 + x^2}$ as $\int_0^\infty e^{-(1+x^2)u} \, du$. This leads to the double integral $e^{-\gamma} = \dfrac{1}{\pi} \displaystyle\int_0^\infty e^{-u} \left(\int_{-\infty}^\infty e^{i\gamma x} e^{-ux^2} \, dx \right) du$ which after evaluation of the inner integral gives (β). This being done we return to $P_y(x)$. We have by (β) and (13):

$$P_y(x) = \frac{1}{\sqrt{\pi}} \int_{\mathbf{R}^n} \left(\int_0^\infty \frac{e^{-u}}{\sqrt{u}} e^{-\pi^2 |t|^2 y^2 / u} \, du \right) e^{-2\pi i t \cdot x} \, dt.$$

* See Bochner [2], Chapter 4.

Then apply (α) with $\delta = \dfrac{\pi y^2}{u}$. We get

$$P_y(x) = \frac{1}{\pi^{\frac{n+1}{2}}} \int_0^\infty e^{-u} e^{-\frac{|x|^2 u}{v^2}} y^{-n} u^{\frac{n-1}{2}} \, du = \frac{y}{(\pi(|x|^2 + y^2))^{\frac{n+1}{2}}} \int_0^\infty e^{-u} u^{\frac{n-1}{2}} \, du,$$

which is the desired formula for the Poisson kernel.

We list the properties of the Poisson kernel that are now more or less evident:

   (i) $P_y(x) > 0$.
   (ii) $\int_{\mathbf{R}^n} P_y(x) \, dx = 1$, $y > 0$; more generally, $P_y^\wedge(x) = e^{-2\pi|x|y}$ by an application of the Fourier inversion formula to (13).
   (iii) $P_y(x)$ is homogeneous of degree $-n$: $P_\varepsilon(x) = P_1(x/\varepsilon)\varepsilon^{-n}$, $\varepsilon > 0$.
   (iv) $P_y(x)$ is a decreasing function of $|x|$, and $P_y(x) \in L^p(\mathbf{R}^n)$, $1 \le p \le \infty$.
   (v) Suppose $f \in L^p(\mathbf{R}^n)$, $1 \le p \le \infty$, then its Poisson integral u, given by (14), is harmonic in $\mathbf{R}_+^{n+1}$. This is a simple consequence of fact that $P_y(x)$ is harmonic in $\mathbf{R}_+^{n+1}$; the latter is immediately derivable from (13).
   (vi) We have the "semi-group property" $P_{y_1} * P_{y_2} = P_{y_1+y_2}$ if $y_1 > 0$ and $y_2 > 0$. This follows immediately from the Fourier transform formula in (ii).

The boundary behavior of Poisson integrals is already described to a significant extent by the following theorem.

THEOREM 1. *Suppose $f \in L^p(\mathbf{R}^n)$, $1 \le p \le \infty$, and let $u(x, y)$ be its Poisson integral. Then:*

   (a) $\sup_{y>0} |u(x, y)| \le Mf(x)$, *where Mf is the maximal function of Chapter I, §1.*
   (b) $\lim_{y\to 0} u(x, y) = f(x)$, *for almost every x.*
   (c) *If $p < \infty$, $u(x, y)$ converges to $f(x)$ in $L^p(\mathbf{R}^n)$ norm, as $y \to 0$.*

The theorem will now be proved in a more general setting, valid for a large class of approximations to the identity.

2.2   Let φ be an integrable function on $\mathbf{R}^n$, and set $\varphi_\varepsilon(x) = \varepsilon^{-n}\varphi(x/\varepsilon)$, $\varepsilon > 0$.

THEOREM 2. *Suppose that the least decreasing radial majorant of φ is integrable; i.e. let $\psi(x) = \sup_{|y| \ge |x|} |\varphi(y)|$, and we suppose $\int_{\mathbf{R}^n} \psi(x) \, dx = A < \infty$. Then with the same A,*

(a) $\sup_{\varepsilon > 0} |(f * \varphi_\varepsilon)(x)| \leq AM(f)(x), f \in L^p(\mathbf{R}^n), 1 \leq p \leq \infty.$

(b) *If in addition* $\int_{\mathbf{R}^n} \varphi(x)\, dx = 1$, *then* $\lim_{\varepsilon \to 0} (f * \varphi_\varepsilon)(x) = f(x)$ *almost everywhere.*

(c) *If* $p < \infty$, *then* $\|f * \varphi_\varepsilon - f\|_p \to 0$, *as* $\varepsilon \to 0$.

We have already considered a special case of this situation in Chapter I, with $\varphi(x) = \dfrac{1}{m(B)} \chi_B$, the characteristic function of the unit ball B, divided by the measure of that ball. The point of the theorem is to reduce matters to this fundamental special case.

We begin with the proof of (c). It is to be remarked that the proof actually holds under the weaker assumption that φ is merely integrable. (Of course the normalization $\int_{\mathbf{R}^n} \varphi\, dx = 1$ is still required.) First we point out that if $f \in L^p(\mathbf{R}^n)$, $p < \infty$, and $\|f(x - y) - f(x)\|_p = \Delta(y)$, then $\Delta(y) \to 0$, as $y \to 0$.* If f_1 is continuous with compact support, the assertion in that case is an immediate consequence of the uniform convergence $f_1(x - y) \to f_1(x)$, as $y \to 0$. In general write $f = f_1 + f_2$, where f_1 is as described and $\|f_2\|_p \leq \delta$; this is possible since such f_1 are dense in L^p, $p < \infty$. Then $\Delta(y) = \Delta_1(y) + \Delta_2(y)$, with $\Delta_1(y) \to 0$, as $y \to 0$, and $\Delta_2(y) \leq 2\delta$. This shows that $\Delta(y) \to 0$ for general $f \in L^p(\mathbf{R}^n)$, $p < \infty$. Now $f * \varphi_\varepsilon - f = \int_{\mathbf{R}^n} [f(x - y) - f(x)] \varphi_\varepsilon(y)\, dy$, because

$$\int_{\mathbf{R}^n} \varphi_\varepsilon(x)\, dx = \int_{\mathbf{R}^n} \varphi(x)\, dx = 1.$$

So

$$\|f * \varphi_\varepsilon - f\|_p \leq \int_{\mathbf{R}^n} \Delta(y) |\varphi_\varepsilon(y)|\, dy = \int_{\mathbf{R}^n} \Delta(\varepsilon y) |\varphi(y)|\, dy \to 0;$$

the latter fact is by the Lebesgue dominated convergence theorem and the fact that $\Delta(\varepsilon y) \to 0$, as $\varepsilon \to 0$. This proves assertion (c) of the theorem. We shall now prove assertion (a). With a slight abuse of notation, let us write $\psi(r) = \psi(x)$, if $|x| = r$; it should cause no confusion since $\psi(x)$ is anyway radial. Now observe that $\displaystyle\int_{r/2 \leq |x| \leq r} \psi(x)\, dx \geq \psi(r) \int_{r/2 \leq |x| \leq r} dx = \psi(r) c r^n$. Therefore the assumption $\psi \in L^1$, (and the fact that $\psi(r)$ is decreasing) proves that $r^n \psi(r) \to 0$, as $r \to 0$, or $r \to \infty$. To prove (a) we need to show that

(16)
$$(f * \psi_\varepsilon)(x) \leq A(Mf)(x)$$

where $f \geq 0, f \in L^p(\mathbf{R}^n), 1 \leq p \leq \infty, \varepsilon > 0$, and $A = \int_{\mathbf{R}^n} \psi(x)\, dx$.

Since the assertion (16) is clearly translation invariant (with respect to

---

* This statement is the continuity of the mapping $y \to f(x - y)$ of $\mathbf{R}^n$ to $L^p(\mathbf{R}^n)$.

f) and also dilation invariant (with respect to ψ), it suffices to show that

(17) $$(f * \psi)(0) \leq A(Mf)(0).$$

In proving (17) we may clearly assume that $Mf(0) < \infty$. Let us write $\lambda(r) = \int_{x \in S^{n-1}} f(rx) \, d\sigma(x)$, and $\Lambda(r) = \int_{|x| \leq r} f(x) \, dx$, so

$$\Lambda(r) = \int_0^r \lambda(t) t^{n-1} \, dt.$$

We have

$$(f * \psi)(0) = \int_{\mathbf{R}^n} f(x) \psi(x) \, dx = \int_0^\infty \lambda(r) \psi(r) r^{n-1} \, dr$$

$$= \lim_{\substack{\varepsilon \to 0 \\ N \to \infty}} \int_\varepsilon^N \lambda(r) \psi(r) r^{n-1} \, dr = \lim_{\substack{\varepsilon \to 0 \\ N \to \infty}} - \int_\varepsilon^N \Lambda(r) \, d\psi(r).$$

The passage to the last equality is by integration by parts. This introduces the error $\Lambda(N)\psi(N) - \Lambda(\varepsilon)\psi(\varepsilon)$; but this term tends to zero as $\varepsilon \to 0$, and $N \to \infty$ in view of our observation regarding ψ and the fact that $\Lambda(r) = \int_{|x| \leq r} f(x) \, dx \leq Vr^n Mf(0)$, where V is the volume of the unit ball. Thus

$$f * \psi(0) = \int_0^\infty \Lambda(r) \, d(-\psi(r)) \leq V Mf(0) \int_0^\infty r^n \, d(-\psi(r)).$$

So (17) and hence (16) is proved.

The almost everywhere convergence (a) is then proved in the familiar way as follows. First, one verifies that if f_1 is continuous and has compact support, $(f_1 * \varphi_\varepsilon)(x) \to f_1(x)$ uniformly as $\varepsilon \to 0$. Next one deals with the case $f \in L^p(\mathbf{R}^n)$, $1 \leq p < \infty$, by writing $f = f_1 + f_2$ with f_1 as described and with the L^p norm of f_2 small. The argument then follows closely that given in §1.5, Chapter I, after equation (6). Thus we get that $\lim_{\varepsilon \to 0} f_\varepsilon(x)$ exists almost everywhere and equals $f(x)$. To deal with the remaining case, that of bounded f, we fix any ball B, and set ourselves the task of showing that $\lim_{\varepsilon \to 0} (f * \varphi_\varepsilon)(x) = f(x)$ for almost every $x \in B$. Let B_1 be any other ball which strictly contains B, and let δ be the distance from B to the complement of B_1. Let $f_1(x) = \begin{cases} f(x), & x \in B \\ 0, & x \notin B \end{cases}$; $f(x) = f_1(x) + f_2(x)$. Then $f_1(x) \in L^1(\mathbf{R}^n)$, and so the appropriate conclusion holds for it. However for $x \in B$,

$$|(f * \varphi_\varepsilon)(x)| = \left| \int f(x - y) \varphi_\varepsilon(y) \, dy \right| \leq \int_{|y| \geq \delta > 0} |f(x - y)| \, \varphi_\varepsilon(y) \, dy$$

$$\leq \|f\|_\infty \int_{|y| \geq \delta/\varepsilon} |\varphi(y)| \, dy \to 0,$$

$$\text{as} \quad \varepsilon \to 0.$$

Theorem 2 is then completely proved. Theorem 2 then applies directly to prove theorem 1, because of properties (i)–(iv) of the Poisson kernel; in this case $\varphi(x) = \psi(x) = P_1(x)$.

There are also some variants of the result of Theorem 2 which, of course, apply equally well to Poisson integrals. The first is an easy adaptation of the argument already given, and is stated without proof.

COROLLARY. *Suppose f is continuous and bounded on $\mathbf{R}^n$. Then $(f * \varphi_\varepsilon)(x) \to f(x)$ uniformly on compact subsets of $\mathbf{R}^n$.*

In particular this shows that if f is a given bounded and continuous function in $\mathbf{R}^n$, we can find a function $u(x, y)$ which is continuous on the closure of $\mathbf{R}^{n+1}_+$, harmonic in the interior, and whose restriction to the boundary is the given f. Thus Dirichlet's problem is resolved in this case.

The second variant is somewhat more difficult. It is the analogue for finite Borel measures in place of integrable functions, and is outlined in §4.1.

2.3 Conjugate harmonic functions. We shall now tie together the Riesz transforms and the theory of harmonic functions, more particularly Poisson integrals. Since we are interested here mainly in the formal aspects we shall restrict ourselves to the L^2 case. (The L^p case and related results are stated in §4.3. and §4.4.)

THEOREM 3. *Let f and $f_1, \ldots, f_n$ all belong to $L^2(\mathbf{R}^n)$, and let their respective Poisson integrals be $u_0(x, y) = P_y * f$, $u_1(x, y) = P_y * f_1, \ldots$, $u_n(x, y) = P_y * f_n$. Then a necessary and sufficient condition that*

$$(18) \qquad f_j = R_j(f), \qquad j = 1, \ldots, n,$$

is that the following generalized Cauchy-Riemann equations hold:

$$(19) \qquad \begin{cases} \displaystyle\sum_{j=0}^{n} \frac{\partial u_j}{\partial x_j} = 0, \\ \displaystyle\frac{\partial u_j}{\partial x_k} = \frac{\partial u_k}{\partial x_j}, \qquad j \neq k, \quad \text{with} \quad x_0 = y. \end{cases}$$

It is to be noted that, at least locally, the system (19) is equivalent with the existence of a harmonic function H (of the $n + 1$ variables), so that

$$u_j = \frac{\partial H}{\partial x_j}, j = 0, 1, 2, \ldots, n.$$

The theorem is one of that class whose proof is nearly obvious but whose statement is nevertheless of some interest.

Suppose $f_j = R_j(f)$, then $\hat{f}_j(t) = \dfrac{it_j}{|t|} \hat{f}(t)$, and so by (12)

$$u_j(x, y) = \int_{\mathbf{R}^n} \hat{f}(t) \frac{it_j}{|t|} e^{-2\pi i t \cdot x} e^{-2\pi |t| y} \, dt, \quad j = 1, \ldots, n.$$

The equations (19) can then be immediately verified by differentiation under the integral sign, which is justified by the rapid convergence of the integrals in question.

Conversely, let $u_j(x, y) = \int_{\mathbf{R}^n} \hat{f}_j(t) e^{-2\pi i t \cdot x} e^{-2\pi |t| y} \, dt, j = 0, 1, \ldots, n.$ Then the fact that $\dfrac{\partial u_0}{\partial x_j} = \dfrac{\partial u_j}{\partial x_0} = \dfrac{\partial u_j}{\partial y}$, $j = 1, \ldots, n$, shows that

$$-2\pi i t_j \hat{f}_0(t) e^{-2\pi |t| y} = -2\pi |t| f_j(t) e^{-2\pi |t| y};$$

therefore $\hat{f}_j(t) = \dfrac{it_j}{|t|} \hat{f}_0(t)$, and so

$$f_j = R_j(f_0) = R_j(f), \quad j = 1, \ldots, n.$$

The theorem indicates that it should be of interest to study harmonic functions satisfying the system (19) in analogy with complex function theory. We shall return to this point of view in Chapters V, VII, and VIII.

2.4 A digression. We shall now digress from our main topic in order to return to a point left open in our treatment of singular integrals in §4.6 of Chapter II. The situation there was as follows: We considered the kernel $K(x) = \dfrac{\Omega(x)}{|x|^n}$, where Ω was homogeneous of degree 0, and its restriction to the unit sphere satisfied the cancellation property (24) and the smoothness property (25) (see page 39). We were concerned with the almost everywhere existence of

$$\lim_{\varepsilon \to 0} T_\varepsilon(f) = \lim_{\varepsilon \to 0} \int_{|y| \geq \varepsilon} \frac{\Omega(y)}{|y|^n} f(x - y) \, dy.$$

We used, without proof, the following lemma.

LEMMA. *If* $T^* f(x) = \sup\limits_{\varepsilon > 0} |T_\varepsilon(f)|(x)$, *then*

$$\|T^* f\|_p \leq A_p \|f\|_p, \quad 1 < p < \infty.$$

Let $T(f)(x) = \lim\limits_{\varepsilon \to 0} T_\varepsilon(f)(x)$, where the limit is taken in the L^p norm. Its existence is guaranteed by theorem 3 in §4.2 of Chapter II. We shall prove the lemma by showing that

$$T^*(f)(x) \le M(Tf)(x) + CM(f)(x).$$

Let φ be a smooth non-negative function on $\mathbf{R}^n$, which is supported in the unit ball, has integral equal to one, and which is also radial and decreasing in $|x|$. Consider

$$K_\varepsilon(x) = \begin{cases} \dfrac{\Omega(x)}{|x|^n}, & \text{if } |x| \ge \varepsilon \\ 0 & \text{if } |x| < \varepsilon \end{cases}$$

This leads us to another function Φ defined by

(20) $$\varphi * K - K_1 = \Phi,$$

where $\varphi * K = \lim_{\varepsilon \to 0} \varphi * K_\varepsilon = \lim_{\varepsilon \to 0} \int_{|x-y| \ge \varepsilon} K(x-y)\varphi(y)\,dy.$

We shall need to prove that the smallest decreasing radial majorant of Φ is integrable (so as to apply Theorem 2). In fact if $|x| < 1$, $\Phi = \varphi * K$ which is $\int_{\mathbf{R}^n} K(y)\varphi(x-y)\,dy$ or $\int_{\mathbf{R}^n} K(y)[\varphi(x-y) - \varphi(x)]\,dx$ and hence is bounded on account of the smoothness of φ. When $1 \le |x| \le 2$, then $\Phi(x) = K * \varphi - K(x)$ which is again bounded by the same reason. Finally when $|x| \ge 2$,

$$\Phi(x) = \int_{\mathbf{R}^n} K(x-y)\varphi(y)\,dy - K(x) = \int_{|y| \le 1} [K(x-y) - K(x)]\varphi(y)\,dy$$

so $|\Phi(x)| \le C' \dfrac{\omega(c/|x|)}{|x|^n}$, by the estimate in §4.2 of Chapter II. Since $\omega(\delta)$

is increasing and $\displaystyle\int_0^1 \frac{\omega(\delta)\,d\delta}{\delta} < \infty$, we have proved here our assertion

about Φ. From (20) it follows, because the singular integral operator $\varphi \to \varphi * K$ commutes with dilations, that

(21) $$\varphi_\varepsilon * K - K_\varepsilon = \Phi_\varepsilon, \quad \text{with} \quad \Phi_\varepsilon(x) = \varepsilon^{-n}\Phi(x/\varepsilon).$$

Now we claim that for any $f \in L^p(\mathbf{R}^n)$, $1 < p < \infty$,

(22) $$(\varphi_\varepsilon * K) * f(x) = T(f) * \varphi_\varepsilon(x),$$

where the identity holds for every x. In fact we notice first that

(23) $$(\varphi_\varepsilon * K_\delta) * f(x) = T_\delta(f) * \varphi_\varepsilon(x), \quad \text{for every} \quad \delta > 0$$

because both sides of (20) are equal for each x to the absolutely convergent double integral $\int_{z \in \mathbf{R}^n} \int_{|y| \ge \delta} K(y)f(z-y)\varphi_\varepsilon(x-z)\,dz\,dy$. Moreover $\varphi_\varepsilon \in L^q(\mathbf{R}^n)$, with $1 < q < \infty$ and $1/p + 1/q = 1$, so $\varphi_\varepsilon * K_\delta \to \varphi_\varepsilon * K$ in

L^q norm, and $T_\delta(f) \to T(f)$ in L^p norm, as $\delta \to 0$. This proves (22), and so by (21)

$$T_\varepsilon(f) = (Tf) * \varphi_\varepsilon - f * \Phi_\varepsilon.$$

Passing to the supremum over ε and applying Theorem 2, part (a), we get our asserted majorization for T^*f. The L^p estimates for $f \to Tf$ and the maximal function M then prove the lemma.

3. *Higher Riesz transforms and spherical harmonics*

3.1   We return to our subject proper, the consideration of special transformations of the form

(24)
$$Tf(x) = \lim_{\varepsilon \to 0} \int_{|y| \geq \varepsilon} \frac{\Omega(y)}{|y|^n} f(x - y)\, dy,$$

where Ω is homogeneous of degree 0 and its integral over S^{n-1} vanishes. We have already considered the example $\Omega_j(y) = c\dfrac{y_j}{|y|}$, $j = 1, \ldots, n$. For $n = 1$, $\Omega(y) = c$ sign y, and this is the only possible case. To study the matter further for $n > 1$ we recall the expression (see (6), p. 57),

$$m(x) = \int_{S^{n-1}} \Gamma(x \cdot y)\Omega(y)\, d\sigma(y), \qquad |x| = 1$$

where m is the multiplier arising from the transformation (24).

We have already remarked that the mapping $\Omega \to m$ commutes with rotations. We shall therefore consider the functions on the sphere S^{n-1} (more particularly the space $L^2(S^{n-1})$) from the point of view of its decomposition under the action of rotations. As is well known, this decomposition is in terms of the spherical harmonics, and it is with a brief review of their properties that we begin.

We fix our attention, as always, on $\mathbf{R}^n$, and we shall consider polynomials in $\mathbf{R}^n$ which are also harmonic. Thus we shall define $\mathcal{H}_k$ to be the linear space of homogeneous polynomials of degree k which are harmonic: the *solid spherical harmonics of degree k*. It will be convenient to restrict these polynomials to the surface of the unit sphere S^{n-1}, and there to define the standard inner product,

$$(P, Q) = \int_{S^{n-1}} P(x)\overline{Q(x)}\, d\sigma(x).$$

We can then affirm:

3.1.1 The finite dimensional spaces $\{\mathscr{H}_k\}_{k=0}^{\infty}$ are mutually orthogonal. In fact if $P \in \mathscr{H}_k$, and $Q \in \mathscr{H}_j$ then

$$(k - j) \int_{S^{n-1}} P\bar{Q} \, d\sigma(x) = \int_{S^{n-1}} \left(\bar{Q} \frac{\partial P}{\partial \nu} - P \frac{\partial \bar{Q}}{\partial \nu} \right) d\sigma(x)$$

$$= \int_{|x| \leq 1} [\bar{Q} \, \Delta P - P \, \Delta \bar{Q}] \, dx = 0$$

by Green's theorem, where $\dfrac{\partial}{\partial \nu}$ denotes differentiation with respect to the outward normal, and $\Delta = \sum\limits_{j=1}^{n} \dfrac{\partial^2}{\partial x_j^2}$ is the Laplacean.

3.1.2 If P is any homogeneous polynomial of degree k (not necessarily harmonic) then $P = P_1 + |x|^2 P_2$, where P_1 is homogeneous of degree k, P_2 is homogeneous of degree $k - 2$, and P_1 is harmonic. To prove this we argue as follows. Let $\mathscr{P}_k$ denote the linear space of all homogeneous polynomials of degree k. We write $P(x) = \sum a_\alpha x^\alpha$, where $\alpha = (\alpha_1, \alpha_2, \ldots, \alpha_n)$, $\alpha_1 + \alpha_2 \cdots + \alpha_n = k$ and $x^\alpha = x_1^{\alpha_1} x_2^{\alpha_2} \cdots x_n^{\alpha_n}$. To each such polynomial corresponds its dual object, the differential operator

$$P\left(\frac{\partial}{\partial x}\right) = \sum a_\alpha \left(\frac{\partial}{\partial x}\right)^\alpha, \quad \text{where} \quad \left(\frac{\partial}{\partial x}\right)^\alpha = \left(\frac{\partial}{\partial x_1}\right)^{\alpha_1} \cdots \left(\frac{\partial}{\partial x_n}\right)^{\alpha_n}. \quad \text{On} \quad \mathscr{P}_k \quad \text{we}$$

define a positive inner product $\langle P, Q \rangle = P\left(\dfrac{\partial}{\partial x}\right) \bar{Q}$. Notice that two distinct monomials x^α and $x^{\alpha'}$ are orthogonal with respect to it, and $\langle P, P \rangle = \sum |a_\alpha|^2 \alpha!$ where $\alpha! = (\alpha_1!)(\alpha_2!) \cdots (\alpha_n!)$.

Let $|x|^2 \mathscr{P}_{k-2}$ be the subspace of $\mathscr{P}_k$ of all polynomials of the form $|x|^2 P_2$, where $P_2 \in \mathscr{P}_{k-2}$. Then its orthogonal complement (with respect to $\langle \cdot, \cdot \rangle$) is exactly $\mathscr{H}_k$. In fact P_1 is in this orthogonal complement if and only if $\langle |x|^2 P_2, P_1 \rangle = 0$ for all P_2. But $\langle |x|^2 P_2, P_1 \rangle = \left(P_2 \left(\dfrac{\partial}{\partial x}\right) \Delta \right) \bar{P}_1 = \langle P_2, \Delta P_1 \rangle$, so ΔP_1 is null and thus $\mathscr{P}_k = \mathscr{H}_k \oplus |x|^2 \mathscr{P}_{k-2}$, which proves the assertion (ii).

3.1.3 Let H_k denote the linear space of restrictions of $\mathscr{H}_k$ to the unit sphere. The elements of H_k are the *surface spherical harmonics* of degree k. Then $L^2(S^{n-1}) = \sum\limits_{k=0}^{\infty} H_k$. The L^2 space is taken with respect to usual measure, and the infinite direct sum is taken in the sense of Hilbert space theory. Since we have already proved the mutual orthogonality of the subspaces H_k, we need only observe that every $f \in L^2(S^{n-1})$ can be approximated in the norm, by finite linear combinations of elements from H_k.

That this is possible can be seen as follows. Use §3.1.2 and apply it again
to P_2, repeating this process. This shows that if P is a polynomial then
$P(x) = P_1(x) + |x|^2 P_2(x) + |x|^4 P_2(x) \ldots$, where each of the P_j are
harmonic polynomials. When we set $|x| = 1$, we see that the restriction of
any polynomial on the unit sphere is a finite linear combination of spherical
harmonics.

Since the restrictions of polynomials are dense in $L^2(S^{n-1})$ in the norm,
the assertion is then established. It may also be restated as follows. If
$f \in L^2(S^{n-1})$, then f has the development

$$(25) \qquad f(x) = \sum_{k=0}^{\infty} Y_k(x), \qquad Y_k \in H_k,$$

where the convergence is in the $L^2(S^{n-1})$ norm, and

$$\int_{S^{n-1}} |f(x)|^2 \, d\sigma(x) = \sum_k \int_{S^{n-1}} |Y_k(x)|^2 \, d\sigma(x).$$

3.1.4   Let Δ_S denote the spherical Laplacean. If $Y_k(x) \in H_k$, then
$\Delta_S Y_k(x) = -k(k + n - 2) Y_k(x)$. In fact if $Y(x)$ is any function defined
on the sphere, then $\Delta_S Y(x)$ equals the restriction of the ordinary Laplacean applied to $Y(x)$, but where $Y(x)$ is now defined in the neighborhood
of this sphere by considering it as homogeneous of degree 0. Thus we must
calculate $\Delta(|x|^{-k} P_k(x))$, where $P_k \in \mathscr{H}_k$. But this is

$$|x|^{-k} \Delta P_k + P_k \Delta(|x|^{-k}) + 2 \sum_{j=1}^{n} \frac{\partial}{\partial x_j} |x|^{-k} \frac{\partial}{\partial x_j} P_k.$$

If we carry out the required differentiation, and use the fact that

$$\sum_{j=1}^{n} x_j \frac{\partial}{\partial x_j} P_k = k P_k$$

(Euler's theorem for homogeneous functions), then we obtain our
assertion.

3.1.5   Suppose f has the development (25). Then f (after correction on a
set of measure zero, if necessary) is indefinitely differentiable on S^{n-1} if
and only if

(26) $\int_{S^{n-1}} |Y_k(x)|^2 \, d\sigma(x) = O(k^{-N})$,   as $k \to \infty$, for each fixed N.

To prove this write (25) as $f(x) = \sum_{k=0}^{\infty} a_k Y_0^k(x)$, where the Y_0^k are
normalized;
our assertion is then equivalent with $a_k = O(k^{-N/2})$, as $k \to \infty$. If f is of
class C^2 then an application of Green's theorem shows that

$$\int_{S^{n-1}} \Delta_S f \, \overline{Y_k^0} \, d\sigma = \int_{S^{n-1}} f \Delta_S \overline{Y_k^0} \, d\sigma.$$

Thus if f is indefinitely differentiable,

$$\int_{S^{n-1}} (\Delta_S^r) f \cdot \overline{Y_k^0} \, d\sigma = \int_{S^{n-1}} f(\Delta_S^r \overline{Y_k^0}) \, d\sigma = a_k[-k(k+n-2)]^r,$$

by §3.1.4. So $a_k = O(k^{-2r})$ for every r and therefore (26) holds. To prove the converse we note that (26) implies not only that $f \in L^2(S^{n-1})$, but that for every positive integer r, $(\Delta_S)^r f$, (when suitably defined) also belongs to $L^2(S^{n-1})$. That this implies that f can be corrected on a set of measure zero so as to become indefinitely differentiable will not be proved here; the rather technical argument will be given in Appendix C.

3.2 Thus, after this rapid review of some of the fundamental facts concerning spherical harmonics, we return to the study of special singular integral transforms. First we deal with the interelation of spherical harmonics and the Fourier transform which is in reality the study of the decomposition of the space $L^2(\mathbf{R}^n)$ under the simultaneous action of rotations and the Fourier transform. At its source is the beautiful identity of Hecke.

THEOREM 4. *Suppose $P_k(x)$ is a homogeneous harmonic polynomial of degree k. Then*

(27)
$$\mathscr{F}(P_k(x)e^{-\pi|x|^2}) = i^k P_k(x)e^{-\pi|x|^2}.$$

The identity to be proved can be rewritten as

(28)
$$\int_{\mathbf{R}^n} P_k(x) \exp[-\pi|x|^2 + 2\pi ix \cdot y] \, dx = i^k P_k(y)e^{-\pi|y|^2}.$$

It is clear however that the left side of (28) equals $Q(y)e^{-\pi|y|^2}$ where Q is a polynomial which we see when we apply the differential operator $P_k\left(\dfrac{\partial}{\partial y}\right)$ to both sides of the identity

$$\int_{\mathbf{R}^n} \exp[-\pi|x|^2 + 2\pi ix \cdot y] \, dx = \exp(-\pi|y|^2).$$

The problem is therefore to show that $Q(y) = P_k(iy)$. But $Q(y) = \int_{\mathbf{R}^n} P_k(x) \exp -\pi\{(x_1 - iy_1)^2 + (x_2 - iy_2)^2 \cdots (x_n - iy_n)^2\} \, dx$. However this integral is equal to $\int_{\mathbf{R}^n} P_k(x + iy)e^{-\pi|x|^2} \, dx$, after a shift of the contours of integration in $\mathbf{C}^n$, which is justified by the analyticity of $P_k(x) \exp[-\pi \sum_{j=1}^{n} x_j^2]$, and its rapid decrease. For the same reason we can replace iy by y. This gives $Q(y/i) = \int_{\mathbf{R}^n} P_k(x + y)e^{-\pi|x|^2} \, dx$. Now P_k is harmonic, so its mean value over any sphere centered at y has the value

$P_k(y)$; the factor $e^{-\pi|x|^2}$ is constant on such spheres, while its total integral over $\mathbf{R}^n$ is 1. Thus $Q(y/i) = P_k(y)$, which proves the theorem.

The theorem implies the following generalization of itself, whose interest is that it links the various components of the decomposition of $L^2(\mathbf{R}^n)$, for different n.

If f is a radial function, we write $f = f(r)$, where $|x| = r$.

COROLLARY. *Let $P_k(x)$ be a homogeneous harmonic polynomial of degree k in $\mathbf{R}^n$. Suppose that f is radial and $P_k(x)f(r) \in L^2(\mathbf{R}^n)$. Then the Fourier transform of $P_k(x)f(r)$ is also of the form $P_k(x)g(r)$, with g a radial function. Moreover the induced transformation $f \to g$, $\mathscr{F}_{n,k}(f) = g$, depends essentially only on $n + 2k$. More precisely, we have Bochner's relation*

$$(29) \qquad \mathscr{F}_{n,k} = i^k \mathscr{F}_{n+2k,0}$$

Proof. Consider the Hilbert space of radial functions

$$\mathscr{R} = \left\{ f(r) : \|f\|^2 = \int_0^\infty |f(r)|^2 \, r^{2k+n-1} \, dr < \infty \right\},$$

with the indicated norm. Fix now $P_k(x)$, and assume that P_k is normalized, i.e., $\int_{S^{n-1}} |P_k(x)|^2 \, d\sigma(x) = 1$. Then there is an obvious unitary correspondence between the elements f of $\mathscr{R}$ and the elements $f(|x|)P_k(x)$ in $L^2(\mathbf{R}^{n+2k})$, and $f(|x|)$ in $L^2(\mathbf{R}^{n+2k})$ respectively. What we have to prove is that

$$(30) \qquad (\mathscr{F}_{n,k}f)(r) = i^k \mathscr{F}_{n+2k,0}(f)(r),$$

for each $f \in \mathscr{R}$. First, if $f(r) = e^{-\pi r^2}$, then (30) is an immediate consequence of the theorem (see (27)). Now consider next $e^{-\pi\delta r^2}$ for a fixed $\delta > 0$. Because of the homogeneity of P_k and the interplay of dilations with the Fourier transform (see (1) in §1.2), we get successively,

$$\mathscr{F}(P_k(x)e^{-\pi\delta|x|^2}) = \delta^{-k/2}\mathscr{F}(P_k(\delta^{\frac{1}{2}}x)e^{-\pi\delta|x|^2})$$

$$= i^k \, \delta^{-k/2-n/2} P_k(x/\delta^{\frac{1}{2}})e^{-\pi|x|^2/\delta} = i^k \, \delta^{-k-n/2} P_k(x)e^{-\pi|x|^2/\delta}.$$

This shows that $\mathscr{F}_{n,k}(e^{-\pi\delta r^2}) = i^k\delta^{-k-n/2}e^{-\pi r^2/\delta}$, and so proves (30) for $f(r) = e^{-\pi\delta r^2}$, $\delta > 0$.

To conclude the proof of the corollary it suffices to see that the linear combination of $\{e^{-\pi\delta r^2}\}_{0<\delta<\infty}$, are dense in $\mathscr{R}$. Suppose the contrary. Then there exists a non-zero $g \in \mathscr{R}$, so that $\int_0^\infty e^{-\pi\delta r^2}g(r)r^{2k+n-1}\,dr = 0$, for all $\delta > 0$. Making the change of variables $r^2 \to r$ brings us back to the Fourier-Laplace transform, and by a very well known argument we can show that $g \equiv 0$, concluding the proof of the corollary.

3.3 We come now to what has been our main goal in our discussion of spherical harmonics.

THEOREM 5. *Let $P_k(x)$ be a homogeneous harmonic polynomial of degree k, $k \geq 1$. Then the multiplier corresponding to the transformation (24) with the kernel $\dfrac{P_k(x)}{|x|^{k+n}}$ is*

$$\gamma_k \frac{P_k(x)}{|x|^k} \text{ , where } \gamma_k = i^k \pi^{n/2} \frac{\Gamma\left(\dfrac{k}{2}\right)}{\Gamma\left(\dfrac{k+n}{2}\right)},$$

Notice that if $k \geq 1$, $P_k(x)$ is orthogonal to the constants on the sphere (§3.1.1), and so its mean value over the sphere is zero.

The statement of the theorem can be interpreted as

$$\tag{31} \left(\frac{P_k(x)}{|x|^{k+n}}\right)^{\wedge} = \gamma_k \frac{P_k(x)}{|x|^k}.$$

As such it will be derived from the following closely related fact,

$$\tag{32} \left(\frac{P_k(x)}{|x|^{k+n-\alpha}}\right)^{\wedge} = \gamma_{k,\alpha} \frac{P_k(x)}{|x|^{k+\alpha}} \text{ with } \gamma_{k,\alpha} = i^k \pi^{n/2-\alpha} \frac{\Gamma(k/2 + \alpha/2)}{\Gamma(k/2 + n/2 - \alpha/2)}.$$

LEMMA. *The identity* (32) *holds in the sense that*

$$\tag{33} \int_{\mathbf{R}^n} \frac{P_k(x)}{|x|^{k+n-\alpha}} \hat{\varphi}(x) \, dx = \gamma_{k,\alpha} \int_{\mathbf{R}^n} \frac{P_k(x)}{|x|^{k+\alpha}} \varphi(x) \, dx$$

for every φ which is sufficiently rapidly decreasing at ∞, and whose Fourier transform has the same property. It is valid for all integral k and for $0 < \alpha < n$.

Observe that both the left and right side of (32) are locally integrable in the range $0 < \alpha < n$, and so the integrals in (33) both converge absolutely.

It should be pointed out that both the theorem, and the lemma from which it is deduced, are in effect special cases of the general law (29); however here the context of L^2 is replaced by that of "generalized functions." Still other generalizations of (29) present themselves, but we shall not pursue this point further.

3.4 We turn to the proof of the lemma. We have already observed in §3.2 that $\mathscr{F}(P_k(x)e^{-\pi\delta|x|^2}) = i^k \delta^{-k-n/2} P_k(x)e^{-\pi|x|^2/\delta}$, so we have

$$\int_{\mathbf{R}^n} P_k(x)e^{-\pi\delta|x|^2} \hat{\varphi}(x) \, dx = i^k \, \delta^{-k-n/2} \int_{\mathbf{R}^n} P_k(x)e^{-\pi|x|^2/\delta} \varphi(x) \, dx,$$

if $\delta > 0$.

We now integrate both sides of the above with respect to δ, after having multiplied the equation by a suitable power of δ, ($\delta^{\beta-1}$, $\beta = \dfrac{k+n-\alpha}{2}$, to be precise).

If we use the fact that $\int_0^\infty e^{-\pi\delta|x|^2}\delta^{\beta-1}\,d\delta = (\pi|x|^2)^{-\beta}\Gamma(\beta)$, if $\beta > 0$, we get $\Gamma\!\left(\dfrac{k+n-\alpha}{2}\right)\pi^{-(k+n-\alpha)/2}\displaystyle\int_{\mathbf{R}^n}\dfrac{P_k(x)}{|x|^{k+n-\alpha}}\,\hat{\varphi}(x)\,dx$ for the integral on the left side. The corresponding integration for the right side gives

$$i^k\Gamma\!\left(\frac{k+\alpha}{2}\right)\pi^{-(k+\alpha)/2}\int_{\mathbf{R}^n|x|^{k+\alpha}}\frac{P_k(x)}{}\,\varphi(x)\,dx$$

which leads to identity (33). It is to be observed that when $0 < \alpha < n$ and both φ and $\hat{\varphi}$ decrease sufficiently rapidly (the estimates $|\varphi(x)| \le A(1+|x|)^{-n}$, and $|\hat{\varphi}(x)| \le A(1+|x|)^{-n}$ suffice), then the double integrals that occur in the above manipulation converge absolutely. Thus the formal argument just given establishes the lemma.

To prove the theorem we make the assumption that $k \ge 1$, and we restrict φ further by supposing that $\hat{\varphi}$ is also smooth (the differentiability of $\hat{\varphi}$ near the origin will suffice). Then,

$$(34)\qquad \lim_{\substack{\alpha\to 0\\ \alpha>0}}\int_{\mathbf{R}^n}\frac{P_k(x)}{|x|^{k+n-\alpha}}\,\hat{\varphi}(x)\,dx = \lim_{\varepsilon\to 0}\int_{|x|\ge\varepsilon}\frac{P_k(x)}{|x|^{k+n}}\,\hat{\varphi}(x)\,dx.$$

In fact, since the integral of P_k over any sphere centered at the origin is zero, then

$$\int_{\mathbf{R}^n}\frac{P_k(x)}{|x|^{k+n-\alpha}}\,\hat{\varphi}(x)\,dx = \int_{|x|\le 1}\frac{P_k(x)}{|x|^{k+n-\alpha}}\,[\hat{\varphi}(x)-\hat{\varphi}(0)]\,dx$$

$$+\int_{|x|\ge 1}\frac{P_k(x)}{|x|^{k+n-\alpha}}\,\hat{\varphi}(x)\,dx.$$

Passing to the limit as $\alpha \to 0$ we get for the first integral on the right side,

$$\int_{|x|\le 1}\frac{P_k(x)}{|x|^{k+n}}\,[\hat{\varphi}(x)-\hat{\varphi}(0)]\,dx = \lim_{\varepsilon\to 0}\int_{\varepsilon\le|x|\le 1}\frac{P_k(x)}{|x|^{k+n}}\,\hat{\varphi}(x)\,dx,$$

which proves our assertion (34). Finally, let f be any sufficiently smooth function with compact support, and for fixed x, set $f(x-y) = \hat{\varphi}(y)$. Then since $(\hat{\varphi})^{\wedge}(y) = \varphi(-y)$, we see that $\varphi(y) = \hat{f}(y)e^{-2\pi ix\cdot y}$, and so our assertion is in this case

$$(35)\qquad \lim_{\varepsilon\to 0}\int_{|y|\ge\varepsilon}\frac{P_k(y)}{|y|^{k+n}}\,f(x-y)\,dy = \gamma_k\int_{\mathbf{R}^n}\frac{P_k(y)}{|y|^k}\,\hat{f}\,(y)e^{-2\pi ix\cdot y}\,dy.$$

Now by the definition of the multiplier m, we have

$$\lim_{\varepsilon \to 0} \int_{|y| \geq \varepsilon} \frac{P_k(y)}{|y|^{k+n}} f(x - y) \, dy = \int_{\mathbf{R}^n} m(y) \hat{f}(y) e^{-2\pi i x \cdot y} \, dy$$

where the convergence of both integrals is in the L^2 sense. Because the type of f just described is dense in L^2, we get $m(y) = \gamma_k \dfrac{P_k(y)}{|y|^k}$, which establishes the theorem.

For fixed k, $k \geq 1$, the (finite-dimensional) linear space of operators (24), where $\Omega(y) = \dfrac{P_k(y)}{|y|^k}$ and the P_k range over the homogeneous harmonic polynomials of degree k, form a natural generalization of the Riesz transforms; the latter arise in the special case $k = 1$. Those for $k > 1$, we call the higher Riesz transforms;* they can also be characterized by their invariance properties (see §4.8).

3.5 We now consider two classes of transformations, defined on $L^2(\mathbf{R}^n)$ (which can later also be defined on $L^p(\mathbf{R}^n)$, $1 < p < \infty$). The first class consists of all transformations of the form

$$(36) \qquad T(f) = c \cdot f + \lim_{\varepsilon \to 0} \int_{|y| \geq \varepsilon} \frac{\Omega(y)}{|y|^n} f(x - y) \, dy.$$

c is a constant; Ω is a homogeneous function of degree 0, which is indefinitely differentiable on the sphere S^{n-1}, and whose mean value on that sphere is zero. The second class is given by those transformations T for which

$$(37) \qquad (Tf)^{\wedge}(y) = m(y) \hat{f}(y)$$

where the multiplier m is homogeneous of degree 0 and is indefinitely differentiable on the sphere.

THEOREM 6. *The two classes of transformations, defined by* (36) *and* (37) *respectively, are identical.*

Suppose first that T is of the form (36). Then according to the theorem in §4.2 of Chapter 2, (see also the formula (6) of the present chapter), T is of the form (37) with m homogeneous of degree 0 and

$$(38) \qquad m(x) = c + \int_{S^{n-1}} \Gamma(x \cdot y) \Omega(y) \, d\sigma(y), \qquad |x| = 1.$$

---

* We refer to k as the *degree* of the higher Riesz transform.

Now write the spherical harmonic developments

$$\Omega(y) = \sum_{k=1}^{\infty} Y_k(y), \qquad m(x) = \sum_{k=0}^{\infty} \tilde{Y}_k(x) \qquad \text{and}$$

(39)

$$\Omega_N(x) = \sum_{k=1}^{N} Y_k(x), \qquad m_N(x) = \sum_{k=0}^{N} \tilde{Y}_k(x).$$

Then by the theorem we have just proved, if $\Omega = \Omega_N$, then $m(x) = m_N(x)$, with

$$\tilde{Y}_k(x) = \gamma_k Y_k(x), \qquad 1 \le k.$$

But $m_M(x) - m_N(x) = \int_{S^{n-1}} \Gamma(x \cdot y)[\Omega_M(y) - \Omega_N(y)]\, d\sigma(y)$. Moreover

$$\sup_{x \in S^{n-1}} |m_M(x) - m_N(x)| \le \left(\sup_x \int_{S^{n-1}} |\Gamma(x \cdot y)|^2\, d\sigma(y) \right)^{\frac{1}{2}}$$

$$\times \left(\int_{S^{n-1}} |\Omega_M - \Omega_N|^2\, d\sigma(y) \right)^{\frac{1}{2}} \to 0,$$

as $M, N \to 0$, since

$$\sup_x \int_{S^{n-1}} |\Gamma(x \cdot y)|^2\, d\sigma(y) = \int_{S^{n-1}} |\Gamma(x \cdot y)|^2\, d\sigma(y)$$

$$= c_1 + c_2 \int_0^{\pi} |\log|\cos\theta||^2 (\sin\theta)^{n-2}\, d\theta$$

$$< \infty$$

in view of the fact that $\Gamma(t) = \dfrac{\pi i}{2} \operatorname{sign} t + \log 1/|t|$. This shows that

$$m(x) = c + \sum_{k=1}^{\infty} \gamma_k Y_k(x).$$

Now by the indefinite differentiability of Ω we have that

$$\int_{S^{n-1}} |Y_k(x)|^2\, d\sigma(x) = O(k^{-N})$$

as $k \to \infty$ for every fixed N. However by the explicit form of γ_k, we see that $\gamma_k \approx k^{-n/2}$, so $m(x)$ is also indefinitely differentiable on the unit sphere.

Conversely, suppose $m(x)$ is indefinitely differentiable on the unit sphere and let its spherical harmonic development be as in (39). Set $c = \tilde{Y}_0$, and $Y_k(x) = \dfrac{1}{\gamma_k} \tilde{Y}_k(x)$. Then $\Omega(y)$, given by (39), has mean value zero in the sphere, and is again indefinitely differentiable there. But as we have just seen the multiplier corresponding to this transformation is m; so the theorem is proved.

As an application of this theorem and a final illustration of the singular integral transforms we shall give the generalization of the estimates for partial derivatives given in §1.3.

Let $P(x)$ be a homogeneous polynomial of degree k in $\mathbf{R}^n$. We shall say that P is *elliptic* if $P(x)$ vanishes only at the origin. For any polynomial P we consider also its corresponding differential polynomial. Thus if $P(x) = \sum a_\alpha x^\alpha$, we write $P\left(\dfrac{\partial}{\partial x}\right) = \sum a_\alpha \left(\dfrac{\partial}{\partial x}\right)^\alpha$ where $\left(\dfrac{\partial}{\partial x}\right)^\alpha = \left(\dfrac{\partial}{\partial x_1}\right)^{\alpha_1} \cdots \left(\dfrac{\partial}{\partial x}\right)^{\alpha_k}$, and with the monomials $x^\alpha = x_1^{\alpha_1} \cdots x_n^{\alpha_n}$ (which are of degree $|\alpha| = \alpha_1 + \alpha_2 \cdots + \alpha_n$).

COROLLARY. *Suppose P is a homogeneous elliptic polynomial of degree k. Let $\left(\dfrac{\partial}{\partial x}\right)^\alpha$ be any differential monomial of degree k. Assume f is k times continuously differentiable with compact support. Then we have the a priori estimate*

$$(40) \qquad \left\| \left(\frac{\partial}{\partial x}\right)^\alpha f \right\|_p \le A_p \left\| P\left(\frac{\partial}{\partial x}\right) f \right\|_p, \qquad 1 < p < \infty.$$

To prove this we note, as in §1.3, the following relation between the Fourier transform of $\left(\dfrac{\partial}{\partial x}\right)^\alpha f$ and $P\left(\dfrac{\partial}{\partial x}\right) f$,

$$P(y)\left[\left(\frac{\partial}{\partial x}\right)^\alpha f \right]^{\wedge}(y) = (-2\pi i y)^\alpha \left(P\left(\frac{\partial}{\partial x}\right) f \right)^{\wedge}(y).$$

Since $P(y)$ is non-vanishing except at the origin, $\dfrac{y^\alpha}{P(y)}$ is homogeneous of degree 0 and is indefinitely differentiable on the unit sphere. Thus

$$\left(\frac{\partial}{\partial x}\right)^\alpha f = T\left(P\left(\frac{\partial}{\partial x}\right) f \right),$$

where T is one of the transformations of the type given by (37). By Theorem 6, T is also given by (36) and hence by the results of Chapter II, we get the estimate (40). An extension of this result is indicated in §7.9 of the next chapter.

4. *Further results*

4.1 Our purpose is to show that certain results for $L^1(\mathbf{R}^n)$ may be extended to the finite Borel measures on $\mathbf{R}^n$, i.e. $\mathscr{B}(\mathbf{R}^n)$:

(a) Let $d\mu \in \mathscr{B}(\mathbf{R}^n)$, and $M(d\mu)(x) = \sup \dfrac{1}{m(B(x, r))} \displaystyle\int_{B(x,r)} |d\mu|$. Then

$$m\{x : M(d\mu)(x) > \alpha\} \le \frac{A}{\alpha} \int_{\mathbf{R}^n} |d\mu|.$$

The argument is the same as in the case of integrable functions.
(b) If $d\mu$ is purely singular, then

$$\lim_{r \to 0} \frac{1}{m(B(x, r))} \int_{B(x,r)} d\mu = 0 \text{ for almost every } x.$$

Hint: Write $d\mu = d\mu_1 + d\mu_2$ where $d\mu_1$ is supported on a closed set F of measure zero and $\|d\mu_2\| \le \delta$. Then

$$\lim_{r \to 0} \frac{1}{m(B(x, r))} \int_{B(x,r)} d\mu_1 = 0 \text{ for every } x \notin F.$$

A more general result of this type holds for any approximation of the identity of the type occurring in Theorem 2, in particular for Poisson integrals.

(c) Let $T_\varepsilon(d\mu)(x) = \displaystyle\int_{|x-y| \ge \varepsilon} \frac{\Omega(x - y)}{|x - y|^n} d\mu(y)$, where Ω satisfies the conditions of Theorems 3 and 4 of Chapter II. Then $\lim\limits_{\varepsilon \to 0} T_\varepsilon(d\mu)(x)$ exists almost everywhere.

See e.g. Zygmund [8]; Calderón and Zygmund [1], for part (c).

4.2 Let $u(x, y)$ be harmonic in $\mathbf{R}^{n+1}_+$.
(a) If $1 \le p \le \infty$, $u(x, y)$ is the Poisson integral of an $L^p(\mathbf{R}^n)$ function if and only if $\sup\limits_{y>0} \|u(x, y)\|_p < \infty$.

(b) $u(x, y)$ is the Poisson integral of a measure in $\mathscr{B}(\mathbf{R}^n)$ if and only if

$$\sup_{y>0} \|u(x, y)\|_1 < \infty.$$

See e.g. Stein and Weiss [2]. See also §1.2.1 in Chapter VII.

4.3 Suppose $f \in L^2(\mathbf{R}^n)$, $f_j = R_j(f)$, and $u_j(x, y)$ is the Poisson integral of f_j. Then

$$u_j(x, y) = \int_{\mathbf{R}^n} Q^{(j)}_y(t) f(x - t) \, dt,$$

where

$$Q^{(j)}_y(x) = c_n \frac{x_j}{(|x|^2 + y^2)^{\frac{n+1}{2}}}.$$

4.4 This result, as well as that of Theorem 3 in §2.3, generalizes to $L^p(\mathbf{R}^n)$, $1 < p < \infty$. For details see Horváth [1]; see also §3.2 in Chapter VII. The case $n = 1$ is treated in Titchmarch [1], Chapter 5.

4.5 It is worthwhile to observe the following easily proved facts:
(a) Let $\mathscr{A}$ be the algebra of operators on $L^2(\mathbf{R}^n)$ which is (algebraically) generated by the Riesz transforms $R_1, R_2, \ldots, R_n$. Then every higher Riesz transform belongs to $\mathscr{A}$.

(b) The closure of $\mathscr{A}$ (in the strong operator topology) is identical with the algebra of bounded transformations on $L^2(\mathbf{R}^n)$ which commute with translations and dilations.

4.6 In §4.6 to 4.8 we shall assume that $n \geq 3$. (The cases $n = 1, 2$ would need minor modifications.) We let $SO(n)$ denote the group of proper rotations in $\mathbf{R}^n$, and $SO(n - 1)$ the subgroup of those rotations leaving the direction along the x_1 axis fixed. Then, for every k, the subspace of the polynomials $P_k(x) \in \mathscr{H}_k$, which are fixed by $SO(n - 1)$, i.e. for which $P_k(\rho^{-1}x) = P_k(x)$, $\rho \in SO(n - 1)$, is exactly one-dimensional. See *Fourier Analysis*, Chapter IV.

4.7 Let V be a finite dimensional Hilbert space and $\rho \to R_\rho$ a continuous homomorphism from $SO(n)$ to the group of unitary transformations on V. The couple (R_ρ, V) is called a *representation* of $SO(n)$. It is *irreducible* if there is no non-trivial subspace of V invariant under the R_ρ, $\rho \in SO(n)$. Two representations $(R_\rho^{(1)}, V_1)$ and $(R_\rho^{(2)}, V_2)$ are *equivalent* if there is a unitary correspondence U, $U : V_1 \leftrightarrow V_2$, so that $U^{-1}R_\rho^{(2)}U = R_\rho^{(1)}$.
(a) Let $V = \mathscr{H}_k$ (the linear space of homogeneous harmonic polynomials of degree k). Define

$$(R_\rho P(x)) = P(\rho^{-1}x), \quad \rho \in SO(n), \, P \in \mathscr{H}_k.$$

This representation is irreducible.
(b) An irreducible representation (R_ρ, V) of $SO(n)$ is equivalent to one obtained from the spherical harmonics, as above, if and only if there exists a non-zero $v \in V$ so that

$$R_\rho(v) = v, \quad \text{all } \rho \in SO(n - 1).$$

For the general theory of representations of the rotation group see Weyl [1], Boerner [1]. §4.7 can be deduced from §4.6 by using the Frobenius reciprocity theorem for compact groups. For the reciprocity theorem see Weil [1].

4.8 Let (R_ρ, V) be an irreducible representation of $SO(n)$ as in §4.7 above. Suppose $f \to T(f)$ is a bounded linear transformation from $L^2(\mathbf{R}^n)$ to $L^2(\mathbf{R}^n, V)$; thus T takes complex-valued functions to functions which take their values in V.
(a) Suppose T commutes with translations and dilations and transforms according to (R_ρ, V) in the sense that

$$\rho T \rho^{-1}(f) = R_\rho Tf.$$

Then $T \equiv 0$ unless (R_ρ, V) is equivalent to a representation obtained from spherical harmonics as in §4.7(a).
(b) If (R_ρ, V) arises from the spherical harmonics of degree k, then T is determined up to a constant multiple. In particular if $\beta_1, \beta_2, \ldots, \beta_N$ is a basis for the linear functionals on V, then each $\beta_j(Tf)$ is a higher Riesz transform of degree k, if $k \geq 1$ (when $k = 0$, T is a constant multiple of the identity).

4.9 The following observation will be useful later: Suppose $u(x, y)$ is the Poisson integral of f, $f \in L^p(\mathbf{R}^n)$. Then $\displaystyle\sup_{y>0} \left| y \frac{\partial u}{\partial x_j}(x, y) \right| \leq A(Mf)(x)$.

Hint: Apply Theorem 2 to the case $\varphi(x) = \dfrac{\partial}{\partial x_j} [P_1(x)]$.

A more general result of this kind is that $\sup_{y>0} |y^{|\alpha|} Du(x, y)| \leq A_\alpha M f(x)$, where D is any differential monomial in x and y of total degree $|\alpha|$.

Notes

Sections 1 and 2. The connection of singular integrals (of the "Riesz transform" type) with estimates like those in §1.3 have a long history. See e.g. Friedrichs [1], for $p = 2$; also Calderón and Zygmund [1] for the case of general p. The identification of Riesz transforms with conjugate harmonic functions (in §2.3) goes back to Horváth [1]. For the Fourier analysis of the n-dimensional Poisson integral see Bochner [1], [2], and Bochner and Chandrasekharan [1]. The relation with the maximal function, which generalizes the classical result of Hardy and Littlewood, is in K. T. Smith [1]. The argument in §2.4 comes from Calderón and Zygmund [1].

Section 3. The main results are Theorem 4 and its corollary. The first is implicit in the work of Hecke [1] and was made explicit by Bochner, who also deduced the corollary; see Bochner [2]; also Calderón and Zygmund [5], and Calderón [3]. A more elaborate treatment of some of these topics, in particular spherical harmonics and the connection with Bessel functions, may be found in *Fourier Analysis*, Chapter IV. In comparing the present formulae with those in *Fourier Analysis* one should keep in mind that the Fourier transform defined here corresponds to the inverse Fourier transform in *Fourier Analysis*.

The germinal idea of the calculus of singular integrals in terms of their "symbols" is given in §3.5, although the latter notion is not explicitly defined there. For further details see Mihlin [1], and Calderón and Zygmund [5]. Later developments involving partial differential operators are in Calderón [3], [5], Seeley [1], Kohn and Nirenberg [1], Unterberger and Bokobza [1], and Hörmander [2]. A sketch of the history of the subject may be found in Seeley [2]. The reader is urged to consult it for the work of earlier writers; of particular note, in this connection, are the contributions of Giraud.

CHAPTER IV

The Littlewood-Paley Theory
and Multipliers

The Littlewood-Paley theory of one-dimensional Fourier series, and its applications, represents one of the most far-reaching advances of that subject. The theory originally proceeded along three main lines, each interesting in its own right:

(i) The auxiliary g-function which, aside from its applications, illustrates the principle that often the most fruitful way of characterizing various analytical situations (such as finiteness of L^p norms, existence of limits almost everywhere, etc.) is in terms of appropriate quadratic expressions.

(ii) The "dyadic" decomposition of a function in terms of its Fourier analysis.

(iii) The multiplier theorem of Marcinkiewicz which gives very useful sufficient conditions for L^p multipliers.

This theory was developed in the main between 1930 and 1939 by Littlewood and Paley, Zygmund, and Marcinkiewicz, but it depended on complex function theory and so its full thrust was limited to the case of one-dimension. The n-dimensional theory is more recent and was inspired in part by the real-variable techniques presented in Chapters I and II.

We should, however, not want the reader to be left with an over-simplified picture of the above. Thus it was realized early that significant n-dimensional results could be deduced from the one-dimensional theory. Moreover, the n-dimensional theory is only partly successful in comparison with the one-dimensional case, and much remains to be done in the general context. (The latter point is taken up again §6.2.)

There are by now several possible approaches to the main results which we present, but we have purposely not chosen the shortest and most direct way; we hope, however, that the longer route we shall follow will be more instructive. In this way the reader will have a better opportunity to examine all the working parts of the complex mechanism detailed below.

1. *The Littlewood-Paley g-function*

1.1   The g-function is a (non-linear) operator which allows one to give a useful characterization of the L^p norm of a function on $\mathbf{R}^n$ in terms of the behavior of its Poisson integral. This characterization will be used not only in this chapter, but also in the succeeding chapter dealing with function spaces. The g-function is defined as follows. Let $f \in L^p(\mathbf{R}^n)$ and write $u(x, y)$ for its Poisson integral

$$u(x, y) = \int_{\mathbf{R}^n} P_y(t) f(x - t)\, dt$$

as defined in Chapter III, §2. We let Δ denote the Laplace operator in $\mathbf{R}^{n+1}_+$, that is $\Delta = \dfrac{\partial^2}{\partial y^2} + \sum_{j=1}^{n} \dfrac{\partial^2}{\partial x_j}$; ∇ is the corresponding gradient, $|\nabla u(x, y)|^2 = \left|\dfrac{\partial u}{\partial y}\right|^2 + |\nabla_x u(x, y)|^2$, where $|\nabla_x u(x, y)|^2 = \sum_{j=1}^{n} \left|\dfrac{\partial u}{\partial x_j}\right|^2$. With these notations we define $g(f)(x)$, by

(1)
$$g(f)(x) = \left(\int_0^\infty |\nabla u(x, y)|^2\, y\, dy \right)^{1/2}.$$

The basic result for g is the following.

THEOREM 1.   *Suppose $f \in L^p(\mathbf{R}^n)$, $1 < p < \infty$. Then $g(f)(x) \in L^p(\mathbf{R}^n)$, and*

(2)
$$A'_p \, \|f\|_p \leq \|g(f)\|_p \leq A_p \, \|f\|_p.$$

1.2   It is best to begin with the simple case $p = 2$. With $f \in L^2(\mathbf{R}^n)$, we have $\|g(f)\|_2^2 = \int_{y=0}^{\infty} \int_{\mathbf{R}^n} y |\nabla u(x, y)|^2\, dx\, dy$. The double integral may be treated either by Green's theorem (as we shall see later in §2.1) or by Plancherel's formula, when we integrate first with respect to x. In fact, in view of the identity

$$u(x, y) = \int_{\mathbf{R}^n} \hat{f}(t) e^{-2\pi i t \cdot x} e^{-2\pi |t| y}\, dt$$

we have

$$\frac{\partial u}{\partial y} = \int_{\mathbf{R}^n} -2\pi |t| \hat{f}(t) e^{-2\pi i t \cdot x} e^{-2\pi |t| y}\, dt,$$

and

$$\frac{\partial u}{\partial x_j} = \int_{\mathbf{R}^n} -2\pi i t_j \hat{f}(t) e^{-2\pi i t \cdot x} e^{-2\pi |t| y}\, dt.$$

Thus

$$\int_{\mathbf{R}^n} |\nabla u(x, y)|^2 \, dx = \int_{\mathbf{R}^n} 8\pi^2 |t|^2 |\hat{f}(t)|^2 \, e^{-4\pi |t| y} \, dt, \qquad y > 0$$

and so

$$\|g(f)\|_2^2 = \int_{\mathbf{R}^n} |\hat{f}(t)|^2 \left\{ 8\pi^2 |t|^2 \int_0^\infty e^{-4\pi |t| y} \, y \, dt \right\} = (1/2) \|\hat{f}\|_2^2.$$

Hence,

(3) $$\|g(f)\|_2 = 2^{-1/2} \|f\|_2$$

It may be appropriate here to introduce the following two "partial" g-functions, one dealing with the y differentiation and the other with the x differentiations,

$$g_1(f)(x) = \left(\int_0^\infty \left| \frac{\partial u}{\partial y}(x, y) \right|^2 y \, dy \right)^{1/2},$$

(4)

$$g_x(f)(x) = \left(\int_0^\infty |\nabla_x u(x, y)|^2 y \, dy \right)^{1/2}$$

Note that $g^2 = g_1^2 + g_x^2$, and what is more interesting, the proof of (3) also shows that

$$\|g_1(f)\|_2 = \|g_x(f)\|_2 = \tfrac{1}{2} \|f\|_2.$$

The whole theory could be based just as well on g_1 or g_x instead of g, and anyway the three are closely related by the Riesz transforms (see §7.1).

1.3 The L^p inequalities, when $p \neq 2$, will be obtained as a corollary of the theory of singular integrals in the context of Hilbert space-valued functions, as given in §5 of Chapter II. We define the Hilbert spaces $\mathscr{H}_1$ and $\mathscr{H}_2$ which are to be considered now. $\mathscr{H}_1$ is the one-dimensional Hilbert space of complex numbers. To define $\mathscr{H}_2$ we define first $\mathscr{H}_2^0$ as the L^2 space on $(0, \infty)$ with measure $y \, dy$, i.e.

$$\mathscr{H}_2^0 = \left\{ f : |f|^2 = \int_0^\infty |f(y)|^2 \, y \, dy < \infty \right\}$$

we let $\mathscr{H}_2$ be the direct sum of $n + 1$ copies of $\mathscr{H}_2^{(0)}$; so the elements of $\mathscr{H}_2$ can be represented as $(n + 1)$ component vectors whose entries belong to $\mathscr{H}_2^0$. Since $\mathscr{H}_1$ is the same as the complex numbers, then $B(\mathscr{H}_1, \mathscr{H}_2)$ is of course identifiable with $\mathscr{H}_2$. Now let $\varepsilon > 0$, and keep it temporarily fixed.

Define

$$K_\varepsilon(x) = \left(\frac{\partial P_{y+\varepsilon}(x)}{\partial y}, \frac{\partial P_{y+\varepsilon}(x)}{\partial x_1}, \ldots, \frac{\partial P_{y+\varepsilon}(x)}{\partial x_k} \right)$$

Notice that for each fixed x, $K_\varepsilon(x) \in \mathscr{H}_2$. This is the same as saying that

$$\int_0^\infty \left|\frac{\partial P_{y+\varepsilon}(x)}{\partial y}\right|^2 y\, dy < \infty \quad \text{and} \quad \int_0^\infty \left|\frac{\partial P_{y+\varepsilon}(x)}{\partial x_j}\right|^2 y\, dy < \infty, \text{ for } j = 1, \dots, n.$$

However it is easily seen from the explicit formula (on p. 61) for the Poisson kernel that both $\dfrac{\partial P_y}{\partial y}$ and $\dfrac{\partial P_y}{\partial x}$ are bounded by $\dfrac{A}{(|x|^2 + y^2)^{\frac{n+1}{2}}}$.

So* $|K_\varepsilon(x)|^2 = A^2(n+1) \displaystyle\int_0^\infty \frac{y\, dy}{(|x|^2 + (y+\varepsilon)^2)^{n+1}} \le A_\varepsilon$, and $\le A\, |x|^{-2n}$.

Thus

(5) $|K_\varepsilon(x)| \in L^2(\mathbf{R}^n).$

Similarly

$$\left|\frac{\partial K_\varepsilon(x)}{\partial x_j}\right|^2 \le A\int_0^\infty \frac{y\, dy}{(|x|^2 + (y+\varepsilon))^{n+2}} \le A\int_0^\infty \frac{y\, dy}{(|x|^2 + y^2)^{n+2}} = A'\, |x|^{-2n-2}.$$

Therefore,

(6) $\left|\dfrac{\partial K_\varepsilon(x)}{\partial x_j}\right| \le A/|x|^{n+1},$

with A independent of ε.

Now consider the operator T_ε defined by

$$T_\varepsilon(f)(x) = \int_{\mathbf{R}^n} K_\varepsilon(t) f(x - t)\, dt$$

The functions f are complex-valued (take their value in $\mathscr{H}_1$), but the $T_\varepsilon f(x)$ take their values in $\mathscr{H}_2$. Observe that

(7) $|T_\varepsilon(f)(x)| = \left(\displaystyle\int_0^\infty |\nabla u(x, y + \varepsilon)|^2\, y\, dy\right)^{1/2} \le g(f)(x).$

Hence $\|T_\varepsilon f(x)\|_2 \le 2^{-1/2} \|f\|_2$, if $f \in L^2(\mathbf{R}^n)$, by (3); therefore

(8) $|\hat{K}_\varepsilon(x)| \le 2^{-1/2},$

(which could also be verified directly).

Because of (5), (6), and (8) and in view of Theorem 5, Chapter II, we can apply the Hilbert space version of Theorem 1, Chapter II. The conclusion is $\|T_\varepsilon(f)\|_p \le A_p \|f\|_p$, $1 < p < \infty$ with A_p independent of ε. By (7), for each x, $|T_\varepsilon(f)(x)|$ increases to $g(f)(x)$, as $\varepsilon \to 0$, so we obtain finally

(9) $\|g(f)\|_p \le A_p \|f\|_p, \qquad 1 < p < \infty$

---

* Notice that here the symbol $|K_\varepsilon(x)|$ denotes the norm in $\mathscr{H}_2$ of $K_\varepsilon(x)$, for each x.

1.4 We should have liked to derive the converse inequalities,

$$(10) \qquad A'_p \|f\|_p \leq \|g(f)\|_p, \qquad 1 < p < \infty,$$

directly from (9) and the corollary in §5.3, Chapter II. But this would have required a preliminary argument, since the operator corresponding to g has been obtained as a limit, and this limit is not a principal value arising from the truncation of kernels, (see (7)); in fact, the limiting approach used here is a little more natural, since principal values are not really relevant in this context. Nevertheless, matters can be settled directly and without any difficulty. Take g_1 instead of g. Then the equality $\|g_1(f)\|_2 = (1/2) \|f\|_2$, for $f \in L^2(\mathbf{R}^n)$, leads by polarization to the identity

$$4 \int_{\mathbf{R}^n} \int_0^\infty y \frac{\partial u_1}{\partial y} (x, y) \overline{\frac{\partial u_2}{\partial y}} (x, y) \, dy \, dx = \int_{\mathbf{R}^n} f_1(x) \overline{f_2}(x) \, dx,$$

where $f_1, f_2 \in L^2(\mathbf{R}^n)$, and where u_j are the Poisson integrals of f_j, $j = 1, 2$. This identity, in turn, leads to the inequality

$$4 \left| \int_{\mathbf{R}^n} f_1(x) \overline{f_2}(x) \, dx \right| \leq \int_{\mathbf{R}^n} g_1(f_1)(x) g_1(f_2)(x) \, dx$$

Suppose now in addition that $f_1 \in L^p(\mathbf{R}^n)$ and $f_2 \in L^q(\mathbf{R}^n)$ with $\|f_2\|_q \leq 1$, and $1/p + 1/q = 1$. Then by Hölders inequality and the result (9),

$$(11) \qquad \left| \int_{\mathbf{R}^n} f_1(x) \overline{f_2}(x) \, dx \right| \leq \tfrac{1}{4} \|g_1(f_1)\|_p \|g_1(f_2)\|_q \leq \tfrac{1}{4} A_q \|g_1(f_1)\|_p$$

Now take the supremum in (11) as f_2 ranges over all function in $L^2 \cap L^q$, with $\|f_2\|_q \leq 1$. We obtain therefore the desired result (10), with $A'_p = 4/A_q$, but where f is restricted to be in $L^2 \cap L^p$. The passage to the general case is provided by an easy limiting argument. Let f_m be a sequence of functions in $L^2 \cap L^p$, which converge in L^p norm to f, (a general element of L^p). Notice that $|g(f_m)(x) - g(f_n)(x)| \leq g(f_n - f_m)(x)$. So $g(f_n)$ converges in L^p norm to $g(f)$, and we obtain the inequality (10) for f as a result of the corresponding inequalities for f_n. We have incidentally also proved the following, which we state as a corollary.

COROLLARY. *Suppose $f \in L^2(\mathbf{R}^n)$, and $g_1(f)(x) \in L^p(\mathbf{R}^n)$, $1 < p < \infty$. Then $f \in L^p(\mathbf{R}^n)$, and $A'_p \|f\|_p \leq \|g_1(f)\|_p$.*

1.5 There are some very simple variants of the above that should be pointed out:

(i) The results hold also with $g_x(f)$ instead of $g(f)$. The direct inequality $\|g_x(f)\|_p \leq A_p \|f\|_p$ is of course a consequence of the one for g. The converse inequality is then proved in the same way as that for g_1.

(ii) For any integer k, $k > 1$, define

$$g_k(f)(x) = \left(\int_0^\infty \left| \frac{\partial^k u}{\partial y^k}(x, y) \right|^2 y^{2k-1} \, dy \right)^{1/2}.$$

Then the L^p inequalities hold for g_k as well. Both (i) and (ii) are stated more systematically in §7.2, below.

(iii) For later purposes it will be useful to note that for each x, $g_k(f)(x) \geq A_k g_1(f)(x)$ where the bound A_k depends only on k.

It is easily verified from the Poisson integral formula that if $f \in L^p(\mathbf{R}^n)$, $1 \leq p \leq \infty$, then

$$\frac{\partial^k u(x, y)}{\partial y^k} \to 0 \text{ for each } x, \quad \text{as} \quad y \to \infty.$$

Thus

$$\frac{\partial^k u(x, y)}{\partial y^k} = - \int_y^\infty \frac{\partial^{k+1} u(x, s)}{\partial s^{k+1}} s^k \frac{ds}{s^k}.$$

By Schwarz's inequality, therefore,

$$\left| \frac{\partial^k u(x, y)}{\partial y^k} \right|^2 \leq \int_y^\infty \left| \frac{\partial^{k+1} u(x, s)}{\partial s^{k+1}} \right|^2 s^{2k} \, ds \left(\int_1^\infty s^{-2k} \, ds \right).$$

Hence $(g_k(f, x))^2 \leq \dfrac{1}{2k - 1} (g_{k+1}(f, x))^2$, and the assertion is proved by induction on k.

2. The function g_λ^*

2.1 The proof that was given for the L^p inequalities for the g-function did not in any essential way depend on the theory of harmonic functions, despite the fact that this function was defined in terms of the Poisson integral. In effect, all that was really used is the fact that the Poisson kernels are suitable approximations to the identity. (See the remarks at the end of the previous section, as well as §7.2.)

There is however another approach, which can be carried out without recourse to the theory of singular integrals, but which leans heavily on characteristic properties of harmonic functions. We present it here (more precisely, we present that part which deals with $1 < p \leq 2$, for the inequality (9)), because its ideas can be adapted to other situations where the methods of Chapter II are not applicable. Everything will be based on the following three observations.

LEMMA 1.   *Suppose u is harmonic and strictly positive. Then*

(12) $$\Delta(u)^p = p(p - 1)u^{p-2} |\nabla u|^2.$$

LEMMA 2. *Suppose $F(x, y)$ is continuous in $\bar{\mathbf{R}}_+^{n+1}$, is of class C^2 in $\mathbf{R}_+^{n+1}$, and suitably small at infinity. Then*

$$(13) \qquad \int_{\mathbf{R}_+^{n+1}} y \, \Delta F(x, y) \, dx \, dy = \int_{\mathbf{R}^n} F(x, 0) \, dx$$

LEMMA 3. *If $u(x, y)$ is the Poisson integral of f; then*

$$(14) \qquad \sup_{y > 0} |u(x, y)| \le (Mf)(x).$$

The proof of Lemma 1 is a straightforward exercise in differentiation. Its main interest is that the right-hand side of (12) does not involve any of the second derivatives of u.

To prove Lemma 2, we use Green's theorem

$$\int_D (u \, \Delta v - v \, \Delta u) \, dx \, dy = \int_{\partial D} \left(u \frac{\partial v}{\partial \nu} - v \frac{\partial u}{\partial \nu} \right) d\sigma$$

where $D = B_r \cap \mathbf{R}_+^{n+1}$, with B_r the ball of radius r in $\mathbf{R}^{n+1}$ centered at the origin. We take $v = F$, and $u = y$. Then we will obtain our result (13) if

$$\int_D y \, \Delta F(x, y) \, dx \, dy \to \int_{\mathbf{R}_+^{n+1}} y \, \Delta F(x, y) \, dx \, dy$$

and

$$\int_{\partial D_0} \left(y \frac{\partial F}{\partial \nu} - \frac{\partial y}{\partial \nu} F \right) d\sigma \to 0, \quad \text{as} \quad r \to 0.$$

Here ∂D_0 is the spherical part of the boundary of D. This will certainly be the case, if for example $\Delta F \ge 0$, and $|F| \le O((|x| + y)^{-n-\varepsilon})$ and $|\nabla F| = O((|x| + y)^{-n-1-\varepsilon})$, as $|x| + y \to \infty$, for some $\varepsilon > 0$.

The third lemma is of course a majorization with which the reader is by now familiar. (See part (a) of the theorem in §2.1, Chapter III).

Once these facts have been set down, the proof of the inequality $\|g(f)\|_p \le A_p \|f\|_p$, $1 < p \le 2$, can be accomplished in a few strokes.

Suppose first that $f \ge 0$, is indefinitely differentiable and has compact support. An examination of the Poisson kernel shows that the Poisson integral u of f is strictly positive in $\mathbf{R}_+^{n+1}$, and the majorizations $u(x, y) = O(|x| + y)^{-n}$ and $|\nabla u| = O(|x| + y)^{-n-1}$, as $|x| + y \to \infty$ are valid. We have

$$g(f, x)^2 = \int_0^\infty y |\nabla u(x, y)|^2 \, dy = \frac{1}{p(p - 1)} \int_0^\infty y u^{2-p} \, \Delta(u)^p \, dy$$

$$\le \frac{1}{p(p - 1)} [(Mf)(x)]^{2-p} \int_0^\infty y \, \Delta(u)^p \, dy$$

first using Lemma 1, then Lemma 3, and the hypothesis $1 < p \leq 2$. We can write this as

(15) $$g(f, x) \leq C_p(Mf(x))^{(2-p)/2}(I(x))^{1/2},$$

where

$$I(x) = \int_0^\infty y \, \Delta u^p \, dy.$$

However,

(16) $$\int_{\mathbf{R}^n} I(x) \, dx = \int_{\mathbf{R}^{n+1}_+} y \, \Delta u^p \, dx \, dy = \int_{\mathbf{R}^n} u^p(x, 0) \, dx = \| f \|_p^p,$$

by Lemma 2. This immediately gives the desired result for $p = 2$. Suppose now $1 < p < 2$. By (15)

$$\int_{\mathbf{R}^n} (g(f, x))^p \, dx \leq C_p^p \int_{\mathbf{R}^n} (Mf)(x)^{p(2-p)/2}(I(x))^{p/2} \, dx$$

$$\leq C_p^p \left(\int_{\mathbf{R}^n} (Mf(x))^p \, dx \right)^{1/r'} \left(\int_{\mathbf{R}^n} I(x) \, dx \right)^{1/r},$$

where we have used Hölder's inequality with exponents r and r'

$$1/r + 1/r' = 1, \qquad (1 < r < 2),$$

which is made possible by the fact that $\left(\dfrac{2 - p}{2} \right) pr' = p$, and $rp/2 = 1$, if $r = 2/p$.

By (16) the last factor of the equation is $\| f \|_p^{p/r}$; the next to the last factor is majorized by $C_p' \| f \|_p^{p/r'}$, according to the maximal theorem of Chapter I. Inserting these two estimates gives $\| g(f) \|_p \leq A_p \| f \|_p$, $1 < p \leq 2$, whenever f is a positive function which is indefinitely differentiable and of compact support. For general $f \in L^p(\mathbf{R}^n)$ (which we assume for simplicity to be real-valued), write $f = f^+ - f^-$ as its decomposition into positive and negative part; then we need only approximate in norm f^+ and f^-, each by a sequences of positive indefinitely differentiable functions with compact support. We omit the routine details that are needed to complete the proof.

2.2 It is unfortunate that the elegant argument just given is not valid for $p > 2$. There is, however, a more intricate variant of the same idea which does work for the case $p > 2$, but we do not intend to reproduce it here.*

We shall, however, use the ideas above to obtain a significant generalization of the inequality for the g-functions. We have in mind the inequalities for the positive function g_λ^* defined as follows,

(17) $$(g_\lambda^*(f)(x))^2 = \int_0^\infty \int_{t \in \mathbf{R}^n} \left(\frac{y}{|t| + y} \right)^{\lambda n} |\nabla u(x - t, y)|^2 \, y^{1-n} \, dt \, dy.$$

* See the literature cited at the end of this chapter, and also the argument in §3.3.2, Chapter VII.

2.3 Before going any further, we shall make a few comments that will help to clarify the meaning of the complicated expression (17).

First, $g_\lambda^*(f)(x)$ will turn out to be a pointwise majorant of $g(f)(x)$. To understand this situation better we have to introduce still another quantity, which is roughly midway between g and g_λ^*. It is defined as follows. Let Γ be fixed proper cone in $\mathbf{R}_+^{n+1}$ with vertex at the origin and which contains $(0, 1)$ in its interior. The exact form of Γ will not really matter, but for the sake of definiteness let us choose for Γ the right-circular cone:

$$\Gamma = \{(t, y) \in \mathbf{R}_+^{n+1} : |t| < y, y > 0\}$$

For any $x \in \mathbf{R}^n$, let $\Gamma(x)$ be the cone Γ translated so that its vertex is at x. Now define the positive function $S(f)(x)$ by

$$
\begin{aligned}
(18) \qquad [S(f)(x)]^2 &= \int_{\Gamma(x)} |\nabla u(t,y)|^2 \, y^{1-n} \, dy \, dt \\
&= \int_\Gamma |\nabla u(x-t, y)|^2 \, y^{2-n} \, dy \, dt
\end{aligned}
$$

We assert, as we shall momentarily prove, that

$$(19) \qquad g(f)(x) \le CS(f)(x) \le C_\lambda g_\lambda^*(f)(x).$$

What interpretation can we put on the inequalities relating these three quantities? A hint is afforded by considering three corresponding approaches to the boundary for harmonic functions.

(a) With $u(x, y)$ the Poisson integral of $f(x)$, the simplest approach to the boundary point $x \in \mathbf{R}^n$ is obtained by letting $y \to 0$, (with x fixed). This is the perpendicular approach, and for it the appropriate limit exists almost everywhere, as we already know.

(b) Wider scope is obtained by allowing the variable point (t, y) to approach $(x, 0)$ through any cone $\Gamma(x)$, (where vertex is x). This is the *non-tangential* approach which will be so important for us later (in Chapters VII and VIII). As the reader may have already realized, the relation of the S-function to the g-function is in some sense analogous to the relation between the non-tangential and the perpendicular approaches; we should add that the S-function is of decisive significance in its own right, but we shall not pursue that matter now.*

(c) Finally the widest scope is obtained by allowing the variable point (t, y) to approach $(x, 0)$ in an arbitrary manner, i.e. the unrestricted approach. The function g_λ^* has the analogous role: it takes into account the unrestricted approach for Poisson integrals.

Notice that $g_\lambda^*(x)$ depends on λ. For each x, the smaller λ the greater $g_\lambda^*(x)$, and this behavior is such that the L^p boundedness of g_λ^* depends

* See Chapter VII.

critically on the correct relation between p and λ. This last point is probably the main interest in g_λ^*, and is what makes its study more difficult than g (or S).

After these various heuristic and imprecise indications let us return to firm ground. The only thing for us to prove here is the assertion (19). The inequality $CS(f)(x) \leq C_\lambda g_\lambda^*(f)(x)$ is obvious, since the integral (17) majorizes that part of the integral taken only over Γ, and

$$\left(\frac{y}{|t| + y}\right)^{\lambda n} \geq (1/2)^{\lambda n}$$

there. The non-trivial part of the assertion is:

$$g(f)(x) \leq CS(f)(x).$$

It suffices to prove this inequality for $x = 0$. Let us denote by B_y the ball in $\mathbf{R}_+^{n+1}$ centered at $(0, y)$ and tangent to the boundary of the cone Γ; the radius of B_y is then proportional to y. Now the partial derivatives $\dfrac{\partial u}{\partial y}$ and $\dfrac{\partial u}{\partial x_k}$ are, like u, harmonic functions. Thus by the mean-value theorem

$$\frac{\partial u}{\partial y}(0, y) = \frac{1}{m(B_y)} \int_{B_y} \frac{\partial u}{\partial y}(x, s)\, dx\, ds$$

(where $m(B_y)$ is the $n + 1$ dimensional measure of B_y, i.e. $m(B_y) = cy^{n+1}$ for an appropriate constant c). By Schwarz's inequality

$$\left|\frac{\partial u(0, y)}{\partial y}\right|^2 \leq \frac{1}{m(B_y)} \int_{B_y} \left|\frac{\partial u}{\partial y}(x, s)\right|^2 dx\, ds.$$

If we integrate this inequality we obtain

$$\int_0^\infty y \left|\frac{\partial u(0, y)}{\partial y}\right|^2 dy \leq \int_0^\infty c^{-2} y^{-n} \left(\int_{B_y} \left|\frac{\partial u}{\partial y}(x, s)\right|^2 dx\, ds\right) dy$$

However $(x, s) \in B_y$ clearly implies that $c_1 s \leq y \leq c_2 s$, for two positive constants c_1 and c_2. Thus, apart from a multiplicative factor, the last integral is majorized by

$$\int_\Gamma \left(\int_{c_1 s}^{c_2 s} y^{-n}\, dy\right) \left|\frac{\partial u}{\partial y}(x, s)\right|^2 dx\, ds$$

This is another way of saying that,

$$\int_0^\infty y \left|\frac{\partial u}{\partial y}(0, y)\right|^2 dy \leq c' \int_\Gamma \left|\frac{\partial u(x, y)}{\partial y}\right|^2 y^{1-n}\, dx\, dy.$$

The same is true for the derivatives $\dfrac{\partial u}{\partial x_j}$, $j = 1, \ldots n$, and adding the corresponding estimates proves our assertion.

2.4   We are now in a position to state the result concerning g_λ^*.

THEOREM 2.   *Let λ be a parameter which is greater than 1.*
Suppose $f \in L^p(\mathbf{R}^n)$. Then
(a) *For every $x \in \mathbf{R}^n$, $g(f)(x) \leq C_\lambda g_\lambda^*(f)(x)$.*
(b) *If $1 < p < \infty$, and $p > 2/\lambda$, then*

$$(20) \qquad \|g_\lambda^*(f)\|_p \leq A_{p,\lambda} \|f\|_p.$$

Part (a) of the theorem has already been proved.

The inequalities for $p \geq 2$ will turn out to be rather easy consequences of the corresponding inequalities of the g-function. This we shall now see. For the case $p \geq 2$, only the assumption $\lambda > 1$ is relevant.

Let ψ denote a positive function on $\mathbf{R}^n$; we claim that

$$(21) \qquad \int_{\mathbf{R}^n} (g_\lambda^*(f)(x))^2 \psi(x)\, dx \leq A_\lambda \int_{\mathbf{R}^n} (g(f)(x))^2 (M\psi)(x)\, dx.$$

The left-side of (21) equals

$$\int_{y=0}^{\infty} \int_{t \in \mathbf{R}^n} y\, |\nabla u(t, y)|^2 \left[\int_{x \in \mathbf{R}^n} \psi(x)[|t - x| + y]^{-\lambda n} y^{\lambda n}\, dx \right] dt\, dy$$

so to prove (21) we must show that

$$(22) \qquad \sup_{y > 0} \int_{\mathbf{R}^n} \psi(x)[|t - x| + y]^{-\lambda n} y^{\lambda n} y^{-n}\, dx \leq A_\lambda M(\psi)(t).$$

However we know by Theorem 2, §2.2, of Chapter III, that

$$\sup_{\varepsilon > 0} (\psi * \varphi_\varepsilon)(t) \leq A M(\psi)(t)$$

for appropriate φ, with $\varphi_\varepsilon(x) = \varepsilon^{-n} \varphi(x/\varepsilon)$. Here we have in fact $\varphi(x) = (1 + |x|)^{-\lambda n}$, $\varepsilon = y$, and so with $\lambda > 1$ the hypotheses of that theorem are satisfied. This proves (22) and thus also (21).

The case $p = 2$ follows immediately from (21) by inserting in this inequality the function $\psi = 1$, and using the L^2 result for g. Suppose now $2 < p$; let us set $1/q + 2/p = 1$, and take the supremum of the left side over all $\psi \geq 0$, such that $\psi \in L^q(\mathbf{R}^n)$ and $\|\psi\|_q \leq 1$. The left side of (21) then gives $\|g_\lambda^*(f)\|_p^2$; Hölder's inequality yields as an estimate for the right side:

$$A_\lambda \|g(f)\|_p^2 \|M\psi\|_q.$$

However by the inequalities for the g-function, $\|g(f)\|_p \leq A'_p \|f\|_p$; and by the theorem of the maximal function $\|M\psi\|_q \leq A''_q \|\psi\|_q = A''_q$, since $q > 1$, if $p < \infty$. If we substitute these in the above we get the result:

$$\|g^*_\lambda(f)\|_p \leq A_{p,\lambda} \|f\|_p, \qquad 2 \leq p < \infty, \qquad \lambda > 1.$$

2.5   The inequalities for $p < 2$ will be proved by an adaptation of the reasoning used in §2.1 for g. Lemmas 1 and 2 will be equally applicable in the present situation, but we need a more general version of Lemma 3, in order to majorize the unrestricted approach to the boundary of a Poisson integral.

It is at this stage where results which depend critically on the L^p class first make their appearance. Matters will depend on a variant of the maximal function which we define as follows. Let $\mu \geq 1$, and write $M_\mu(f)(x)$ for

$$(23) \qquad M_\mu(f)(x) = \left(\sup_{r>0} \frac{1}{m(B(x,r))} \int_{B(x,r)} |f(y)|^\mu \, dy\right)^{1/\mu}$$

Then $M_1(f)(x) = M(f)(x)$, and $M_\mu(f)(x) = (M\,|f|^\mu)(x))^{1/\mu}$. From the theorem of the maximal function it then immediately follows that

$$(23') \qquad \|M_\mu(f)\|_p \leq A_{p,\mu} \|f\|_p, \quad \text{for} \quad p > \mu.$$

This inequality fails for $p \leq \mu$, as in the special case $\mu = 1$.

2.5.1   The substitute for Lemma 3 is as follows

LEMMA 4.   *Let* $f \in L^p(\mathbf{R}^n)$, $p \geq \mu$, $\mu \geq 1$; *if* $u(x, y)$ *is the Poisson integral of* f, *then*

$$(24) \qquad |u(x - t, y)| \leq A\left(1 + \frac{|t|}{y}\right)^n M(f)(x),$$

and more generally

$$(24') \qquad |u(x - t, y)| \leq A_\mu\left(1 + \frac{|t|}{y}\right)^{n/\mu} M_\mu(f)(x).$$

We begin by deducing (24).

One notices that (24) is unchanged by the dilatation $(x, t, y) \to (x\delta, t\delta, y\delta)$; it is then clear that it suffices to prove (24) with $y = 1$.

Setting $y = 1$ in the Poisson kernel, we have $P_1(x) = \dfrac{c_n}{(1 + |x|^2)^{\frac{n+1}{2}}}$, and $u(x - t, 1) = f(x) * P_1(x - t)$, for each t. Theorem 2 of Chapter III (in §2.2) shows that $|u(x - t, 1)| \leq A_t(Mf)(x)$, where $A_t = \int Q_t(x) \, dx$, and

$Q_t(x)$ is the smallest decreasing radial majorant of $P_1(x - t)$, i.e.

$$Q_t(x) = c_n \cdot \sup_{|x'| \geq |x|} \left\{ \frac{1}{(1 + |x' - t|^2)^{(n+1)/2}} \right\}$$

For $Q_t(x)$ we have the easy estimates, $Q_t(x) \leq c_n$ for $|x| \leq 2|t|$ and $Q_t(x) \leq A'(1 + |x|^2)^{\frac{-n-1}{2}}$, for $|x| \geq 2|t|$, from which it is obvious that $A_t \leq A(1 + |t|)^n$ and hence (24) is proved.

Since $u(x - t, y) = \int_{s \in R^n} P_y(s) f(x - t - s)\, ds$, and $\int_{R^n} P_y(s)\, ds = 1$, we have $|u(x - t, y)|^\mu \leq \int_{s \in R^n} P_y(s) |f(x - t - s)|^\mu\, ds = U(x - t, y)$, where U is the Poisson integral of $|f|^\mu$. Apply (24) to U; this gives

$$|u(x - t, y)| \leq A^{1/\mu}(1 + |t|/y)^{n/\mu}(M(|f|^\mu)(x))^{1/\mu}$$
$$= A_\mu(1 + |t|/y)^{n/\mu} M_\mu(f)(x),$$

and the Lemma is established.

2.5.2   We shall now complete the proof of the inequality (20) for the case $1 < p < 2$, with the restriction $p > 2/\lambda$.

Let us observe that we can always find a μ, $1 \leq \mu < p$, so that if we set $\lambda' = \lambda - \dfrac{2 - p}{\mu}$, then one still has $\lambda' > 1$. In fact if $\mu = p$, then $\dfrac{\lambda - 2}{\mu} - p > 1$ since $\lambda > 2/p$; this inequality can then be maintained by a small variation of μ. With this choice of μ we have by Lemma 4

$$(25) \qquad |u(x - t, y)| \left(\frac{y}{y + |t|} \right)^{n/\mu} \leq A M_\mu(f)(x)$$

We now proceed as in §2.1, where we treated the function g.

$$(26) \quad (g_\lambda^*(f)(x))^2 = \frac{1}{p(p - 1)} \int_{R_+^{n+1}} y^{1-n} \left(\frac{y}{y + |t|} \right)^{\lambda n} u^{2-p} |\Delta u^p|\, dt\, dy$$

$$\leq A^{2-p}(M_\mu(f)(x))^{2-p} I^*(x),$$

where

$$I^*(x) = \int_{R_+^{n+1}} y^{1-n} \left(\frac{y}{y + |t|} \right)^{\lambda' n} \Delta u^p(x - t, y)\, dt\, dy$$

It is clear that

$$\int_{R^n} I^*(x)\, dx = \int_{R_+^{n+1}} \int_{x \in R^n} y^{1-n} \left(\frac{y}{y + |t - x|} \right)^{\lambda' n} \Delta u^p(t, y)\, dx\, dt\, dy$$

$$= C_{\lambda'} \int_{R_+^{n+1}} y\, \Delta u^p(t, y)\, dt\, dy.$$

The last step follows from the fact that

$$y^{-n} \int_{\mathbf{R}^n} \left(\frac{y}{y + |x|}\right)^{\lambda' n} dx = \int_{\mathbf{R}^n} \frac{dx}{(1 + |x|)^{\lambda' n}} = C_{\lambda'} < \infty, \quad \text{if} \quad \lambda' > 1.$$

So, by Lemma 2

$$\text{(27)} \qquad\qquad \int_{\mathbf{R}^n} I^*(x)\, dx = C_{\lambda'} \|f\|_p^p.$$

Thus (26) takes the role of (15) and (27) that of (16).

The proof is then concluded as in §2.1 if we make use of the L^p bounds for $M_\mu(f)$ in (23'), instead of those for $M(f)$.

3. *Multipliers (first version)*

3.1   The first application of the theory of the functions g and g_λ^* will be in the study of multipliers. The theorem presented below (Theorem 3) will be a "preliminary" version of the multiplier theorem. A "final" form will be presented in §6, where a comparison of the two variants will also be made.

Let m be a bounded measurable function on $\mathbf{R}^n$. One can then define a linear transformation T_m, whose domain is $L^2(\mathbf{R}^n) \cap L^p(\mathbf{R}^n)$, by the following relation between Fourier transforms

$$\text{(28)} \qquad\qquad (T_m f)^\wedge(x) = m(x)\hat{f}(x), \qquad f \in L^2 \cap L^p.$$

We shall say that m is *a multiplier for L^p* $(1 \leq p \leq \infty)$ if whenever $f \in L^2 \cap L^p$ then $T_m f$ is also in L^p, (notice it is automatically in L^2), and T_m is bounded, that is,

$$\text{(29)} \quad \|T_m(f)\|_p \leq A \|f\|_p, \qquad f \in L^2 \cap L^p, \quad \text{(with A independent f)}.$$

The smallest A for which (29) holds will be called the norm of the multiplier. Observe that if (29) is satisfied, and $p < \infty$, then T_m has a unique bounded extension to L^p, which again satisfies the same inequality. We shall also write T_m for this extension.

We denote by $\mathscr{M}_p$ the class of multipliers with the indicated norm. It is clearly a Banach algebra under pointwise multiplication.

We begin with some examples. The observation that the operators T_m commute with translations, together with the propositions in §1.2 and §1.4 of Chapter II, lead directly to the following.

Example (i). $\mathscr{M}_2$ is the class of all bounded measurable functions and the multiplier norm is identical with the $L^\infty(\mathbf{R}^n)$ norm.

Example (ii). $\mathscr{M}_1$ is the class of Fourier transforms of elements of $\mathscr{B}(\mathbf{R}^n)$, (the finite Borel measures), and the norm of $\mathscr{M}_1$ is identical with the norm of $\mathscr{B}(\mathbf{R}^n)$.

The theory of singular integrals of Chapters II and III allows us to assert the following:

Example (iii).   Suppose m is homogeneous of degree 0. If either m is indefinitely differentiable on the sphere, or more generally, if m is representable in the form of equation (26) of Chapter II, (up to an additive constant), then $m \in \mathcal{M}_p$, $1 < p < \infty$.

We now return to some general considerations.

A basic duality property (which we have already used in somewhat different terms in §2.5, Chapter II) and which reflects the duality of L^p spaces, is contained in the proposition that follows.

PROPOSITION.   *Suppose* $1/p + 1/p'$, $1 \leq p \leq \infty$, *then* $\mathcal{M}_p = \mathcal{M}_{p'}$ *with an identity of norms.*

Proof.   Let σ denote the involution $\sigma(f)(x) = \bar{f}(-x)$. As is immediately verified $\sigma^{-1} T_m \sigma = T_{\bar{m}}$; therefore if m belongs to $\mathcal{M}$, so does $\bar{m}$; moreover $\bar{m}$ has the same norm as m. Now by Plancherel's formula,

$$\int_{\mathbf{R}^n} T_m f \bar{g}\, dx = \int_{\mathbf{R}^n} m(x)\hat{f}(x)\overline{\hat{g}(x)}\, dx = \int_{\mathbf{R}^n} \hat{f}(x)\overline{\bar{m}\hat{g}(x)}\, dx$$

$$= \int_{\mathbf{R}^n} f \overline{T_{\bar{m}}g}\, dx, \quad \text{whenever } f, g \in L^2(\mathbf{R}^n)$$

Assume in addition that $f \in L^{p'}(\mathbf{R}^n)$, $g \in L^p(\mathbf{R}^n)$, and $\|g\|_p \leq 1$. Then

$$\left| \int_{\mathbf{R}^n} T_m f \bar{g}\, dx \right| \leq \|f\|_{p'} \|T_m g\|_p \leq A \|f\|_{p'},$$

where A is the norm of the multiplier m (or $\bar{m}$) in $\mathcal{M}$. Taking the supremum over all indicated g, gives

$$\|T_m f\|_{p'} \leq A \|f\|_{p'}$$

Therefore m belongs to $\mathcal{M}_{p'}$, and its $\mathcal{M}_{p'}$ norm is no larger than its $\mathcal{M}_p$ norm; since the situation is symmetric in p and p', the two norms are identical.

We have already pointed out that if m is a multiplier (in $\mathcal{M}_p$), then the transformation T_m, which is bounded in $L^p(\mathbf{R}^n)$, commutes with translations. The converse also holds: Suppose that T is a bounded linear transformation on $L^p(\mathbf{R}^n)$, $p < \infty$, which commutes with translations; then there exists an $m \in \mathcal{M}_p$ so that $T_m = T$. The proof of this fact will be outlined in §7.3 below.

After these clarifying comments about multipliers we should warn the reader that the deeper structure of the class of multipliers $\mathcal{M}_p$ (except in the "trivial" cases corresponding to $p = 1, 2,$ or ∞) is still to a large extent unknown, even in the context of $\mathbf{R}^1$. What we shall obtain below, however, is an important sufficient condition, which incidentally contains to a large extent the results cited in example (iii).

3.2 THEOREM 3.   *Suppose that $m(x)$ is of class C^k in the complement of the origin of $\mathbf{R}^n$, where k is an integer $> n/2$. Assume also that for every differential monomial* $\left(\dfrac{\partial}{\partial x}\right)^{\alpha}$, $\alpha = (\alpha_1, \alpha_2, \ldots \alpha_n)$, *with* $|\alpha| = \alpha_1 + \alpha_2 \cdots$, $+ \alpha_n$, *we have*

$$
(30) \qquad \left| \left(\frac{\partial}{\partial x}\right)^{\alpha} m(x) \right| \le B \, |x|^{-|\alpha|}, \quad \text{whenever} \quad |\alpha| \le k.
$$

Then $m \in \mathcal{M}_p$, $1 < p < \infty$; that is $\| T_m f \|_p \le A_p \| f \|_p$.

The proof will show that the bound A_p will depend only on B, p, and n.

The proof of the theorem leads to a generalization of its statement. This we formulate as a corollary.

COROLLARY.   *The assumption (30) can be replaced by the weaker assumptions,*

$$
|m(x)| \le B'
$$

$$
(31) \qquad \sup_{0 < R < \infty} R^{2|\alpha|+n} \int_{R \le |x| \le 2R} \left| \left(\frac{\partial}{\partial x}\right)^{\alpha} m(x) \right|^2 dx \le B', \qquad |\alpha| \le k.
$$

We mention now two illustrations of the relevance of the theorem:

Example (1). $m(x) = |x|^{it}$, where t is a real number. This example has connection with the Riesz potentials of §1, Chapter V. See also §6.12 in Chapter II.

Example (2). $m(x)$ is homogeneous of degree 0, and is of class C^k on the unit sphere. (See also §3.5 in Chapter III.)

The theorem (and corollary) will be a consequence of the following lemma. Its statement illuminates at the same time the nature of the multiplier transformations considered here, and the role played by the g-functions and their variants.

LEMMA.   *Under the assumptions of Theorem 3 (or its corollary), let us set for each $f \in L^2(\mathbf{R}^n)$*

$$
F(x) = (T_m f)(x).
$$

Then

$$
(32) \qquad g_1(F, x) \le B_\lambda g^*_\lambda(f, x), \quad \text{where} \quad \lambda = 2k/n.
$$

Thus in view of the lemma, the g-functions and their variants are the characterizing expressions which deal at once with all the multipliers considered. On the other hand, the fact that the relation (32) is pointwise shows that to a large extent the mapping T_m is "semi-local."

The theorem is deduced from the lemma as follows. Our assumption on k is such that $\lambda > 1$. Thus Theorem 2 shows us that

$$\|g_\lambda^*(f, x)\|_p \leq A_{\lambda, p} \|f\|_p, \qquad 2 \leq p < \infty \quad \text{if} \quad f \in L^2 \cap L^p.$$

However by Theorem 1, (see the corollary in §1.4), $A_p' \|F\|_p \leq \|g_1(F, x)\|_p$; therefore

$$\|F\|_p = \|T_m f\|_p \leq A_p \|f\|_p, \quad \text{if} \quad 2 \leq p < \infty \quad \text{and} \quad f \in L^2 \cap L^p.$$

That is, $m \in \mathcal{M}_p$, $2 \leq p < \infty$. By duality, the proposition in §3.1, we have also $m \in \mathcal{M}_p$, $1 < p \leq 2$, which gives the assertion of the theorem.

3.3 We shall now prove the lemma.

Let $u(x, y)$ denote the Poisson integral of f, and $U(x, y)$ the Poisson integral of F. Then with $\hat{\ }$ denoting the Fourier transform with respect to the x variable, we have

$$\hat{u}(x, y) = e^{-2\pi|x|y}\hat{f}(x),$$

and

$$\hat{U}(x, y) = e^{-2\pi|x|y}\hat{F}(x) = e^{-2\pi|x|y}m(x)\hat{f}(x).$$

Define $M(x, y) = \int_{\mathbf{R}^n} e^{-2\pi i x \cdot t} e^{-2\pi|t|y}m(t)\, dt$. Then clearly $\hat{M}(x, y) = e^{-2\pi|x|y}m(x)$, and so

$$\hat{U}(x, y_1 + y_2) = \hat{M}(x, y_1)\hat{u}(x, y_2), \qquad y = y_1 + y_2, y_1 > 0.$$

This can be written as

$$U(x, y_1 + y_2) = \int_{\mathbf{R}^n} M(t, y_1)u(x - t, y_2)\, dt.$$

We differentiate this relation k times with respect to y_1 and once with respect to y_2, and set $y_1 = y_2 = y/2$. This gives us the identity

$$(33) \qquad U^{(k+1)}(x, y) = \int_{\mathbf{R}^n} M^{(k)}(t, y/2)u^{(1)}(x - t, y/2)\, dt.$$

(The superscripts denote differentiation with respect to y.)

3.3.1 With the aid of this identity it will not be difficult to prove the lemma. The assumptions (30) (or (31)) on m need be translated in terms of $M(x, y)$. The result is:

$$(34) \qquad |M^{(k)}(t, y)| \leq B' y^{-n-k}$$

$$(34') \qquad \int_{\mathbf{R}^n} |t|^{2k} |M^{(k)}(t, y)|^2\, dt \leq B' y^{-n}.$$

In fact, by the definition of M, it follows that

$$|M^{(k)}(x, y)| \le B(2\pi)^k \int_{\mathbf{R}^n} |t|^k e^{-2\pi|t|y} \, dt = B' \int_0^\infty r^k e^{-2\pi ry} r^{n-1} \, dr = B'' y^{-n-k},$$

which is (34).

To prove (34′) let us show more particularly that

$$\int_{\mathbf{R}^n} |t^\alpha M^{(k)}(t, y)|^2 \, dt \le B' y^{-n},$$

whenever $\alpha = (\alpha_1, \alpha_2, \ldots, \alpha_n)$, so that $\alpha_1 + \alpha_2 \cdots + \alpha_n = k$, with $t^\alpha = t_1^{\alpha_1} t_2^{\alpha_2} \cdots t_n^{\alpha_n}$.

By Plancherel's theorem

$$\|t^\alpha M^{(k)}(t, y)\|_2 = \left\| (2\pi)^{2k} \left(\frac{\partial}{\partial x} \right)^\alpha (|x|^k m(x) e^{-2\pi|x|y}) \right\|_2$$

But

$$y^{2r} \int_{\mathbf{R}^n} |x|^{2r} e^{-4\pi|x|y} \, dx \le C y^{-n}, \qquad \text{for} \qquad 0 \le r,$$

and by the hypothesis (30) and Leibniz's rule $\left| \left(\frac{\partial}{\partial x} \right)^\alpha (|x|^k m(x)) \right| \le B' |x|^{k-|\alpha|}$, with $|\alpha| \le k$. So using Leibniz's rule again to evaluate

$$\left(\frac{\partial}{\partial x} \right)^\alpha (|x|^k m(x) e^{-2\pi|x|y}),$$

we get

$$\|t^\alpha M^{(k)}(t, y)\|_2^2 \le B' y^{-n}, \qquad |\alpha| = k$$

which proves the assertion (34′).

3.3.2 Return to the identity (33), and for each y divide the range of integration into two parts, $|t| \le y/2$ and $|t| \ge y/2$. In the first range use the estimate (34) on $M^{(k)}$ and in the second range use the estimate (34′). This together with Schwarz's inequality gives immediately

$$|U^{(k+1)}(x, y)|^2 \le A y^{-n-2k} \int_{|t| \le y/2} |U^{(1)}(x - t, y/2)|^2 \, dt$$
$$+ A y^{-n} \int_{|t| > y/2} \frac{|U^{(1)}(x - t, y/2)|^2 \, dt}{|t|^{2k}} = I_1(y) + I_2(y).$$

Now

$$(g_{k+1}(F, x))^2 = \int_0^\infty |U^{(k+1)}(x, y)|^2 y^{2k+1} \, dy = \sum_{j=1}^2 \int_0^\infty I_j(y) y^{2k+1} \, dy.$$

However

$$\int_0^\infty I_1(y) y^{2k+1}\, dy \leq B \int_{|t| \leq y/2} |U^{(1)}(x - t, y/2)|^2 y^{-n+1}\, dt\, dy$$

$$\leq B' \int_\Gamma |\nabla u(x - t, y)|^2 y^{-n+1}\, dt\, dy$$

$$= B'(S(F, x))^2 \leq B_\lambda g_\lambda^*(F, x).$$

Similarly

$$\int_0^\infty I_2(y) y^{2k+1}\, dy \leq B' \int_{|t| \geq y} y^{-n+2k+1} |t|^{-2k} |\nabla u(x - t, y)|^2\, dt\, dy$$

$$\leq B'' g_\lambda^*(F, x), \qquad \text{with} \qquad n\lambda = 2k.$$

This shows that $g_{k+1}(F, x) \leq B_\lambda g_\lambda^*(f, x)$. However by §1.5 (see remark (iii)), we know that $g_1(F, x) \leq A_k g_{k+1}(F, x)$. Thus the proof of the lemma is concluded, and with it that of Theorem 2.

It is to be noted that the proof of the corollary is the same as that of the theorem, except for one slight change: In the lemma, the estimate $y^{2r} \int_{\mathbf{R}^n} |x|^{2r} e^{-4\pi|x|y}\, dx \leq Cy^{-n}$, must be replaced by the estimate

$$y^{2r} \int_{\mathbf{R}^n} |x|^{2r} |m_0(x)|^2 e^{-4\pi|x|y}\, dx \leq C'y^{-n},$$

whenever m_0 satisfies the inequality

$$\sup_{0 < R < \infty} R^{-n} \int_{R \leq |x| \leq 2R} |m_0(x)|^2\, dx \leq 1.$$

4. *Application of the partial sums operators*

4.1 We shall now develop the second main tool in the Littlewood-Paley theory, (the first being the usage of the functions g and g^*). It is here already that the n-dimensional theory is so much more restricted than the one-dimensional case, but we postpone further discussion of this point until §4.3.

Let ρ denote an arbitrary *rectangle* in $\mathbf{R}^n$. By rectangle we shall mean (in the rest of this chapter) a possibly infinite rectangle with sides parallel to the axes, i.e. the Cartesian product of n intervals. For each rectangle ρ denote by S_ρ the "partial sum operator," that is the multiplier operator with $m = \chi_\rho =$ characteristic function of the rectangle ρ. So

(35) $$S_\rho(f)^\wedge = \chi_\rho \cdot \hat{f}, \qquad f \in L^2(\mathbf{R}^n) \cap L^p(\mathbf{R}^n).$$

For this operator we have the following theorem.

THEOREM 4.

$$\|S_\rho(f)\|_p \le A_p \, \|f\|_p, \qquad f \in L^2 \cap L^p$$

if $1 < p < \infty$. *The constant* A_p *does not depend on the rectangle* ρ.

We shall need however a more extended version of the theorem which arises when we replace complex-valued functions by functions taking their values in a Hilbert space.

Let $\mathscr{H}$ be the sequence Hilbert space, $\mathscr{H} = \{(c_j)_{j=1}^\infty : (\sum_j |c_j|^2)^{\frac{1}{2}} = |c| < \infty\}$. Then we can represent a function $f \in L^p(\mathbf{R}^n, \mathscr{H})$, as sequences $f(x) = (f_1(x), \ldots f_n(x), \ldots)$, where each f_j is complex-valued and $|f(x)| = (\sum_{j=1}^\infty |f_j(x)|^2)^{\frac{1}{2}}$. Let $\mathfrak{R}$ be a sequence of rectangle, $\mathfrak{R} = \{\rho_j\}_{j=1}^\infty$. Then we can define the operator $S_{\mathfrak{R}}$, mapping $L^2(\mathbf{R}^n, \mathscr{H})$ to itself, by the rule

(36) $\quad S_{\mathfrak{R}}(f) = (S_{\rho_1}(f_1), \ldots, S_{\rho_j}(f_j), \ldots)$, where $f = (f_1, f_2, \ldots, f_j, \ldots)$.

The generalization of Theorem 4 is then as follows:

THEOREM 4′. *Let* $f \in L^2(\mathbf{R}^n, \mathscr{H}) \cap L^p(\mathbf{R}^n, \mathscr{H})$. *Then*

(37) $$\|S_{\mathfrak{R}}(f)\|_p \le A_p \, \|f\|_p, \qquad 1 < p < \infty$$

where A_p *does not depend on the family* $\mathfrak{R}$ *of rectangles.*

4.2 The theorem will be proved in a series of steps, the first two of which already contain the essence of the matter.

4.2.1 First stage: Here $n = 1$, and the rectangles $\rho_1, \rho_2, \ldots \rho_j, \ldots$ are the semi-infinite intervals $(-\infty, 0)$.

We recall the Hilbert transform $f \to H(f)$, which corresponds to the multiplier i sign x (see Chapter III). Then clearly,

(38) $$S_{(-\infty, 0)} = \frac{I + iH}{2}$$

where I is the identity, and $S_{(-\infty, 0)}$ is the partial sum operator corresponding to the interval $(-\infty, 0)$. Everything will depend on the following lemma.

LEMMA. *Let* $f(x) = (f_1(x), \ldots f_j(x), \ldots) \in L^2(\mathbf{R}^n, \mathscr{H}) \cap L^p(\mathbf{R}^n, \mathscr{H})$. *Set* $\tilde{H}f(x) = (Hf_1(x), \ldots Hf_j(x), \ldots)$. *Then*

(39) $$\|\tilde{H}f\|_p \le A_p \, \|f\|_p \qquad 1 < p < \infty$$

where A_p *is the same constant as in the scalar case, i.e. when* $\mathscr{H}$ *is one-dimensional.*

We use the vector-valued version of the Hilbert transform, as is described more generally in §5 of Chapter II. Let the Hilbert spaces $\mathscr{H}_1$ and $\mathscr{H}_2$ be both identical with $\mathscr{H}$. Take in $\mathbf{R}^1$, $K(x) = I \cdot 1/\pi x$, where I is the identity mapping on $\mathscr{H}$. Then the kernel $K(x)$ satisfies all the assumptions of Theorem 5 and Theorem 3 of Chapter II. Moreover

$$\lim_{\varepsilon \to 0} \int_{|y| \geq \varepsilon} K(y) f(x - y) \, dy = \tilde{H}(f)(x),$$

and so our lemma is proved. (Another proof is indicated in §7.12 below.)

Now if all the rectangles are the intervals $(-\infty, 0)$, then because of (38),

$$S_{\mathfrak{R}} = \frac{I + i\tilde{H}}{2}$$

and so because of the lemma the theorem is proved in this case.

4.2.2 *Second stage:* Here $n = 1$, and the rectangles are the intervals $(-\infty, a_1), (-\infty, a_2), \ldots (-\infty, a_j), \ldots$.

Notice that $(f(x)e^{-2\pi i x \cdot a})^\wedge = \hat{f}(x + a)$, therefore $H(e^{-2\pi i x \cdot a}f)^\wedge = i \operatorname{sign} x \hat{f}(x + a)$, and hence $[e^{-2\pi i x \cdot y}H(e^{-2\pi i x \cdot a}f)]^\wedge = i \operatorname{sign}(x - a)\hat{f}(x)$. From this we see that

(40) $$(S_{(-\infty, a_j)}f_j)(x) = \frac{f_j + ie^{2\pi i x \cdot a_j}H(e^{-2\pi i x \cdot a_j}f_j)}{2}$$

If we now write symbolically $e^{-2\pi i x \cdot a}f$ for $(e^{-2\pi i x \cdot a}f_1, \ldots e^{-2\pi i x \cdot a_j}f_j, \ldots)$, where $f = (f_1, \ldots f_j, \ldots)$, then (40) may be rewritten as

(41) $$S_{\mathfrak{R}}f = \frac{f + ie^{2\pi i x \cdot a}\tilde{H}(e^{-2\pi i x \cdot a}f)}{2}$$

and so the result again follows in this case by the lemma.

4.2.3 *Third stage:* General n, but the rectangles ρ_j are the half-spaces $x_1 < a_j$; i.e. $\rho_j = \{x : x_1 < a_j\}$.

Let $S^{(1)}_{(-\infty, a_j)}$ denote the operator defined on $L^2(\mathbf{R}^n)$, which acts only on the x_1 variable, by the action given by $S_{(-\infty, a_j)}$. We claim that

(42) $$S_{\rho_j} = S^{(1)}_{(-\infty, a_j)}$$

This identity is obvious for L^2 functions of the product form

$$f'(x_1)f''(x_2, \ldots, x_n);$$

since their linear span is dense in L^2 the identity (42) is established.

We now use the L^p inequality, which is the result of the previous stage, for each fixed $x_2, x_3, \ldots, x_n$. We raise this inequality to the pth power and integrate with respect to $x_2, \ldots x_n$. This gives the desired result for the present case. Notice that the result holds as well if the half-space $\{x:x_1 < a_j\}_{j=1}^{\infty}$, is replaced by the half-space $\{x:x_1 > a_j\}_{j=1}^{\infty}$; or if the role of the x_1 axis is taken by the x_2 axis, etc.

4.2.4   Final stage: Observe that every general finite rectangle of the type considered is the intersection of $2n$ half-spaces, each half-space having its boundary hyperplane perpendicular to one of the axes of $\mathbf{R}^n$. Thus a $2n$-fold application of the result of the third stage proves the theorem, where the family $\mathfrak{R}$ is made up of finite rectangles. Since the bounds obtained do not depend on the family $\mathfrak{R}$, we can pass to the general case where $\mathfrak{R}$ contains possibly infinite rectangles by an obvious limiting argument.

4.3   Some problems. We wish to make some remarks about the limitations of Theorems 4 and 4′. When $n = 1$, the theorems deal with the partial sum operators taken with respect to intervals. Not much more can be wished for in the one-dimensional case since intervals are the only "regular" sets in $\mathbf{R}^1$: they are the only convex sets, the only connected sets, etc.

However when $n > 1$, the situation changes radically. The rectangles with sides parallel to the axes are now only *very special* sets, and the fact that we consider only those is a serious limitation of the generality of the theorems. Thus what we have proved is only an n-fold superposition of the one-dimensional results, and is not genuinely an n-dimensional result.

To clarify the situation, we wish to describe two particular test problems for an essentially n-dimensional theory. These problems are interesting in themselves, but the solution of each would surely have many further consequences.

PROBLEM A.   *Let B be the unit ball in* $\mathbf{R}^n$. *Can we replace the rectangle ρ by the ball B in Theorem 4?*

It is known that the answer can be affirmative only in the range $\dfrac{2n}{n+1} < p < \dfrac{2n}{n-1}$, but there is no positive result, except when $p = 2$. See §7.7 and §7.8 below.

PROBLEM B.   *Can the rectangles of Theorem 4′ be replaced by rectangles that are each arbitrarily rotated?*

It can be shown that the positive solution of Problem A would imply the resolution of Problem B for the same p. It can also be shown that the answer to Problem B is in the negative for p outside the interval $\dfrac{2n}{n+1} \leq p \leq \dfrac{2n}{n-1}$.*

4.4 We state here the continuous anologue of Theorem 4′. Let $(\Gamma, d\gamma)$ be an abstract measure space, and consider the Hilbert space $\mathcal{H}$ of square integrable functions on Γ, i.e. $\mathcal{H} = L^2(\Gamma, d\gamma)$. The elements

$$f \in L^p(\mathbf{R}^n, \mathcal{H})$$

are the complex-valued functions $f(x, \gamma) = f_\gamma(x)$ on $\mathbf{R}^n \times \Gamma$, which are jointly measurable, and for which $(\int_{\mathbf{R}^n} (\int_\Gamma |f(x, \gamma)|^2 \, d\gamma)^{p/2} \, dx)^{1/p} = \|f\|_p < \infty$, (if $p < \infty$). In analogy with §4.1 let $\Re = \{\rho_\gamma\}_{\gamma \in \Gamma}$, and suppose that the mapping $\gamma \to \rho_\gamma$ is a measurable function from Γ to rectangles; that is, the numerical-valued functions which assign to each γ the components of the vertices of ρ_γ are all measurable.

Suppose $f \in L^2(\mathbf{R}^n, \mathcal{H})$. Then we define $F = S_\Re f$ by the rule

$$F(x, \gamma) = S_{\rho_\gamma}(f_\gamma)(x), \qquad (f_\gamma(x) = f(x, \gamma)).$$

THEOREM 4″.

(42) $$\|S_\Re f\|_p \leq A_p \|f\|_p, \qquad 1 < p < \infty$$

for $f \in L^2(\mathbf{R}^n, \mathcal{H}) \cap L^p(\mathbf{R}^n, \mathcal{H})$, where the bound A_p does not depend on the measure space (Γ, γ), or on the function $\gamma \to \rho_\gamma$.

The proof of this theorem is an exact repetition of the argument given for Theorem 4′. The reader may, if he wishes, also obtain it from Theorem 4′ by a limiting argument.

5. *The dyadic decomposition*

5.1 We shall now consider a cannonical decomposition of $\mathbf{R}^n$ into rectangles. First, in the case of $\mathbf{R}^1$ we decompose it as the union of the "disjoint" intervals (that is, whose interiors are disjoint) $[2^k, 2^{k+1}]$, $-\infty < k < \infty$, and $[-2^{k+1}, -2^k]$, $-\infty < k < \infty$. This double collection of intervals, one collection for the positive half-line, the other for the negative half-line, will be the dyadic decomposition of $\mathbf{R}^1$. (Strictly speaking, the origin is left out; but for the sake of simplicity of terminology we still refer to it as the decomposition of $\mathbf{R}^1$.) Having obtained this decomposition of $\mathbf{R}^1$, we take the corresponding product decomposition for $\mathbf{R}^n$. Thus we write $\mathbf{R}^n$ as the union of "disjoint" rectangles, which rectangles are products of the intervals which occur for the dyadic decomposition of each of the axes. This is the *dyadic decomposition* of $\mathbf{R}^n$.

---

* Y. Meyer, personal communication.

The family of resulting rectangles will be denoted by Δ. We recall the partial sum operator S_ρ, defined in (35) for each rectangle. Now in an obvious sense, (e.g. L^2 convergence)

$$\sum_{\rho \in \Delta} S_\rho = \text{Identity}.$$

Also in the L^2 case the different blocks, $S_\rho(f)$, $\rho \in \Delta$, behave as if they were independent; they are of course mutually orthogonal. To put the matter precisely: The L^2 norm of f can be given exactly in terms of the L^2 norms of the $S_\rho f$, that is

$$(43) \qquad \sum_{\rho \in \Delta} \|S_\rho f\|_2^2 = \|f\|_2^2,$$

(and this is true for *any* decomposition of $\mathbf{R}^n$). For the general L^p case not as much can be hoped for, but the following important theorem can nevertheless be established.

THEOREM 5.   *Suppose* $f \in L^p(\mathbf{R}^n)$, $1 < p < \infty$. *Then*

$$(\sum_{\rho \in \Delta} |S_\rho f(x)|^2)^{\frac{1}{2}} \in L^p(\mathbf{R}^n), \text{ and the ratio } \|(\sum_{\rho \in \Delta} |S_\rho f(x)|^2)^{\frac{1}{2}}\|_p / \|f\|_p$$

is contained between two bounds (*independent of* f).

5.2   The Rademacher functions provide a very useful device in the study of L^p norms in terms of quadratic expressions. These functions, $r_0(t), r_1(t), \dots, r_m(t), \dots$ are defined on the interval $(0, 1)$ as follows: $r_0(t) = +1$, for $0 \le t \le 1/2$, and $r_0(t) = -1$ for $1/2 < t \le 1$; r_0 is extended outside the unit interval by periodicity, that is $r_0(t + 1) = r_0(t)$. In general $r_m(t) = r_0(2^m t)$. The sequence of Rademacher functions are orthonormal (and in fact mutually independent) over $[0, 1]$. For our purposes their importance arises from the following fact. Suppose $\sum_0^\infty |a_m|^2 < \infty$ and set $F(t) = \sum_{m=0}^\infty a_m r_m(t)$. Then $F(t) \in L^p[0, 1]$ for every $p < \infty$, and for $p < \infty$

$$(44) \qquad A_p \|F\|_p \le \|F\|_2 = \left(\sum_{m=0}^\infty |a_m|^2 \right)^{\frac{1}{2}} \le B_p \|F\|_p$$

for two positive constants A_p and B_p.

Thus for functions which can be expanded in terms of the Rademacher functions, all the L^p norms, $p < \infty$, are comparable.

We shall also need the n-dimensional form of (44). We consider the unit cube Q in $\mathbf{R}^n$, $Q = \{t = (t_1, t_2, \dots t_n) : 0 \le t_j \le 1\}$. Let m be an n-tuple of non-negative integers $m = (m_1, m_2, \dots, m_n)$. Define $r_m(t) = r_{m_1}(t_1) r_{m_2}(t_2) \cdots r_{m_n}(t_n)$. Write $F(t) = \sum a_m r_m(t)$. With

$$\|F\|_p = \left(\int_Q |F(t)|^p \, dt \right)^{1/p},$$

we also have (44), whenever $\sum |a_m|^2 < \infty$. The proof of these facts is not overly long, but it will be best not to digress at this point. For this reason we postpone it until later, and present it in an appendix.*

5.3 We come now to the proof of the theorem itself. It will be presented in several steps.

5.3.1 We show here that it suffices to prove the inequality

$$(45) \qquad \left\|\left(\sum_{\rho \in \Delta} |S_\rho f(x)|^2\right)^{\frac{1}{2}}\right\|_p \leq A_p \|f\|_p, \quad 1 < p < \infty$$

for $f \in L^2(\mathbf{R}^n) \cap L^p(\mathbf{R}^n)$. To see this let $g \in L^2(\mathbf{R}^n) \cap L^q(\mathbf{R}^n)$, $1/p + 1/q = 1$, and consider the identity:

$$\sum_{\rho \in \Delta} \int_{\mathbf{R}^n} S_\rho(f) \overline{S_\rho(g)} \, dx = \int_{\mathbf{R}^n} f \bar{g} \, dx$$

which follows from (43) by polarization. By Schwarz's inequality and then Hölder's inequality

$$\left| \int_{\mathbf{R}^n} f \bar{g} \, dx \right| \leq \int_{\mathbf{R}^n} \left(\sum_\rho |S_\rho f|^2\right)^{\frac{1}{2}} \left(\sum_\rho |S_\rho g|^2\right)^{\frac{1}{2}} dx$$

$$\leq \left\|\left(\sum_\rho |S_\rho f|^2\right)^{\frac{1}{2}}\right\|_p \left\|\left(\sum_\rho |S_\rho g|^2\right)^{\frac{1}{2}}\right\|_q .$$

Taking the supremum over all such g with the additional restriction that $\|g\|_q \leq 1$, gives $\|f\|_p$ for the left side of the above inequality. The right side is majorized by $\|(\sum |S_\rho f|^2)^{\frac{1}{2}}\|_p A_q$, since we assume (45) for all p, (in particular q). Thus we have also

$$(46) \qquad B_p \|f\|_p \leq \left\|\left(\sum_\rho |S_\rho f|^2\right)^{\frac{1}{2}}\right\|_p .$$

To dispose of the additional assumption that $f \in L^2$, for $f \in L^p$ take $f_j \in L^2 \cap L^p$ so that $\|f_j - f\|_p \to 0$; use the inequalities (45) and (46) for f_j and $f_j - f_{j'}$; after a simple limiting argument we get (45) and (46) for f as well.

5.3.2 Here we shall prove the inequality (45) for $n = 1$. We shall need first to introduce a little more notation. We let Δ_1 be the family of dyadic intervals in $\mathbf{R}^1$, as explained in §5.1; we can enumerate them as $I_0, I_1, \ldots,$ $I_m \ldots$ (the order is here immaterial). For each $I \in \Delta_1$ we consider the partial sum operator S_I, and a modification of it that we now define. Let

* Appendix D.

φ be a fixed function* of class C^1 with the following properties:

$$\begin{cases} \varphi(x) = 1 & \text{if} \quad 1 \leq x \leq 2 \\ \varphi(x) = 0 & \text{if} \quad x \leq 1/2, \quad \text{or} \quad x \geq 4. \end{cases}$$

Suppose I is any dyadic interval, and assume that it is of the form $[2^k, 2^{k+1}]$. Define $\tilde{S}_I$ by

(47) $$(\tilde{S}_I f)^\wedge(x) = \varphi(2^{-k}x)\hat{f}(x) = \varphi_I(x)\hat{f}(x).$$

That is, $\tilde{S}_I$, like S_I, is a multiplier transformation where the multiplier is equal to one on the interval I; but unlike S_I, the multiplier of $\tilde{S}_I$ is smooth.

A similar definition is made for $\tilde{S}_I$ when $I = [-2^{k+1}, -2^k]$. We observe that

(48) $$S_I \tilde{S}_I = S_I,$$

since S_I has as multiplier the characteristic function of I.

Now for each $t \in [0, 1]$, consider the multiplier transformation

$$\tilde{T}_t = \sum_{m=0}^{\infty} r_m(t)\tilde{S}_{I_m}$$

That is, $\tilde{T}_t$ is for each t the multiplier transformation whose multiplier is $m_t(x)$, with

(49) $$m_t(x) = \sum_m r_m(t)\varphi_{I_m}(x).$$

By the definition of φ_{I_m} it is clear that for any x at most three terms in the sum (49) can be non-zero. Moreover we also see easily that

(50) $$|m_t(x)| \leq B, \qquad \left|\frac{dm_t}{dx}(x)\right| \leq B/|x|$$

where B is independent of t. Thus by the multiplier theorem (Theorem 3 in §3)

(51) $$\|\tilde{T}_t f\|_p \leq A_p \|f\|_p, \quad \text{for} \quad f \in L^2 \cap L^p$$

and with A_p independent of t. From this it follows obviously that

$$\left(\int_0^1 \|\tilde{T}_t(f)\|_p^p \, dt \right)^{1/p} \leq A_p \|f\|_p.$$

However

$$\int_0^1 \|\tilde{T}_t(f)\|_p^p \, dt = \int_0^1 \int_{\mathbf{R}^1} |\sum r_m(t)(\tilde{S}_{I_m}f)(x)|^p \, dx \, dt$$

$$\geq A_p' \int_{\mathbf{R}^1} \left(\sum_m |\tilde{S}_{I_m}f(x)|^2 \right)^{p/2} dx,$$

---

* It is kept fixed in the rest of this argument.

by the property (44) of the Rademacher functions. Thus we have

$$(52) \qquad \left\| \left(\sum_m |\tilde{S}_{I_m}(f)|^2 \right)^{\frac{1}{2}} \right\|_p \leq B_p \, \|f\|_p.$$

Now apply the general theorem about partial sums, Theorem 4′, with $\Re = \Delta_1$ here; and using (48) we get

$$(53) \qquad \left\| \left(\sum_m |S_{I_m}(f)|^2 \right)^{\frac{1}{2}} \right\|_p \leq C_p \, \|f\|_p,$$

which is the one-dimensional case of the inequality (45), and this is what we had set out to prove.

5.3.3   We are still in the one-dimensional case, and we write T_t for the operator

$$T_t = \sum_m r_m(t) S_{I_m}.$$

Our claim is that

$$(54) \qquad \|T_t(f)\|_p \leq A_p \, \|f\|_p, \qquad 1 < p < \infty,$$

with A_p independent of t, and $f \in L^2 \cap L^p$.

Write $T_t^N = \sum_{m=0}^N r_m(t) S_{I_m}$, and it suffices to show that (54) holds, with T_t^N in place of T_t (and with A_p independent of N and t). Since each S_{I_m} is a bounded operator on L^2 and L^p, we have that $T_t^N f \in L^2 \cap L^p$ and so we can apply (46) to it, which has already been proved (in the case $n = 1$). So

$$B_p \, \|T_t^N f\|_p \leq \left\| \left(\sum^N |S_{I_m} f|^2 \right)^{\frac{1}{2}} \right\|_p \leq C_p \, \|f\|_p,$$

using also (53). Letting $N \to \infty$, we get (54).

5.3.4   We now turn to the n-dimensional case and define $T_{t_1}^{(1)}$, as the operator T_{t_1} acting only on the x_1 variable. Then by the inequality (54) we get

$$(55) \qquad \int_{\mathbf{R}^1} |T_{t_1}^{(1)} f(x_1, x_2, \ldots x_n)|^p \, dx_1 \leq A_p^p \int_{\mathbf{R}^1} |f(x_1, \ldots x_n)|^p dx_1$$

for almost every fixed $x_2, x_3, \ldots x_n$, since $x_1 \to f(x_1, x_2, \ldots x_n) \in L^2(\mathbf{R}^1) \cap L^p(\mathbf{R}^1)$ for almost every fixed $x_2, \ldots x_n$, if $f \in L^2(\mathbf{R}^n) \cap L^p(\mathbf{R}^n)$. If we integrate (55) with respect to $x_2, \ldots x_n$ we then obtain

$$(56) \qquad \|T_{t_1}^{(1)} f\|_p \leq A_p \, \|f\|_p, \qquad f \in L^2 \cap L^p,$$

with A_p independent of t_1. The same inequality of course holds with x_1 replaced by x_2, or x_3, etc.

5.3.5   We come now to the final step of the proof. We first describe the additional notation we shall need. With Δ representing the collection of dyadic rectangles in $\mathbf{R}^n$, we write any $\rho \in \Delta$, as $\rho = I_{m_1} \times I_{m_2}, \ldots \times I_{m_n}$ where $I_0, I_1, \ldots I_m, \ldots$ represents the (arbitrary) enumeration of the dyadic intervals used above. Thus if $m = (m_1, m_2, \ldots m_n)$, (with each $m_j \geq 0$), we write $\rho_m = I_{m_1} \times I_{m_2} \cdots \times I_{m_n}$.

We now apply the operator $T_{t_1}^{(1)}$ for the x_1 variable, and successively its analogues for x_2, x_3, etc. The result is

(57) $\|T_t(f)\|_p \leq A_p^n \|f\|_p.$

Here

$$T_t = \sum_{\rho_m \in \Delta} r_m(t) S_{\rho_m}$$

with $r_m(t) = r_{m_1}(t_1) \cdots r_{m_n}(t_n)$ as described in §5.2. The inequality holds uniformly for each $(t_1, t_2, \ldots t_n)$ in the unit cube Q.

We raise this inequality to the pth power and integrate it with respect to t, making use of the properties of the Rademacher functions cited in (44). We then get, as in the analogous proof of (52), that

$$\left\| \left(\sum_{\rho_m \in \Delta} |S_{\rho_m} f|^2 \right)^{\frac{1}{2}} \right\|_p \leq A_p \|f\|_p$$

if $f \in L^2(\mathbf{R}^n) \cap L^p(\mathbf{R}^n)$. This together with the remarks of §5.3.1 concludes the proof of Theorem 5.

6. *The Marcinkiewicz multiplier theorem*

6.1   We now present the second version of the multiplier theorem. This form is to a large extent the synthesis of the ideas developed in §4 and 5, and as such is one of the most important results of the whole theory. For the sake of clarity we state first the one-dimensional case.

THEOREM 6.   *Let m be a bounded function on $\mathbf{R}^1$, which is of bounded variation on every finite interval not containing the origin. Suppose*

(a) $|m(x)| \leq B, \ -\infty < x < \infty$

(b) $\int_I |dm(x)| \leq B$, *for every dyadic interval I.*

Then $m \in \mathcal{M}_p$, $1 < p < \infty$; and more precisely, if $f \in L^2 \cap L^p$

$$\|T_m(f)\|_p \leq A_p \|f\|_p,$$

where A_p depends only on B and p.

To present the general theorem we consider $\mathbf{R}^1$ as divided into its two half-lines, $\mathbf{R}^2$ as divided into its four quadrants, and generally $\mathbf{R}^n$ as divided into its 2^n "octants." Thus the first octant in $\mathbf{R}^n$ will be the open "rectangle" of those x all of whose coordinates are strictly positive. We shall assume that $m(x)$ is defined on each such octant and is there continuous together with its partial derivatives up to and including order n. Thus m may be left undefined on the set of points where one or more coordinate variables vanishes.

For every $k \leq n$, we regard $\mathbf{R}^k$ embedded in $\mathbf{R}^n$ in the following obvious way: $\mathbf{R}^k$ is the subspace of all points of the form $(x_1, x_2, \ldots, x_k, 0, \ldots 0)$.

THEOREM 6'. *Let m be a bounded function on $\mathbf{R}^n$ of the type described. Suppose also*

(a) $|m(x)| \leq B$

(b) *for each $0 < k \leq n$*

$$\sup_{x_{k+1}, \ldots, x_n} \int_\rho \left| \frac{\partial^k m}{\partial x_1 \, \partial x_2 \cdots \partial x_k} \right| dx_1 \cdots dx_k \leq B$$

as ρ ranges over dyadic rectangles of $\mathbf{R}^k$. (If $k = n$ the "sup" sign is omitted.)

(c) *The condition analogous to (b) is valid for every one of the $n!$ permutations of the variables $x_1, x_2, \ldots, x_n$.*

Then $m \in \mathcal{M}_p$, $1 < p < \infty$; and more precisely, if $f \in L^2 \cap L^p$, $\|T_m f\|_p \leq A_p \|f\|_p$, where A_p depends only on B, p, and n.

6.2 Comments. Before we come to the proof of the theorem we need to clarify certain matters of a technical nature; also the relation of this theorem with the first multiplier theorem, treated in §3.

6.2.1 Theorem 6 appears stronger than Theorem 6' for the case $n = 1$, because the hypotheses of the second theorem require that $m(x)$ is continuously differentiable away from the origin, while Theorem 6 requires only that m is of bounded variation in intervals separated from the origin. However, in reality, both results are equally strong, since whenever m satisfies the hypotheses of Theorem 6, we can find a sequence $\{m_j(x)\}$ (with $m_j(x)$ in fact indefinitely differentiable away from the origin), for which the bounds of Theorem 6' hold uniformly in j, and so that $m_j(x) \to m(x)$ almost everywhere. It then follows that $T_{m_j} \to T_m$, and in this way the assertion can be established. We leave the details to the interested reader.

6.2.2 Another pedantic remark is the following. As the reader may have surmised, there is nothing indispensable about the role of the powers of 2 in the definition of the dyadic rectangles in Theorems 5, 6, and 6'. The dyadic rectangles could in fact be replaced by other rectangles; for instance, those whose vertices have coordinates $\{-\lambda_k\}_{k=-\infty}^{k=\infty}$, and $\{\lambda_k\}_{k=-\infty}^{k=\infty}$, instead of $\{-2^k\}_{k=-\infty}^{k=\infty}$ and $\{2^k\}_{k=-\infty}^{k=\infty}$, with $\lambda_{k+1}/\lambda_k \geq r > 1$, all k. However the conclusions obtained this way are no stronger than in the dyadic case. (See also §7.10 below.)

6.2.3 It is more interesting to compare Theorem 6', with our first multiplier theorem, Theorem 3 and its corollary. It is clear that for $n = 1$ Theorem 6' is the stronger. However for $n \geq 2$, they overlap and neither includes the other. This difference for $n \geq 2$ is also illustrated by simple invariance considerations. Thus the class of multipliers treated by Theorem 3 is invariant under dilations, $m(x) \to m(\varepsilon x)$, $\varepsilon > 0$, and also under rotations, $m(x) \to m(\rho^{-1}x)$. The set of multipliers of Theorem 6' is not invariant under rotations, but is however invariant under a larger group of dilations, $m(x) \to m(\varepsilon \circ x)$, where $(\varepsilon \circ x) = (\varepsilon_1 x_1, \varepsilon_2 x_2, \ldots, \varepsilon_n x_n)$, $x = (x_1, \ldots, x_n)$ and the ε_j are *independent* non-zero quantities.

6.2.4 Nevertheless, in various applications Theorem 6' seems to be the more useful of the two. For example for those multipliers which arise typically in elliptic differential equations (see Chapter III, §1.3, §3.5, and also §7.9 of this chapter) both theorems apply equally well.

However the multiplier $\dfrac{x_1}{x_1 + i(x_2^2 + x_3^2 \cdots + x_n^2)}$ which appears in parabolic equations falls under the scope of Theorem 6' only. The same can be said of the multiplier

$$\frac{|x_1|^{\alpha_1}|x_2|^{\alpha_2}\cdots|x_n|^{\alpha_n}}{(x_1^2 + x_2^2 \cdots + x_n^2)^{\alpha/2}}, \qquad \alpha = \alpha_1 + \alpha_2 \cdots + \alpha_n, \quad \text{with} \quad \alpha_j > 0,$$

which is not untypical of a class arising in connection with the study of spaces of fractional potentials. (Compare with §3.2 in Chapter V.)

6.2.5 Finally, both theorems have very definite shortcomings. This is already so because of the matters raised in §4.3. What seems to be needed is a more far-reaching theory which gives sufficient conditions for a multiplier to belong to some $\mathscr{M}_p$, $p \neq 2$, without implying also that it belongs to all $\mathscr{M}_p$, $1 < p < \infty$. Few tools seems to be available for this difficult task. The only thing that readily suggests itself is a possible development of the ideas centering about the function g_λ^*.

6.2.6 The limitations of Theorems 3 and 6' may be further illustrated by the following remark. Consider the characteristic function of an arbitrary polyhedron in $\mathbf{R}^n$. By the use of the same considerations as that of Theorem 4 we may show that it is a multiplier for L^p, $1 < p < \infty$. But this simple example does not fall under the scope of either Theorem 3 or Theorem 6'.

6.3 Proof. It will be best to prove Theorem 6' in the case $n = 2$. This case is already completely typical of the general situation, and in doing only it we can avoid some notational complications.

Let $f \in L^2(\mathbf{R}^2) \cap L^p(\mathbf{R}^2)$, and write $F = T_m f$, that is $F(x)^\wedge = m(x) \hat{f}(x)$.

Let Δ denote the dyadic rectangles, and for each $\rho \in \Delta$, write $f_\rho = S_\rho f$, $F_\rho = S_\rho F$, thus $F_\rho = T_m f_\rho$.

In view of the theorem of dyadic decompositions (Theorem 5) it suffices to show that

$$(58) \qquad \left\| \left(\sum_{\rho \in \Delta} |F_\rho|^2 \right)^{\frac{1}{2}} \right\|_p \leq C_p \left\| \left(\sum_{\rho \in \Delta} |f_\rho|^2 \right)^{\frac{1}{2}} \right\|_p$$

The rectangles in Δ come in four sets, those in the first, the second, the third, and fourth quadrants respectively. In estimating the left side of (58) consider the rectangles of each quadrant separately, and assume from now on that our rectangles belong to the first quadrant.

We will express F_ρ in terms of an integral involving f_ρ and the partial sum operators. That this is possible is the essential idea of the proof.

Fix ρ and assume $\rho = \{(x_1 x_2) : 2^k \leq x_1 \leq 2^{k+1}, 2^l \leq x_2 \leq 2^{l+1}\}$. Then for $(x_1, x_2) \in \rho$, we have the identity

$$m(x_1, x_2) = \int_{2^k}^{x_1} \int_{2^l}^{x_2} \frac{\partial^2 m(t_1 t_2)}{\partial t_1 \, \partial t_2} \, dt_1 \, dt_2 + \int_{2^k}^{x_1} \frac{\partial}{\partial t_1} m(t_1, 2^l) \, dt_1$$
$$+ \int_{2^l}^{x_2} \frac{\partial}{\partial t_2} m(2^k, t_2) \, dt_2 + m(2^k, 2^l).$$

Now let S_t denote the multiplier transformation corresponding to the rectangle $2^k < x_1 < t_1$, $2^l < x_2 < t_2$. Similarly let $S_{t_1}^{(1)}$ denote the multiplier corresponding to the rectangle $2^k < x_1 < t_1$, similarly for $S_{t_2}^{(2)}$. Thus in fact $S_t = S_{t_1}^{(1)} \cdot S_{t_2}^{(2)}$. Then the above equation is obviously

$$(59) \quad S_\rho T_m = \int_{2^l}^{2^{l+1}} \int_{2^k}^{2^{k+1}} S_t \frac{\partial^2 m}{\partial t_1 \, \partial t_2} \, dt_1 \, dt_2 + \int_{2^k}^{2^{k+1}} S_{t_1}^1 \frac{\partial}{\partial t_1} m(t_1, 2^l) \, dt_1$$
$$+ \int_{2^l}^{2^{l+1}} \cdots + m(2^k, 2^l) S_\rho.$$

Now use the fact that $S_\rho T_m f = F_\rho$, and $S_{t_1}^{(1)} S_\rho = S_{t_1}^{(1)}$, $S_{t_2}^{(2)} S_\rho = S_{t_2}^{(2)}$, $S_t S_\rho = S_t$, together with Schwarz's inequality and the assumptions of the theorem. This gives

$$|F_\rho|^2 \le B' \left\{ \iint_\rho |S_t(f_\rho)|^2 \left| \frac{\partial^2 m}{\partial t_1 \, \partial t_2} \right| dt_1 \, dt_2 + \int_{I_1} |S_{t_1}^{(2)}(f_\rho)|^2 \left| \frac{\partial(m(t_1, 2^l)}{\partial t_1} \right| dt_1 \right.$$

$$(60) \qquad + \int_{I_2} |S_{t_2}^{(2)}(f_\rho)|^2 \left| \frac{\partial}{\partial t_2} \, m(2^k, t_2) \right| dt_2 + |f_\rho| \right\}$$

$$= \mathfrak{I}_\rho^1 + \mathfrak{I}_\rho^2 + \mathfrak{I}_\rho^3 + \mathfrak{I}_\rho^4, \quad \text{with} \quad \rho = I_1 \times I_2.$$

To estimate $\|(\sum_\rho |F_\rho|^2)^{\frac{1}{2}}\|_p$, we estimate separately the contributions of each of the four terms on the right side of (60) by the use of Theorem 4" in §4.4. To apply that theorem in the case of $\mathfrak{I}_\rho^1$ we take for Γ the first quadrant, and $d\gamma = \left| \dfrac{\partial^2 m(t_1, t_2)}{\partial t_1 \partial t_2} \right| dt_1 \, dt_2$; the functions $\gamma \to \rho_\gamma$ are constant on the dyadic rectangles. Since for every rectangle,

$$\int_\rho d\gamma = \int_\rho \left| \frac{\partial^2 m(t_1, t_2)}{\partial t_1 \, \partial t_2} \right| dt_1 \, dt_2 \le B, \text{ then } \|(\sum_\rho |\mathfrak{I}_\rho^1|^2)^{\frac{1}{2}}\|_p \le C_\rho \|(\sum |f_\rho|^2)^{\frac{1}{2}}\|_p.$$

Similarly for $\mathfrak{I}^2$, $\mathfrak{I}^3$, and $\mathfrak{I}^4$, which concludes the proof.

7. Further results

7.1 Suppose that $R_1, R_2, \dots R_n$ are the Riesz transforms. Then

(a) $(g(f, x))^2 = (g_1(f)(x))^2 + \sum_{j=1}^n (g_1(R_j f)(x))^2$

(b) $g_1^2(f)(x) \le \sum_{j=1}^n (g_x(R_j f)(x))^2$

7.2 (a) Suppose φ continuously differentiable in $\mathbf{R}^n$ and

(i) $|\varphi(x)| \le A(1 + |x|)^{-n+\delta}$,

(ii) $\left| \dfrac{\partial \varphi}{\partial x_j} \right| \le A(1 + |x|)^{-n-\delta}$ for each $j = 1, \dots, n$

(iii) $\displaystyle\int_{\mathbf{R}^n} \left| \dfrac{\partial \varphi}{\partial x_j}(x + t) - \dfrac{\partial \varphi}{\partial x_j}(x) \, dx \right| \le A \, |t|^\delta$ for some $\delta > 0$.

Define $f_\varepsilon(x) = f * \varphi_\varepsilon$, where $\varphi_\varepsilon(x) = \varepsilon^{-n} \varphi(x/\varepsilon)$. Then

$$\left\| \left(\int_0^\infty \varepsilon \left| \frac{\partial f_\varepsilon}{\partial \varepsilon} \right|^2 d\varepsilon \right)^{\frac{1}{2}} \right\|_p \le A_p \, \|f\|_p, \quad f \in L^p, \quad 1 < p < \infty.$$

If, in addition, $\left\| \left(\displaystyle\int_0^\infty \varepsilon \left| \dfrac{\partial f_\varepsilon}{\partial \varepsilon} \right|^2 d\varepsilon \right)^{\frac{1}{2}} \right\|_2 = C \, \|f\|_2, C > 0$, then also the converse

inequality holds. Similarly for $\left(\displaystyle\int_0^\infty \varepsilon \left| \dfrac{\partial f_\varepsilon}{\partial x_k} \right|^2 d\varepsilon \right)^{\frac{1}{2}}$. (See Benedek, Calderón, and

Panzone [1] for closely related results.)

(b) An example in $\mathbf{R}^1$ is given by $\left(\int_0^\infty \dfrac{|F(x+t) + F(x-t) - 2F(x)|^2}{t^3}\, dt\right)^{\frac{1}{2}}$,

where $F(x) = \int_0^x f(t)\, dt$. See Marcinkiewicz [2], Zygmund [1]; also Stein [3], Hörmander [1] for generalizations.

7.3 Let T be a bounded linear transformation of $L^p(\mathbf{R}^n)$ to itself, $1 \le p \le \infty$, which commutes with translations. Then there exists a bounded function m so that $(Tf)^\wedge(x) = m(x)\hat{f}(x)$, whenever $f \in L^2 \cap L^p$.

Outline of proof. (i) Since T commutes with translations $(Tf) * g = T(f * g)$, for appropriate f and g. Thus $Tf * g = f * Tg$.

(ii) Let $1/p + 1/q = 1$, and suppose that f and g both belong to $L^p(\mathbf{R}^n) \cap L^q(\mathbf{R}^n)$. Then the convolutions $Tf * g$ and $f * Tg$ represent continuous functions, and so they are equal at every point, in particular the origin. Hence

$$\int_{\mathbf{R}^n}(Tf)(x)g(-x)\, dx = \int_{\mathbf{R}^n}(Tg)(x)f(-x)\, dx.$$

The usual duality argument then shows that T is bounded on L^q, and by the interpolation theorem (Theorem 4, Chapter I) T is also bounded on L^2. Finally apply the Proposition in §1.4, Chapter II. See also *Fourier Analysis*, Chapter I.

7.4 §7.4–§7.6 give interesting illustrations of multipliers which cannot be treated by the methods of this chapter.

Let $m(x)$ be a function in $\mathbf{R}^n$ which is of the form $m(x) = \varphi_0(x)\dfrac{e^{i|x|^\alpha}}{(1 + |x|^2)^\beta}$, $\alpha > 0$, $\beta > 0$, where φ_0 is a smooth function which vanishes near the origin and is 1 for sufficiently large α. Assume that not both $n = 1$ and $\alpha = 1$.

(a) If $|1/2 - 1/p| < \theta$, then $m(x) \in \mathscr{M}_p$. Here $\theta = 2\beta/\alpha n$.

(b) If $|1/2 - 1/p| > \theta$, then $m(x) \notin \mathscr{M}_p$.

In the exceptional case ($n = 1$, $\alpha = 1$),

(a′) With $\beta = 0$, $m \in \mathscr{M}_p \Leftrightarrow 1 < p < \infty$.

(b′) With $\beta > 0$, $m \in \mathscr{M}_p$, all p.

See Hirschmann [2], Wainger [1], Hörmander [3], Stein [8], and Fefferman [1].

7.5 Suppose that $m(x) \in \mathscr{M}_p(\mathbf{R}^n)$, and is continuous at each point of $\mathbf{R}^k$, $k < n$ ($\mathbf{R}^k$ is considered as a subspace of $\mathbf{R}^n$). Then $m(x)$ restricted to $\mathbf{R}^k$ belongs to $\mathscr{M}_p(\mathbf{R}^k)$. deLeeuw [1].

7.6 Suppose $m(x) = (m_1 * m_2)(x)$, where $m_1 \in L^r(\mathbf{R}^n)$, and $m_2 \in L^{r'}(\mathbf{R}^n)$, with $1/r + 1/r' = 1$. Then $m \in \mathscr{M}_p$, if $|1/2 - 1/p| \le 1/r$, if $2 \le r \le \infty$. See Hahn [1].

7.7 (a) Let χ_B be the characteristic function of the unit ball of $\mathbf{R}^n$. Then $\chi_B \notin \mathscr{M}_p$ if $p \le \dfrac{2n}{n+1}$ or if $p \ge \dfrac{2n}{n-1}$. See Herz [1].

(b) More generally: suppose $m(x)$ is a radial function and suppose $m \in \mathcal{M}_p(\mathbf{R}^n)$. Then if $p < \dfrac{2n}{n+1}$, or $p > \dfrac{2n}{n-1}$, then m is continuous everywhere, except possibly the origin. *Hint*: Suppose $f \in L^p(\mathbf{R}^n)$, $p < \dfrac{2n}{n+1}$ and f is radial. Then f is continuous except possibly at the origin. To prove this assertion use the representation of f in terms of Bessel integrals, as in *Fourier Analysis*, Chapter IV.

7.8   Consider the question of whether the function

$$m_\delta(x) = \begin{cases} (1 - |x|^2)^\delta, & \text{if } |x| \le 1 \\ 0 & , \text{ if } |x| > 1 \end{cases}$$

is a multiplier for $L^p(\mathbf{R}^n)$, i.e. whether $m_\delta \in \mathcal{M}_p$. For $\delta = 0$, this is Problem A discussed in §4.3 above, and also in §7.7. The following positive results are known.

(a) if $\delta > \left(\dfrac{n-1}{2}\right) |1 - 2/p|$, then $m_\delta \in \mathcal{M}_p$. See Stein [1].

(b) This has recently been significantly improved in Fefferman [1]. A particular result of his is that $m_\delta \in \mathcal{M}_p$, if $n = 2$, $\delta > 2(n-1)|1/p - 3/4|$, and $1 \le p < 6/5$. This result is in the nature of best possible, for p in the range $1 \le p < 6/5$. For n dimensions, $n \ge 3$, there are similar results, in the nature of best possible, when $1 \le p < 4n/(3n+1)$.

7.9   Let $P(x)$ be a polynomial in $\mathbf{R}^n$ of degree k. Suppose $P(x)$ is elliptic in the sense that its part of homogeneous degree k is non-vanishing except at the origin. Let f be any k times continuously differentiable function in $\mathbf{R}^n$ with compact support. Then we have the inequality

$$\left\| \left(\frac{\partial}{\partial x}\right)^\alpha f \right\|_p \le A_p \left[\left\| P\left(\frac{\partial}{\partial x}\right)f \right\|_p + \|f\|_p \right], \qquad 1 < p < \infty,$$

as long as $|\alpha| \le k$.

Hint: Let $\varphi(x)$ be a smooth function which vanishes in a neighborhood of the zero set of P, and which is 1 outside a sufficiently large ball. Then $\varphi(x)\dfrac{x^\alpha}{P(x)}$ satisfies the conditions of the multiplier Theorems 3 or 6′; $x^\alpha(1 - \varphi(x))$ is the Fourier transform of an L^1 function, and finally, $x^\alpha f(x) = x^\alpha(1 - \varphi(x)) f(x) + \varphi(x)\dfrac{x^\alpha}{P(x)} \cdot P(x) f(x)$. See also §3.5 of Chapter III, and Agmon, Douglis and Nirenberg [1].

7.10   (a) The conditions (a) and (b) of Theorem 6 are equivalent with the statements

$$|m(x)| \le B'$$

$$\sup_{0 < R} \frac{1}{R} \int_{|x| \le R} |x| \, |dm(x)| \le B'$$

(b) What is the analogous reformulation of the conditions of Theorem 6′?

7.11 The result of Theorem 2 in §2.4 can be strengthened as follows. Let $1 < p < 2$ and $p = 2/\lambda$. Then the mapping $f \to g_\lambda^*(f)$ is of weak-type (p, p). See Fefferman [1]. An earlier result (for an analogous maximal function) is stated in §4.5 of Chapter VII.

7.12 Let T be a bounded operator of $L^p(\mathbf{R}^n)$ to itself, $1 \leq p \leq \infty$. Let $\mathscr{H}$ be any Hilbert space. Then T has a unique "extension" to an operator $T \otimes I$, taking $L^p(\mathbf{R}^n, \mathscr{H})$ to itself, with the property that $(T \otimes I)(\varphi f(x)) = \varphi \cdot Tf(x)$, for any $\varphi \in \mathscr{H}$, and $f \in L^p(\mathbf{R}^n)$. Moreover the norm of $T \otimes I$ on $L^p(\mathbf{R}^n, \mathscr{H})$ is the same as the norm of T on $L^p(\mathbf{R}^n)$. Marcinkiewicz and Zygmund [1], Zygmund [8], Chapter XV.

Notes

Section 1. The classical theory (which used complex methods) is described in Chapters XIV and XV of Zygmund [8]; further historical references will be found there. The theorems for the g-function in n-dimensions are in Stein [3]; further generalizations were given by Hörmander [1], Schwartz [1], and Benedeck, Calderón, and Panzone [1].

Section 2. The function g^* was studied systematically by Zygmund in [1], and the n-dimensional theory by Stein [6] and [10]. The particular approach, described in §2.1 is taken from Stein [10]; a related idea was independently developed by Gasper [1]. This approach is a starting point for various generalizations of the theory, as in Stein [13].

Section 3. The original Marcinkiewicz multiplier theorem is in Marcinkiewicz [4], where it is given in the periodic set-up. Non-periodic variants of this theorem are due to Mihlin [2], Hörmander [1], and Kree [1]. The statement of Theorem 3 is identical with Hörmander's; the present proof however, as it uses comparison with the g and g^* functions is different, and can be adapted in various other circumstances, as in Chapter VII below.

For a general discussion of multipliers, see also Edwards [1].

Sections 4, 5, and 6. The one-dimensional version of Theorem 4′ is in Zygmund [8, Chapter XV]. The more general version given here is in reality a simple consequence of this special case.

The proof given for Theorem 6′ is a simple adaptation of the original periodic argument given in Marcinkiewicz [4]. See also Lizorkin [1] and Kree [1].

CHAPTER V

Differentiability Properties in
Terms of Function Spaces

In this chapter we shall study properties of differentiability and smoothness that can best be described in the context of Banach spaces of functions.

One of the motivations for this study is based on the wide scope of its applications, as a useful tool in a variety of problems in analysis, although much of what we do is in reality suggested by the ideas and methods already developed. In fact such techniques as the interpolation theorem of Marcinkiewicz, the application of harmonic functions, and the Littlewood-Paley g-function, are essential parts of the theory detailed below.

The function spaces we shall treat are the following:

(1) The *Sobolov spaces*, $L_k^p(\mathbf{R}^n)$. These are useful in many questions and consist of all functions on $\mathbf{R}^n$ whose derivatives up to and including order k belong to $L^p(\mathbf{R}^n)$; k is, of course, a non-negative integer.

The two other types of function spaces that will be considered are attempts to "generalize" the Sobolov spaces to the case when k is not integral.

(2) The *potential spaces*, $\mathscr{L}_\alpha^p(\mathbf{R}^n)$, consisting of all "potentials" of order α of L^p functions. When α is integral and $1 < p < \infty$, these spaces are equivalent with the Sobolov spaces.

(3) The *spaces* $\Lambda_\alpha^{p,q}$. These are function spaces defined in terms of the L^p modulus of continuity. As such they represent a more easily defined "generalization" of the spaces $L_k^p(\mathbf{R}^n)$, and because of this are very useful in applications. They are, however, not a genuine generalization of the Sobolov spaces and so a comparison between them and the spaces $\Lambda_\alpha^{p,q}(\mathbf{R}^n)$ and $\mathscr{L}_\alpha^p(\mathbf{R}^n)$ is called for. This comparison may be viewed as one of the central problems treated in this chapter, and it is here where the Littlewood-Paley theory of Chapter IV is applied.

We shall begin by studying the fractional powers of the Laplacean, $(-\Delta)^{\alpha/2}$. This, together with its variant $(I - \Delta)^{\alpha/2}$, represents an important formal device that we shall use.

1. Riesz potentials

1.1 The Fourier transform of a function f which is sufficiently smooth, and small at infinity, and its Laplacean, $\Delta f = \sum_{j=1}^{n} \dfrac{\partial^2 f}{\partial x_j^2}$, are related by

(1) $$(-\Delta f)^\wedge(x) = 4\pi^2 |x|^2 \hat{f}(x).$$

From this it is only one step to replace the exponent 2 in $|x|^2$ by a general exponent β, and thus to define (at least formally) the fractional power of the Laplacean by

(2) $$((-\Delta)^{\beta/2} f)^\wedge = (2\pi |x|)^\beta \hat{f}(x).$$

Of special significance will be the negative powers β in the range, $-n < \beta < 0$. For these there will be a realization of the formal operator (2) as an integral operator. That is, with a slight change of notation we shall have

(3) $$I_\alpha(f) = (-\Delta)^{-\alpha/2}(f), \qquad 0 < \alpha < n$$

where we have defined the *Riesz potentials* by

(4) $$(I_\alpha f)(x) = \frac{1}{\gamma(\alpha)} \int_{\mathbf{R}^n} |x - y|^{-n+\alpha} f(y) \, dy,$$

with

$$\gamma(\alpha) = \pi^{n/2} 2^\alpha \Gamma(\alpha/2) \Big/ \Gamma\left(\frac{n}{2} - \frac{\alpha}{2}\right)$$

The formal manipulations have a precise meaning.

For this purpose it is convenient to use the class $\mathscr{S}$ of functions φ, which are indefinitely differentiable on $\mathbf{R}^n$ and all of whose derivatives remain bounded when multiplied by polynomials.

LEMMA 2. *Let* $0 < \alpha < n$.
(a) *The Fourier transform of the function* $|x|^{-n+\alpha}$ *is the function* $\gamma(\alpha)(2\pi)^{-\alpha} |x|^{-\alpha}$, *in the sense that*

(5) $$\int_{\mathbf{R}^n} |x|^{-n+\alpha} \varphi(x) \, dx = \int_{\mathbf{R}^n} \gamma(\alpha)(2\pi)^{-\alpha} |x|^{-\alpha} \overline{\varphi^\wedge(x)} \, dx,$$

whenever $\varphi \in \mathscr{S}$.
(b) *The identity* $(I_\alpha f)^\wedge = (2\pi |x|)^{-\alpha} \hat{f}(x)$ *holds in the sense that*

$$\int_{\mathbf{R}^n} I_\alpha(f)(x) \overline{g(x)} \, dx = \int_{\mathbf{R}^n} \hat{f}(x)(2\pi |x|)^{-\alpha} \overline{g^\wedge(x)} \, dx$$

whenever $f, g \in \mathscr{S}$.

The first part of the lemma is merely a restatement of the result in Chapter III, §3.3, since $\gamma(\alpha) = \gamma_{0,\alpha}(2\pi)^\alpha$.

Part (b) follows immediately from part (a) by writing

$$\frac{(2\pi)^\alpha}{\gamma(\alpha)} \int_{\mathbf{R}^n} f(x-y)\,|y|^{-n+\alpha}\,dy = \int_{\mathbf{R}^n} \hat{f}(-y)\,|y|^{-\alpha}e^{-2\pi i x \cdot y}\,dy,$$

(which is a rephrasing of (5)) and then integrating both sides of this identity after multiplying through by $\bar{g}(x)$.

We state now two further identities which can be obtained from Lemma 1 and which reflect essential properties of the potentials I_α.

(6) $I_\alpha(I_\beta f) = I_{\alpha+\beta}(f)$, $f \in \mathscr{S}$, $\alpha > 0$, $\beta > 0$, $\alpha + \beta < n$.

(7) $\Delta(I_\alpha f) = I_\alpha(\Delta f) = -I_{\alpha-2}(f)$, $f \in \mathscr{S}$, $n > 3$, $n \geq \alpha \geq 2$.

The deduction of (6) and (7) offer no real difficulties; these are best left to the interested reader to work out.

A simple consequence of (6) is the n-dimensional variant of the *beta integral*,

(8) $$\int_{\mathbf{R}^n} |1 - y|^{-n+\alpha} |y|^{-n+\beta}\,dy = \frac{\gamma(\alpha)\gamma(\beta)}{\gamma(\alpha+\beta)}$$

with $0 < \alpha$, $0 < \beta$, $\alpha + \beta < n$.

1.2 L^p inequality for potentials.   Up to this stage we have considered the Riesz potentials only from a formal point of view; in particular, we have operated only with very smooth functions which are suitably small at infinity. But since the Riesz potentials are integral operators it is natural to inquire about their actions on the spaces $L^p(\mathbf{R}^n)$.

For this reason we formulate the following problem. Given α, $0 < \alpha < n$, for what pairs p and q, is the operator $f \to I_\alpha(f)$ bounded from $L^p(\mathbf{R}^n)$ to $L^q(\mathbf{R}^n)$? That is, when do we have the inequality

(9) $$\|I_\alpha(f)\|_q \leq A \|f\|_p \; ?$$

There is a simple necessary condition, which is merely a reflection of the homogeneity of the kernel $(\gamma(\alpha))^{-1} |y|^{-n+\alpha}$. In fact, consider the dilation operator τ_δ, defined by

$$\tau_\delta(f)(x) = f(\delta x), \qquad \delta > 0.$$

Then clearly

(10) $$\tau_\delta^{-1} I_\alpha \tau_\delta = \delta^{-\alpha} I_\alpha, \qquad \delta > 0.$$

Also

(11) $$\|\tau_\delta(f)\|_p = \delta^{-n/p} \|f\|_p, \qquad \|\tau_\delta^{-1} I_\alpha(f)\|_q = \delta^{n/q} \|I_\alpha(f)\|_q.$$

Thus (9) is possible only if

(12) $$1/q = 1/p - \alpha/n.$$

We shall see below that this condition is also sufficient, save for two exceptional cases.

The two instances arise when $p = 1$, (then $q = n/(n - \alpha)$) and when $q = \infty$, (then $p = n/\alpha$). Let us consider the case $p = 1$. It is not hard to see that the presumed inequality

(13) $$\|I_\alpha(f)\|_{n/(n-\alpha)} \leq A \|f\|_1,$$

cannot hold. If in fact (13) were valid, we could put in the place of f a sequence $\{f_n\}$ of positive integrable functions whose common integral is one and whose supports converge to the origin (an "approximation to the identity"). A simple limiting argument then shows that this implies that

$$\left\| \frac{1}{\gamma(\alpha)} |x|^{-n+\alpha} \right\|_{n/(n-\alpha)} \leq A < \infty,$$

which means

$$\int_{\mathbf{R}^n} |x|^{-n} \, dx < \infty,$$

and this is a contradiction.

The second atypical case occurs when $q = \infty$. Again the inequality of the type (9) cannot hold, and one immediate reason is that this case is dual to the case $p = 1$ just considered. The failure at $q = \infty$ may also be seen directly as follows: Let $f(x) = |x|^{-\alpha}(\log 1/|x|)^{-(\alpha/n)(1+\varepsilon)}$, for $|x| \leq 1/2$, and $f(x) = 0$, for $|x| > 1/2$, where ε is positive but small. Then $f \in L^p(\mathbf{R}^n)$, $p = n/\alpha$, since $\int_{|x| \leq \frac{1}{2}} |x|^{-n}(\log 1/|x|)^{-1-\varepsilon} \, dx < \infty$. However, $I_\alpha(f)$ is essentially unbounded near the origin since

$$I_\alpha(f)(0) = \frac{1}{\gamma(\alpha)} \int_{|x| \leq \frac{1}{2}} |x|^{-n} (\log 1/|x|)^{-(\alpha/n)(1+\varepsilon)} \, dx = \infty,$$

as long as $(\alpha/n)(1 + \varepsilon) \leq 1$.

After these observations we can formulate the positive theorem: The Hardy-Littlewood-Sobolev theorem of *fractional integration.*

THEOREM 1. *Let* $0 < \alpha < n$, $1 \leq p < q < \infty$, $1/q = 1/p - \alpha/n$.
(a) *If* $f \in L^p(\mathbf{R}^n)$, *then the integral* (4), *defining* $I_\alpha(f)$, *converges absolutely for almost every x.*
(b) *If, in addition,* $1 < p$, *then*

$$\|I_\alpha(f)\|_q \leq A_{p,q} \|f\|_p.$$

(c) *If* $f \in L^1(\mathbf{R}^n)$, *then* $m\{x : |I_\alpha| > \lambda\} \leq \left(\dfrac{A \, \|f\|_1}{\lambda}\right)^q$, *for all* λ. *That is,*
the mapping $f \to I_\alpha(f)$ *is of "weak-type"* $(1, q)$, $(1/q = 1 - \alpha/n)$.

1.3   Proof of Theorem 1.   Let us write $K(x) = |x|^{-n+\alpha}$, and we shall consider the transformation $f \to K * f$, instead of $f \to I_\alpha(f)$ from which it differs by a constant multiple. Let us decompose K as $K_1 + K_\infty$, where

$$K_1(x) = K(x) \quad \text{if} \quad |x| \leq \mu, \qquad K_1(x) = 0 \quad \text{if} \quad |x| > \mu$$
$$K_\infty(x) = K(x) \quad \text{if} \quad |x| > \mu, \qquad K_\infty(x) = 0 \quad \text{if} \quad |x| \leq \mu.$$

At this instance μ is a fixed positive constant which need not be specified. We have $K * f = K_1 * f + K_\infty * f$. The integral expressing $K_1 * f$ converges absolutely *almost everywhere* since it represents the convolution of an L^1 function (K_1) with an L^p function. Similarly the integral representing $K_\infty * f$ converges *everywhere* since it is a convolution of a function in L^p, (f), and another in the dual space $L^{p'}$, (K_∞). In fact if $1/p + 1/p' = 1$, then $\|K_\infty\|_{p'}^{p'} = \int_{|x| \geq \mu} |x|^{(-n+\alpha)p'} \, dx < \infty$, since $(-n + \alpha)p' < -n$ is equivalent with $q < \infty$. Thus part (a) of the theorem is proved.

We shall show next, by a similar but more detailed reasoning, that if $1 \leq p < q < \infty$, and $1/q - 1/p - \alpha/n$, then the mapping $f \to K * f$ is of *weak-type* (p, q), in the sense that

$$(14) \quad m\{x : |K * f| > \lambda\} \leq \left(A_{p,q} \frac{\|f\|_p}{\lambda}\right)^q, \quad f \in L^p(\mathbf{R}^n), \quad \text{all} \quad \lambda > 0.$$

We notice first that it suffices to prove the inequality (14) with 2λ in place of λ in the left side of this inequality, and with $\|f\|_p = 1$. Now

$$m\{x : |K * f| > 2\lambda\} \leq m\{x : |K_1 * f| > \lambda\} + m\{x : |K_\infty * f| > \lambda\},$$

since $K * f = K_1 * f + K_\infty * f$. However

$$m\{x : |K_1 * f| > \lambda\} \leq \frac{\|K_1 * f\|_p^p}{\lambda^p} \leq \frac{\|K_1\|_1^p \, \|f\|_p^p}{\lambda^p} = \frac{\|K_1\|_1^p}{\lambda^p}.$$

But

$$\|K_1\|_1 = \int_{|x| \leq \mu} |x|^{-n+\alpha} \, dx = c_1 \mu^\alpha.$$

Next

$$\|K_\infty * f\|_\infty \leq \|K_\infty\|_{p'} \|f\|_p \leq \|K_\infty\|_{p'}.$$

However,

$$\|K_\infty\|_{p'} = \left(\int_{|x| > \mu} (|x|^{-n+\alpha})^{p'} \, dx\right)^{1/p'} = c_2 \mu^{-n/q},$$

and so $\|K_\infty\|_{p'} = \lambda$, if $c_2 \mu^{-n/q} = \lambda$, i.e., if $\mu = c_3 \lambda^{-q/n}$. Choose, therefore, μ to have this value. Then $\|K_\infty * f\|_\infty \leq \lambda$, and so $m\{x : |K_\infty * f| > \lambda\} = 0$.

Finally then

$$m\{x:|K*f| > 2\lambda\} \leq \left(c_1 \frac{\mu^\alpha}{\lambda}\right)^p = c_4 \lambda^{-q} = c_4 \left(\frac{\|f\|_p}{\lambda}\right)^q$$

(since $\|f\|_p = 1$). This is (14), and so the mapping $f \to K*f$ is of weak type (p, q). The special case for $p = 1$ then gives part (c) of the theorem, and part (b) follows by the Marcinkiewicz interpolation theorem. (See Appendix B.)

1.4 Comment. The following retrospective comment about the proof of the theorem is in order. In the proof of Theorem 1 for the operator $f \to K*f$ what was decisive was not the specific structure of the kernel K. The only thing that really mattered was the *distribution function* of K (in the terminology of Chapter I.) A more detailed examination would show that we only needed the fact that $m\{x:|K(x)| \geq \lambda\} \leq A\lambda^{-n/(n-\alpha)}$; that is, the kernel is of "weak type" $n/(n - \alpha)$.

If we had the stronger assumption that $K \in L^{n/(n-\alpha)}$ we would have obtained *a fortiori* the result

$$\|K*f\|_q \leq A \|f\|_p, \quad \text{with} \quad 1/q = 1/p + 1/r - 1, r = n/(n - d).$$

This is essentially the more familiar Young's inequality, which is also valid when $p = 1$ or $q = \infty$. (See Appendix A.)

2. *The Sobolov spaces,* $L_k^p(\mathbf{R}^n)$

We come now to the study of the relation of a function and its partial derivatives. The concept of the partial derivative that will be used is the general notion given us by the theory of distributions, and the appropriate definition is stated in terms of the space $\mathscr{D}$ of all indefinitely differentiable functions on $\mathbf{R}^n$, each with compact support.

Let $\dfrac{\partial^\alpha}{\partial x^\alpha} = \dfrac{\partial^{\alpha_1 + \alpha_2 + \cdots + \alpha_n}}{\partial x_1^{\alpha_1} \partial x_2^{\alpha_2} \cdots \partial x_n^{\alpha_n}}$, be a differential monomial, whose total order is $|\alpha| = \alpha_1 + \alpha_2 + \cdots + \alpha_n$. Suppose we are given two locally integrable functions on $\mathbf{R}^n$, f and g. Then we say that $\dfrac{\partial^\alpha f}{\partial x^\alpha} = g$, (we add the designation "in the weak sense," whenever this is necessary to avoid ambiguities), if

$$(15) \quad \int_{\mathbf{R}^n} f(x) \frac{\partial^\alpha \varphi}{\partial x^\alpha}(x) \, dx = (-1)^{|\alpha|} \int_{\mathbf{R}^n} g(x)\varphi(x) \, dx, \quad \text{for all} \quad \varphi \in \mathscr{D}.$$

Integration by parts shows us that this is indeed the relation that we would expect if f had continuous partial derivatives up to order $|\alpha|$, and $\dfrac{\partial^\alpha f}{\partial x^\alpha} = g$ had the usual meaning.

It is of course not true that every locally integrable function has partial derivatives in this sense: consider, for example, $f(x) = c^{i/|x|^n}$. However when the partial derivatives exist they are determined almost everywhere by the defining relation (15).

For any non-negative integer k, the Sobolov space $L_k^p(\mathbf{R}^n) = L_k^p$ is defined as the space of functions f, with $f \in L^p(\mathbf{R}^n)$ and where all $\dfrac{\partial^\alpha f}{\partial x^\alpha}$ exist and $\dfrac{\partial^\alpha f}{\partial x^\alpha} \in L^p(\mathbf{R}^n)$ in the above sense, whenever $|\alpha| \leq k$. This space of functions can be normed by the expression

$$(16) \qquad \|f\|_{L_k^p} = \sum_{|\alpha| \leq k} \left\| \frac{\partial^\alpha}{\partial x^\alpha} f \right\|_p, \qquad \left(\frac{\partial^0}{\partial x^0} f = f \right).$$

The resulting normed space is complete. The proof of this is as follows. If $\{f_m\}$ is a Cauchy sequence in L_k^p, then for each α, $\left\{ \dfrac{\partial^\alpha}{\partial x^\alpha} f_m \right\}$ is a Cauchy sequence in L^p, $|\alpha| \leq k$. If now $f^{(\alpha)} = \lim\limits_m \dfrac{\partial^\alpha}{\partial x^\alpha} f_m$ (the limit taken in L^p norm), then clearly $\int_{\mathbf{R}^n} f \dfrac{\partial^\alpha}{\partial x^\alpha} \varphi \, dx = (-1)^{|\alpha|} \int_{\mathbf{R}^n} f^{(\alpha)} \varphi \, dx$, for each $\varphi \in \mathcal{D}$ and the assertion is proved.

It is often convenient to use an equivalent characterization of the functions of $L_k^p(\mathbf{R}^n)$, which does not explicitly involve the notion of the weak derivative given in (15).

PROPOSITION 1. *Let* $1 \leq p < \infty$. *Then* $f \in L_k^p$ *if and only if there exists a sequence* $\{f_m\}$, *so that*
  (a) *each* $f_m \in \mathcal{D}$
  (b) $\|f - f_m\|_p \to 0$
  (c) *For each* α, $\left\{ \dfrac{\partial^\alpha f_m}{\partial x^\alpha} \right\}$ *converges in* L^p *norm, for* $|\alpha| \leq k$.

That the conditions (a), (b), and (c) are sufficient is readily obvious. In fact let $f^{(\alpha)} = \lim\limits_{m \to \infty} \dfrac{\partial^\alpha f_m}{\partial x^\alpha}$, $f^{(0)} = f$; then since

$$\int f_m \frac{\partial^\alpha \varphi}{\partial x^\alpha} \, dx = (-1)^{|\alpha|} \int \frac{\partial^\alpha f_m}{\partial x^\alpha} \varphi \, dx,$$

we get

$$\int f \frac{\partial^\alpha \varphi}{\partial x^\alpha} \, dx = (-1)^{|\alpha|} \int f^{(\alpha)} \varphi \, dx;$$

this shows that $f \in L^p_k$.

The converse is more interesting. The argument that is required for its proof is typical of a great deal of similar reasoning which involves the device of *regularization*.

For this purpose let ψ be a *fixed* element of $\mathscr{D}$, with the property that $\int_{\mathbf{R}^n} \psi(x) \, dx = 1$. For every $\varepsilon > 0$, consider $\psi_\varepsilon(x)$ defined by $\psi_\varepsilon(x) = \varepsilon^{-n} \psi(x/\varepsilon)$, and for each $f \in L^p$, set $f_\varepsilon = f * \psi_\varepsilon$. The family $\{f_\varepsilon\}$ is a regularization of f: In the present context this means

(a) $\|f_\varepsilon - f\|_p \to 0$, as $\varepsilon \to 0$

(b) each f_ε is indefinitely differentiable

(c) if f has a partial derivative $\dfrac{\partial^\alpha f}{\partial x^\alpha}$ (in the weak sense), then

$$\frac{\partial^\alpha f_\varepsilon}{\partial x^\alpha} = \left(\frac{\partial^\alpha f}{\partial x_\alpha} \right)_\varepsilon = \frac{\partial^\alpha f}{\partial x^\alpha} * \psi_\varepsilon.$$

(a) is valid under the more general condition that ψ is integrable as we have already seen in Chapter III, §2.2.

(b) Since $f_\varepsilon(x) = \int_{\mathbf{R}^n} f(y) \psi_\varepsilon(x - y) \, dy$, it is clear by differentiation under the integral sign that f_ε is indefinitely differentiable.

(c) Let us carry out this differentiation. Then

$$\frac{\partial^\alpha}{\partial x^\alpha} f_\varepsilon(x) = \int_{\mathbf{R}^n} f(y) \frac{\partial^\alpha}{\partial x^\alpha} (\psi_\varepsilon(x - y)) \, dy$$

$$= (-1)^{|\alpha|} \int_{\mathbf{R}^n} f(y) \frac{\partial^\alpha}{\partial y^\alpha} (\psi_\varepsilon(x - y)) \, dy.$$

For each x, the function $y \to \psi_\varepsilon(x - y)$ is in $\mathscr{D}$, and hence an application of the definition (15) gives

$$\frac{\partial^\alpha}{\partial x^\alpha} f_\varepsilon(x) = \int \left(\frac{\partial^\alpha f}{\partial y^\alpha} \right) \psi_\varepsilon(x - y) \, dt = \left(\frac{\partial^\alpha f}{\partial x^\alpha} \right) * \psi_\varepsilon.$$

We can now apply property (a) to $\dfrac{\partial^\alpha f}{\partial x^\alpha} * \psi_\varepsilon$, and we see that $\dfrac{\partial^\alpha f_\varepsilon}{\partial x^\alpha}$ converges in L^p norm as $\varepsilon \to 0$. Thus the functions $\{f_\varepsilon\}$ give the required approximation, except that they do not each have compact support, and hence a final modification is called for. Let η be a fixed indefinitely differentiable function of compact support with $\eta(0) = 1$ and consider the two-parameter family $\{\eta(\delta x) f_\varepsilon(x)\}$, $\varepsilon > 0$, $\delta > 0$. Choose then ε

first so that $\dfrac{\partial^\alpha f_\varepsilon}{\partial x^\alpha}$ are sufficiently close to their limits. Next, with ε fixed, choose δ sufficiently small so $\dfrac{\partial^\alpha}{\partial x^\alpha}(\eta(\delta x)f_\varepsilon(x))$ are sufficiently close to $\dfrac{\partial^\alpha f_\varepsilon}{\partial x^\alpha}$. Since each $\eta(\delta x)f_\varepsilon(x)$ is indefinitely differentiable with compact support the proposition is then proved.

There is a parallel proposition dealing with case $p = \infty$, but it requires the usual modification since smooth functions are not dense in the $L^\infty(\mathbf{R}^n)$ space.

Alternative characterizations in the case $n = 1$ for all $1 \le p \le \infty$, and in the case of general n for $p = \infty$, may be found in §6.1 and §6.2 below.

2.1.1   As far as the proof of Proposition 1 is concerned, the requirement that ψ have compact support is not absolutely necessary. We could, e.g., have carried out the proof (with a little sacrifice of elegance) by setting $\psi(x) = \dfrac{c_n}{(1 + |x|^2)^{(n+1)/2}}$. Then $f_\varepsilon = f * \psi_\varepsilon$ would equal $u(x, \varepsilon)$, where $u(x, y)$ is the Poisson integral of f. (See Chapter III, §2.)

It is in other problems (see for example, Chapter VI, §3.2.4.) that the regularization with a ψ of *compact* support plays a more essential role. This type of regularization has the property that $f_\varepsilon(x)$ depends only on the values of f in a small neighborhood of x.

2.2   Sobolov's theorem.   The importance of the Sobolov spaces just considered is that in terms of them we can account in a relatively simple way how restrictions on the "size" of partial derivatives imply corresponding restrictions on the functions in question. A general theorem may be formulated as follows.

THEOREM 2.   *Suppose k is a positive integer, and $1/q = 1/p - k/n$.*

(i) *If $q < \infty$ (i.e. $p < n/k$), then $L_k^p(\mathbf{R}^n) \subset L^q(\mathbf{R}^n)$ and the natural inclusion map is continuous.*

(ii) *If $q = \infty$ (i.e. $p = n/k$), then the restriction of an $f \in L_k^p(\mathbf{R}^n)$ to a compact subset of $\mathbf{R}^n$ belongs to $L^r(\mathbf{R}^n)$, for every $r < \infty$.*

(iii) *If $p > n/k$, then every $f \in L_k^p(\mathbf{R}^n)$ can be modified on a set of zero measure so that the resulting function is continuous.*

2.3   To prove the theorem it is required that we find an appropriate way of expressing a function in terms of its partial derivatives. To do this let us proceed in a purely formal way, operating always with functions of the class $\mathscr{S}$ (or $\mathscr{D}$).

With the testing function f we consider its Fourier transform $\hat{f}$. Then the Fourier transform of $\dfrac{\partial f}{\partial x_j}$ is $-2\pi i x_j \hat{f}(x)$. Now recall the Riesz transforms of Chapter III, §1.2. The effect of R_j is multiplication on the Fourier transform side by $\dfrac{ix_j}{|x|}$ (see formula (8)). Thus

$$\left(R_j\left(\frac{\partial}{\partial x_j}f\right)\right)^{\wedge}(x) = 2\pi \frac{x_j^2}{|x|}\hat{f}(x).$$

In view of formula (3) we then get

$$(17) \qquad f = I_1\left(\sum_{j=1}^n R_j\left(\frac{\partial}{\partial x_j}f\right)\right).$$

This identity, which expresses f in terms of its first partial derivatives, contains two elements: the Riesz transforms and the potential of order one. The former are operators which preserve the class $L^p(\mathbf{R}^n)$, and the latter maps $L^p(\mathbf{R}^n)$ to $L^q(\mathbf{R}^n)$, for appropriate p and q (by Theorem 1). This glimpse reveals the essential mechanism behind the present theorem.

There is, however, a simpler approach which is closely related to the identity (17), but avoids the use of the rather deep theory of the Riesz transforms. It is based on the elementary identity

$$(18) \qquad f(x) = \frac{1}{\omega_{n-1}}\sum_{j=1}^n \int_{\mathbf{R}^n}\frac{\partial f}{\partial x_j}(x-y)\cdot\frac{y_j}{|y|^n}\,dy,$$

where ω_{n-1} is the "area" of the sphere S^{n-1}.

The formula (18) is proved as follows. We start with the one-dimensional formula,

$$f(x) = \int_0^\infty f'(x-t)\,dt$$

which is certainly valid whenever f is a testing function. Its n-dimensional analogue is an immediate consequence of itself, namely

$$(19) \qquad f(x) = \int_0^\infty (\nabla f(x-\xi t),\,\xi)\,dt$$

where ξ is any unit vector, and ∇f is the vector with components

$$\left(\frac{\partial f}{\partial x_1},\frac{\partial f}{\partial x_2},\cdots\frac{\partial f}{\partial x_n}\right).$$

We integrate (19) over ξ ranging on the unit sphere. This gives

$$f(x) = \frac{1}{\omega_{n-1}}\int_{\xi\in S^{n-1}}\int_0^\infty (\nabla f(x-\xi t),\,\xi)\,dt\,d\xi.$$

Changing from polar coordinates to rectangular coordinate yields (18). The formula (18) will be applied below. It may be worthwhile, however, to take this opportunity to point out some other identities of the same kind as (17), each of which is transparent on the formal level.

First, suppose we wish to express f in terms of its *second* partial derivatives. Then we can do it in terms of the particular combination

$$\frac{\partial^2 f}{\partial x_1^2} + \frac{\partial^2 f}{\partial x_2^2} + \cdots + \frac{\partial^2 f}{\partial x_n^2} = \Delta f.$$

The identity is

$$f = -I_2(\Delta f)$$

which is just a special case of formula (7). We get in this way a classical formula of potential theory.*

Another useful observation is:

(20)   If $F = I_1(f),$ then $\dfrac{\partial F}{\partial x_j} = -R_j(f).$

In fact according to Lemma 1, with $g = \dfrac{\partial \varphi}{\partial x_j}$, we have

$$\int I_1(f) \frac{\overline{\partial \varphi}}{\partial x_j}\, dx = \int \hat{f}(x)(2\pi\,|x|)^{-1} 2\pi i x_j \overline{\hat{\varphi}(x)}\, dx$$

$$= \int f(x) i \frac{x_j}{|x|}\, \bar{\hat{\varphi}}(x)\, dx = \int R_j(f)\bar{\varphi}\, dx.$$

So $\displaystyle\int I_1(f) \frac{\partial \bar{\varphi}}{\partial x_j}\, dx = \int R_j(f)\bar{\varphi}\, dx$, and therefore (20) is proved, at least when $f \in \mathscr{S}$.

It is also possible to extend (20) to wider classes of functions, as the need may arise (see §6.3 below).

2.4 We prove the theorem first in the case $k = 1$, and where $1 < p$, $q < \infty$. Assume that $f \in \mathscr{D}$. Then the identity (18) shows that

$$|f(x)| \le A \sum_{j=1}^{n} \int_{R^n} \left| \frac{\partial f}{\partial x_j} (x - y) \right| |y|^{-n+1}\, dy.$$

Therefore by Theorem 1, (the case $\alpha = 1$), we get

(21) $\displaystyle \|f\|_q \le A' \sum_{j=1}^{n} \left\| \frac{\partial f}{\partial x_j} \right\|_p,$ $1/q = 1/p - 1/n.$

---

* At least in the case $n \ge 3$. The case $n = 2$ must then be studied separately by a limiting argument, and gives the representation in terms of the logarithmic potential.

Let now f be any function in $L_1^p(\mathbf{R}^n)$. According to Proposition 1, there exists a sequence of elements of $\mathcal{D}$, $\{f_m\}$, so that $f_m \to f$ in L^p norm, and $\dfrac{\partial f_m}{\partial x_j}$ converges in L^p norm. The limit $\displaystyle\lim_{m\to\infty} \dfrac{\partial f_m}{\partial x_j}$ must equal $\dfrac{\partial f}{\partial x_j}$, since

$$\lim_m \int \frac{\partial f_m}{\partial x_j}\, \varphi\, dx = -\lim_m \int f_m \frac{\partial \varphi}{\partial x_j}\, dx = -\int f \frac{\partial \varphi}{\partial x_j}\, dx = \int \frac{\partial f}{\partial x_j}\, \varphi\, dx, \quad \text{for}$$

every $\varphi \in \mathcal{D}$. Substituting in (21) we get

$$\|f_m - f_{m'}\|_q \le A' \left\| \sum_{j=1}^n \frac{\partial f_m}{\partial x_j} - \frac{\partial f_{m'}}{\partial x_j} \right\|_p,$$

and so the sequence f_m also converges in $L^q(\mathbf{R}^n)$ norm, and this limit must also equal f. Thus $f \in L^q(\mathbf{R}^n)$, and

$$\|f\|_q \le A' \sum_{j=1}^n \left\| \frac{\partial f}{\partial x_j} \right\|_p \le A' \|f\|_{L^p_1(\mathbf{R})^n}, \quad f \in L_1^p(\mathbf{R}^n).$$

This shows that $f \in L^q(\mathbf{R}^n)$ and the inclusion mapping of $L_1^p(\mathbf{R}^n)$ into $L^q(\mathbf{R}^n)$ is continuous.

We consider next the situation when $k = 1$, but $q = \infty$ ($p = n/k = n$), or $p > n/k = n$. In both instances the relevant conclusions of Theorem 2 (that is, (ii) and (iii)) are local in character and so we may simplify matters by reducing to the case when f (and therefore its partial derivatives) have compact support. Thus given any fixed compact set, K, let η be a function in $\mathcal{D}$ which is one on that set. If $f \in L_1^p(\mathbf{R}^n)$, consider $\eta \cdot f$.

It will be enough to prove the conclusions (ii) and (iii) for ηf, which incidentally also belongs to $L_1^p(\mathbf{R}^n)$. To see that $\eta f \in L_1^p(\mathbf{R}^n)$ it suffices to verify that the derivative $\dfrac{\partial}{\partial x_j}(\eta f)$, in the weak sense, equals $\dfrac{\partial \eta}{\partial x_j} f + \eta \dfrac{\partial f}{\partial x_j}$. However

$$\int \frac{\partial}{\partial x_j}(\eta f)\varphi\, dx = -\int \eta f \frac{\partial \varphi}{\partial x_j}\, dx$$

$$= -\int f \frac{\partial(\varphi\eta)}{\partial x_j}\, dx + \int \varphi f \frac{\partial \eta}{\partial x_j}\, dx$$

$$= \int \left(\frac{\partial \eta}{\partial x_j} f + \eta \frac{\partial f}{\partial x_j} \right) \varphi\, dx$$

and this assertion is proved.

We start therefore with $f \in L_1^p(\mathbf{R}^n)$ and its approximating sequence $\{f_m\}$ given by Proposition 1. Thus clearly ηf_m is an approximating sequence to ηf. Now choose an R so large that if K_1 is the (compact) support of η, then the set $K_1 - K_1$ is contained in the ball of radius R about the origin.

It then follows again by the identity (18) that

$$(22) \quad |\eta(x)f_m(x)| \leq A \sum_{j=1}^{n} \int_{|y|\leq R} \left| \frac{\partial(\eta(x)f_m)}{\partial x_j}(x-y)\right| |y|^{-n+1}\, dy, \qquad x \in K.$$

We now use Young's inequality which states that if

$$\mathscr{C}(x) = \int_{\mathbf{R}^n} \mathscr{A}(x-y)\mathscr{B}(y)\, dy,$$

then $\|\mathscr{C}\|_r \leq \|\mathscr{A}\|_p \|\mathscr{B}\|_s$, where $1/r = 1/p + 1/s - 1$. We set $\mathscr{A} = A \sum_{j=1}^{n} \left| \frac{\partial}{\partial x_j}(\eta f_m)\right|$; $\mathscr{B}(y) = |y|^{-n+1}$ if $|y| \leq R, \mathscr{B}(y) = 0$ otherwise; and we let s be any exponent $< n/(n-1)$. Notice that then $\|\mathscr{B}\|_s < \infty$, since $\|\mathscr{B}\|_s^s = \int_{|y|\leq R} |y|^{(-n+1)s}\, dy < \infty$, because $(-n+1)s > -n$.

Hence Young's inequality shows that

$$\int_K |f_m|^r\, dx \leq \int |\eta f_m|^r\, dx \leq \|\mathscr{C}\|_r^r \leq A' \|\sum_{j=1}^{n} |\frac{\partial(\eta f_m)}{\partial x_j}|\|_p^r.$$

Similarly,

$$\int_K |f_m - f_{m'}|^r\, dx \leq A' \left\| \sum_{j=1}^{n} \left|\frac{\partial(\eta(f_m - f_{m'}))}{\partial x_j}\right|\right\|_p.$$

So we see that the sequence $\{f_m\}$, which converges to f in the L^p norm, also converges in the L^r norm, when restricted to the set K. Thus f is in L^r, when restricted to K.

In the present case $p = n$, and so the condition $s < n/(n-1)$, is the same as $r < \infty$, since $1/r = 1/p + 1/s - 1$. Therefore the assertion (ii) of the theorem is proved (assuming of course $k = 1$).

The argument for the proof (iii) is very similar to that just carried out, except here we use the estimate

$$\sup_x |\mathscr{C}(x)| \leq \|\mathscr{A}\|_p \|\mathscr{B}\|_{p'}, \qquad 1/p + 1/p' = 1$$

which follows trivially from Hölder's inequality. Notice that if $p > n$, then $\|\mathscr{B}\|_{p'}^{p'} = \int_{|y|\leq R} |y|^{(-n+1)p'}\, dy < \infty$, so we get, in analogy with the above

$$\sup_{x\in K} |f_m(x) - f_{m'}(x)| \leq A' \left\| \sum_{j=1}^{n} \left|\frac{\partial\eta(f_m - f_{m'})}{\partial x_j}\right|\right\|_p.$$

This shows that the continuous functions $\{f_m(x)\}$ converge uniformly on every compact set and hence f may be taken to be continuous.

2.5   The case $p = 1$. With the assumption that $k = 1$, the assertions of the theorem have been completely proved save for the exceptional case

$p = 1$. The argument used so far will of course not work in this case because conclusion (b) of Theorem 1 fails for $p = 1$. A different idea is needed in this circumstance and it is contained in the following elegant inequality:

$$(23) \qquad \|f\|_q \le \left(\prod_{j=1}^{n} \left\|\frac{\partial f}{\partial x_j}\right\|_1\right)^{1/n}, \qquad 1/q = 1 - 1/n, \quad f \in \mathscr{D}.$$

We prove (23) by induction on n. The case $n = 1$ is trivial because it states $\|f\|_\infty \le \|f'\|_1$, which follows immediately from $f(x) = \int_{-\infty}^{x} f'(t)\, dt$.

We assume therefore that the inequality (23) is valid for $n - 1$. For the purposes of the induction, we write $x \in \mathbf{R}^n$, as $x = (x_1, x')$, with $x' \in \mathbf{R}^{n-1}$, $x_1 \in \mathbf{R}^1$. We set

$$I_j(x_1) = \int_{\mathbf{R}^{n-1}} \left|\frac{\partial f}{\partial x_j}(x_1, x')\right| dx', \qquad j = 2, \ldots, n,$$

and

$$I_1(x') = \int_{\mathbf{R}^1} \left|\frac{\partial f}{\partial x_1}(x_1, x')\right| dx_1.$$

Suppose now that q is the index that corresponds to n, $(q = n/(n - 1))$ and q' the index that corresponds to $n - 1$, $(q' = (n - 1)/(n - 2))$. Then by the case $n - 1$, we have

$$(24) \qquad \left(\int_{\mathbf{R}^{n-1}} |f(x_1, x')|^{q'}\, dx'\right)^{1/q'} \le \left(\prod_{j=2}^{n} I_j(x_1)\right)^{1/(n-1)}.$$

Clearly, however, $|f(x)| \le I_1(x')$ (this is the one-dimensional case again!), so $|f|^q \le (I_1(x'))^{1/(n-1)} |f|$, since

$$q = \frac{1}{n-1} + 1.$$

Thus

$$\int_{\mathbf{R}^{n-1}} |f|^q\, dx' \le \int_{\mathbf{R}^{n-1}} (I_1(x'))^{1/(n-1)} |f|\, dx'$$

$$\le \left(\int_{\mathbf{R}^{n-1}} I_1(x')\, dx'\right)^{1/(n-1)} \left(\int_{\mathbf{R}^{n-1}} |f|^{q'}\, dx'\right)^{1/q'}$$

by Hölder's inequality with the conjugate exponents $n - 1$, and q'.

Substituting (24) in the above gives

$$\int_{\mathbf{R}^{n-1}} |f|^q\, dx' \le \left(\int_{\mathbf{R}^{n-1}} I_1(x')\, dx'\right)^{1/(n-1)} \left(\prod_{j=2}^{n} I_j(x_1)\right)^{1/(n-1)}.$$

We integrate this with respect to x_1 and use Hölder's inequality again, to wit,

$$\int_{\mathbf{R}^1} \left(\prod_{j=2}^{n} I_j(x_1) \right)^{1/(n-1)} dx_1 \leq \prod_{j=2}^{n} \left(\int_{\mathbf{R}^1} I_j(x_1) \, dx_1 \right)^{1/(n-1)} = \prod_{j=2}^{n} \left\| \frac{\partial f}{\partial x_j} \right\|_1^{1/(n-1)}.$$

The final result is

$$\int_{\mathbf{R}^n} |f|^q \, dx \leq \left(\prod_{j=1}^{n} \left\| \frac{\partial f}{\partial x_j} \right\|_1 \right)^{1/(n-1)}$$

which is the desired inequality (23), since $q = n/(n-1)$.

If we use the fact that

$$\left(\prod_{j=1}^{n} a_j \right)^{1/n} \leq \frac{1}{n} \sum_{j=1}^{n} a_j, \quad \text{if} \quad a_j \geq 0,$$

then as a consequence of (23) we have

$$\|f\|_q \leq \frac{1}{n} \sum_{j=1}^{n} \left\| \frac{\partial f}{\partial x_j} \right\|_1, \quad f \in \mathscr{D}, \quad 1/q = 1 - 1/n.$$

This result and the reasoning used in §2.4 above, shows that $L^q(\mathbf{R}^n) \subset L_1^1(\mathbf{R}^n)$, and that the inclusion map is continuous.

2.6   To conclude the proof of the theorem we can argue by induction and show that the case of $k \geq 2$ can be reduced to the case $k = 1$. Let us take, for example, the assertion (i) of the theorem. The assumption $f \in L_k^p(\mathbf{R}^n)$ clearly implies that $f \in L_{k-1}^p(\mathbf{R}^n)$ and $\dfrac{\partial f}{\partial x_j} \in L_{k-1}^p(\mathbf{R}^n)$. Hence the case of the theorem for $k - 1$ implies that $f \in L^{q'}(\mathbf{R}^n)$ and $\dfrac{\partial f}{\partial x_j} \in L^{q'}(\mathbf{R}^n)$, where $1/q' = 1/p - \dfrac{(k-1)}{n}$. That is, $f \in L_1^{q'}(\mathbf{R}^n)$. The case $k = 1$ then implies that $f \in L^q(\mathbf{R}^n)$, with $1/q = 1/q' - 1/n = 1/p - \left(\dfrac{k-1}{n} \right) - 1/n = 1/p - k/n$. The corresponding inclusion mappings are also continuous. The cases (ii) and (iii) can be argued similarly.

A final remark is in order. Theorem 2 is valid for $p = 1$, unlike the closely related Theorem 1. However when $q = \infty$, (the case (ii)), it is not true that in general $f \in L^\infty$; (here the situation is again similar to that of Theorem 1). For further details see §6.3 at the end of this chapter.

3. *Bessel potentials*

3.1   The Riesz potentials I_α lead to very elegant and useful formulae, as we have already seen. Nevertheless the present formalism suffers from a shortcoming which may be explained as follows. The importance of the potentials I_α lies above all in their role as "smoothing operators." While the *local* behavior, $(|x| \to 0)$, of the kernels $\dfrac{|x|^{-n+\alpha}}{\gamma(\alpha)}$ is suited to this purpose, the *global* behavior $(|x| \to \infty)$ is less favorable and leads to increasing awkwardness the greater α is.

A way out of this dilemma is by a modification of the Riesz potentials which maintains the essential local behavior but eliminates the irrelevant problems of infinity. There are several roughly equivalent ways of doing this, but the simplest and most natural approach consists in replacing the "non-negative" operator $-\Delta$, by the "strictly positive" operator $I - \Delta$, ($I = $ identity) and defining the *Bessel potentials* $\mathcal{J}_\alpha$ by

$$\mathcal{J}_\alpha = (I - \Delta)^{-\alpha/2}$$

in analogy with

$$I_\alpha = (-\Delta)^{-\alpha/2}.$$

To put matters in logical order, we must begin by deriving the kernel of the Bessel potential, that is the presumed function $G_\alpha(x)$, with the property that $(G_\alpha(x))^\wedge = (1 + 4\pi^2 |x|^2)^{-\alpha/2}$.

The starting point for this derivation is the idea (already used in Chapter III, §3.2) that a "general" function of $|x|$ can be expressed in terms of the $\{e^{-\pi\delta|x|^2}\}_{\delta>0}$. In this instance we have a simple identity, namely

$$(25) \quad (4\pi)^{-\alpha/2}(1 + 4\pi^2 |x|^2)^{-\alpha/2} = \frac{1}{\Gamma(\alpha/2)} \int_0^\infty e^{-\frac{\delta}{4\pi}(1+4\pi^2|x|^2)} \, \delta^{\alpha/2} \frac{d\delta}{\delta}, \quad \alpha > 0$$

which is nothing but a rephrasing of the fact that

$$t^{-a} = \frac{1}{\Gamma(a)} \int_0^\infty e^{-t\delta} \, \delta^a \frac{d\delta}{\delta}$$

with $a = \alpha/2 > 0$.

We therefore set down the following table, where the entries on the right are the Fourier transforms of the corresponding entries on the left, with $a = \alpha/2$.

(i) $e^{-\pi|x|^2}$ $e^{-\pi|x|^2}$

(ii) $e^{-\pi\delta|x|^2}$ $e^{-\pi|x|^2/\delta}\delta^{-n/2}$

(iii) $\displaystyle\int_0^\infty e^{-\pi\delta|x|^2} \, \delta^a \frac{d\delta}{\delta}$ $\displaystyle\int_0^\infty e^{-\pi|x|^2/\delta} \, \delta^{-n/2} \, \delta^a \frac{d\delta}{\delta}$

(iv) $\Gamma(a)(\pi |x|^2)^{-a}$ $\Gamma(n/2 - a)(\pi |x|^2)^{n/2-a}$

By (25) and (ii)

(v) $(1 + 4\pi^2 |x|^2)^{-\alpha/2}$ $\dfrac{1}{(4\pi)^{\alpha/2}} \dfrac{1}{\Gamma(\alpha/2)} \displaystyle\int_0^\infty e^{-\pi|x|^2/\delta} e^{-\delta/4\pi} \, \delta^{(-n+\alpha)/2} \frac{d\delta}{\delta}$

Therefore we *define* $G_\alpha(x)$ by

(26) $\qquad G_\alpha(x) = \dfrac{1}{(4\pi)^{\alpha/2}} \dfrac{1}{\Gamma(\alpha/2)} \displaystyle\int_0^\infty e^{-\pi|x|^2/\delta} e^{-\delta/4\pi} \delta^{(-n+\alpha)/2} \dfrac{d\delta}{\delta}.$

PROPOSITION 2.

(1) *For each* $\alpha > 0$, $G_\alpha(x) \in L^1(\mathbf{R}^n)$.

(2) $G_\alpha{}^\wedge(x) = (1 + 4\pi^2 |x|^2)^{-\alpha/2}$.

Proof.

Since $\int_{\mathbf{R}^n} e^{-\pi|x|^2/\delta}\, dx = \delta^{n/2}$, Fubini's theorem applied to (26) shows

$$\int_{\mathbf{R}^n} G_\alpha(x)\, dx = \frac{1}{(4\pi)^{\alpha/2}\Gamma(\alpha/2)} \int_0^\infty e^{-\delta/4\pi} \delta^{\alpha/2} \frac{d\delta}{\delta} = 1, \qquad \alpha > 0,$$

and so the first conclusion is demonstrated. To prove the second conclusion we use the reasoning schematized by (i)–(v) above. In fact if we let f stand for one of the entries on the left and $\hat{f}$ for the corresponding entry on the right, we have whenever $\varphi \in \mathscr{S}$

(27) $\qquad\qquad \displaystyle\int_{\mathbf{R}^n} f(x)\hat{\varphi}(x)\, dx = \int_{\mathbf{R}^n} \hat{f}(x)\varphi(x)\, dx.$

First we take $f(x) = e^{\pi\delta|x|^2}$, $\hat{f}(x) = e^{-\pi|x|^2/\delta}\delta^{-n/2}$; then $f(x) = e^{-\delta/4\pi}e^{-\pi|x|^2}$, $\hat{f}(x) = e^{-\delta/4\pi}e^{-\pi|x|^2/\delta}\delta^{-n/2}$. With this choice of f and $\hat{f}$ we integrate both sides with respect to $\delta^{\alpha/2}\, d\delta/\delta$ (see (25)). An interchange of the order of integration (validated by Fubini's theorem) then shows that

$$\int_{\mathbf{R}^n} (1 + 4\pi^2 |x|^2)^{-\alpha/2}\varphi(x)\, dx = \int_{\mathbf{R}^n} G_\alpha(x)\hat{\varphi}(x)\, dx.$$

Since $G_\alpha \in L^1(\mathbf{R}^n)$ this shows that $\hat{G}_\alpha(x) = (1 + 4\pi^2 |x|^2)^{-\alpha/2}$.

A result similar to (26) which also follows from the above table is

(28) $\qquad \dfrac{|x|^{-n+\alpha}}{\gamma(\alpha)} = \dfrac{1}{(4\pi)^{\alpha/2}} \dfrac{1}{\Gamma(\alpha/2)} \displaystyle\int_0^\infty e^{-\pi|x|^2/\delta} \delta^{(-n+\alpha)/2} \dfrac{d\delta}{\delta}.$

If we use the fact that $e^{-\delta/4\pi} = 1 + o(e^{-\delta/4\pi})$, $\delta \to 0$, we get, upon comparing (28) with (26), that

(29) $\qquad\qquad G_\alpha(x) = \dfrac{|x|^{-n+\alpha}}{\gamma(\alpha)} + o(|x|^{-n+\alpha}), \quad \text{as} \quad |x| \to 0,$

if $0 < \alpha < n$.

A straightforward examination of the integral defining $G_\alpha(x)$ also shows that

(30) $\qquad G_\alpha(x) = O(e^{-c|x|})$ as $|x| \to \infty$, for some $c > 0$,

so that the kernel G_α is rapidly decreasing as $|x| \to \infty$.

Notice that $e^{-\pi|x|^2/\delta}e^{-\delta/4\pi}$ has as a maximum value $e^{-|x|}$ (which it attains at $\delta = 2\pi|x|$). Also if $|x| \geq 1$ then clearly $e^{-\pi|x|^2/\delta}e^{-\delta/4\pi} \leq e^{-\pi/\delta}e^{-\delta/4\pi}$. Combining the two gives, when $|x| \geq 1$, $e^{-\pi|x|^2/\delta}e^{-\delta/4\pi} \leq e^{-|x|/2}e^{-\pi/2\delta}e^{-\delta/8\pi}$. Now inserting this in the defining formula (30) yields

$$(4\pi)^{\alpha/2}\Gamma(\alpha/2)G_\alpha(x) \leq e^{-|x|/2}\int_0^\infty e^{-\pi/2\delta}e^{-\delta/8\pi}\,\delta^{(-n+\alpha)/2}\,\frac{d\delta}{\delta}$$

which is (30), with $c = 1/2$.

3.1.1 The kernel G_α may also be given another integral representation which shows that it is essentially a Bessel function of the "third kind" (see §6.5 below); that was the original derivation. We shall, however, not need any of the properties of the Bessel functions and so the terminology of "Bessel potential" has for us only a vestigial significance.

3.2 Relation between Riesz and Bessel potentials. It may be surmised from its very definition, and also from the asymptotic relation (29), that there is an intimate connection between the Bessel potentials and the Riesz potentials. This affinity between the two is given precision in the following lemma.

LEMMA 2. *Let $\alpha > 0$.*
(i) *There exists a finite measure μ_α on $\mathbf{R}^n$ so that its Fourier transform $\mu_\alpha^\wedge$ is given by*

$$\mu_\alpha(x) = \frac{(2\pi|x|)^\alpha}{(1 + 4\pi^2|x|^2)^{\alpha/2}}.$$

(ii) *There exist a pair of finite measures ν_α and λ_α on $\mathbf{R}^n$ so that*

$$(1 + 4\pi^2|x|^2)^{\alpha/2} = \nu_\alpha^\wedge(x) + (2\pi|x|)^\alpha\lambda_\alpha^\wedge(x).$$

The first part of the lemma states in effect that the following formal quotient operator is bounded on every $L^p(\mathbf{R}^n)$, $1 \leq p \leq \infty$,

(31)
$$\frac{(-\Delta)^{\alpha/2}}{(I - \Delta)^{\alpha/2}}, \qquad \alpha > 0.$$

The second part states also to what extent the same thing is true of the operator inverse to (31).

To prove (i) we use the expansion

(32)
$$(1 - t)^{\alpha/2} = 1 + \sum_{m=1}^\infty A_{m,\alpha}t^m,$$

which is valid when $|t| < 1$. All the $A_{m,\alpha}$ are of constant sign for m sufficiently large, so $\sum |A_{m,\alpha}| < \infty$, since $(1 - t)^{\alpha/2}$ remains bounded as $t \to 1$, (if $\alpha \geq 0$). Let $t = \dfrac{1}{1 + 4\pi^2 |x|^2}$. Then

(33) $$\left(\frac{4\pi^2 |x|^2}{1 + 4\pi^2 |x|^2}\right)^{\alpha/2} = 1 + \sum_{m=1}^{\infty} A_{m,\alpha}(1 + 4\pi^2 |x|^2)^{-m}.$$

However $G_{2m}(x) \geq 0$ and $\int_{\mathbf{R}^n} G_{2m}(x)e^{2\pi i x \cdot y} \, dx = (1 + 4\pi^2 |y|^2)^{-m}$.

We noticed already that $\int G_{2m}(x) \, dx = 1$ and so $\|G_{2m}\|_1 = 1$.

Thus from the convergence of $\sum |A_{m,\alpha}|$ it follows that if μ_α is defined by

(34) $$\mu_\alpha = \delta_0 + \left(\sum_{m=1}^{\infty} A_{m,\alpha}G_{2m}(x)\right) dx$$

with δ_0 the Dirac measure at the origin, then μ_α represents a finite measure. Moreover by (33),

(35) $$\hat{\mu_\alpha}(x) = \frac{(2\pi |x|)^\alpha}{(1 + 4\pi^2 |x|^2)^{\alpha/2}}.$$

We now invoke the n-dimensional version of Weiner's theorem, to wit: If $\Phi_1 \in L^1(\mathbf{R}^n)$ and $\hat{\Phi_1}(x) + 1$ is nowhere zero, then there exists a $\Phi_2 \in L^1(\mathbf{R}^n)$ so that $(\hat{\Phi_1}(x) + 1)^{-1} = \hat{\Phi_2}(x) + 1$.

For our purposes we then write

$$\Phi_1(x) = \sum_{m=1}^{\infty} A_{m,\alpha}G_{2m}(x) + G_\alpha(x).$$

Then by (35) we see that

$$\hat{\Phi_1}(x) + 1 = \frac{(2\pi |x|)^\alpha + 1}{(1 + 4\pi^2 |x|^2)^{\alpha/2}},$$

which vanishes nowhere. Thus for an appropriate $\Phi_2 \in L^1$,

$$(1 + 4\pi^2|x|^2)^{\alpha/2} = (1+(2\pi|x|)^\alpha)[\hat{\Phi_2}(x) + 1],$$

and so we obtain the desired conclusion with $\nu_\alpha = \lambda_\alpha = \delta_0 + \Phi_2(x) \, dx$.

3.3 Spaces $\mathscr{L}_\alpha^p$. It is now our intention to study more systematically the one-parameter family of operators $\{\mathscr{J}_\alpha\}_\alpha$.

For any $\alpha \geq 0$, and $f \in L^p(\mathbf{R}^n)$, $1 \leq p \leq \infty$, we can define $\mathscr{J}_\alpha(f)$, as $\mathscr{J}_\alpha(f) = G_\alpha * f$, if $\alpha > 0$, and $\mathscr{J}_0(f) = f$. In view of the fact that $\|G_\alpha\|_1 = 1$ $(= \int_{\mathbf{R}^n} G_\alpha(x) \, dx)$, we see that the convolution is in fact well-defined and

(36) $$\|\mathscr{J}_\alpha(f)\|_p \leq \|f\|_p, \qquad 1 \leq p \leq \infty.$$

It is also apparent that

$$(37) \qquad \mathcal{J}_\alpha \cdot \mathcal{J}_\beta = \mathcal{J}_{\alpha+\beta}, \qquad \alpha \geq 0, \beta \geq 0$$

since $G_\alpha * G_\beta = G_{\alpha+\beta}$, as Proposition 2 shows.

The main definition that we wish to make here is that of the *potential spaces*, $\mathcal{L}_\alpha^p$. Symbolically we write

$$(38) \qquad \mathcal{L}_\alpha^p(\mathbf{R}^n) = \mathcal{J}_\alpha(L^p(\mathbf{R}^n)), \qquad 1 \leq p \leq \infty, \ \alpha \geq 0.$$

In other words, $\mathcal{L}_\alpha^p(\mathbf{R}^n)$ is a subspace of $L^p(\mathbf{R}^n)$, consisting of all f which can be written in the form $f = \mathcal{J}_\alpha(g)$, $g \in L^p(\mathbf{R}^n)$. The $\mathcal{L}_\alpha^p$ norm of f is written as $\|f\|_{p,\alpha}$, and is defined to be the L^p norm of g, i.e.,

$$(39) \qquad \|f\|_{p,\alpha} = \|g\|_p, \quad \text{if} \ \ f = \mathcal{J}_\alpha(g).$$

To see that this gives a consistent definition of $\|f\|_{p,\alpha}$ we must observe that if $\mathcal{J}_\alpha(g_1) = \mathcal{J}_\alpha(g_2)$, then $g_1 = g_2$. However, if $\varphi \in \mathcal{S}$,

$$\int \mathcal{J}_\alpha(g_1)\varphi(x) \, dx = \int\int G_\alpha(x - y)g_1(y)\varphi(x) \, dx \, dy = \int g_1 \mathcal{J}_\alpha(\varphi) \, dx,$$

by Fubini's theorem. So $\mathcal{J}_\alpha(g_1) = \mathcal{J}_\alpha(g_2)$ implies

$$\int_{\mathbf{R}^n} (g_1 - g_2) \mathcal{J}_\alpha(\varphi) \, dx = 0, \quad \text{all} \quad \varphi \in \mathcal{S}.$$

Now the mapping $\mathcal{J}_\alpha$ is actually an onto mapping of $\mathcal{S}$ to itself. In fact, suppose $\psi \in \mathcal{S}$ is given, and take $\hat{\varphi}(x) = \hat{\psi}(x)(1 + 4\pi^2 |x|^2)^{-\alpha/2}$. Then since $\hat{\psi} \in \mathcal{S}$, so is $\hat{\varphi}$, and hence $\varphi \in \mathcal{S}$. But $\hat{\psi}(x) = \hat{\varphi}(x)(1 + 4\pi^2 |x|^2)^{\alpha/2}$. Therefore $\psi = \mathcal{J}_\alpha(\varphi)$. This shows $\int (g_1 - g_2)\psi \, dx = 0$ for all $\psi \in \mathcal{S}$ and therefore $g_1 = g_2$.

It is an immediate consequence of the definition and (36) that

$$(40) \qquad \mathcal{L}_\beta^p \subset \mathcal{L}_\alpha^p, \quad \text{and} \quad \|f\|_{p,\alpha} \leq \|f\|_{p,\beta} \quad \text{if} \ \ \beta > \alpha.$$

Also

$(41) \quad$ *$\mathcal{J}_\beta$ is an isomorphism of $\mathcal{L}_\alpha^p$ to $\mathcal{L}_{\alpha+\beta}^p$, if $\alpha \geq 0, \beta \geq 0$.*

After this rehearsal of the routine related to the spaces $\mathcal{L}_\alpha^p$, we come back to matters of greater consequence. We return to the germinal idea, already exploited in §2: that of the connection between potentials and partial derivatives. In the present context it takes the form of a close connection between the scale of potential spaces $\mathcal{L}_\alpha^p(\mathbf{R}^n)$ and the scale of Sobolov spaces $L_k^p(\mathbf{R}^n)$.

THEOREM 3. *Suppose k is a positive integer and $1 < p < \infty$. Then*

$$\mathcal{L}_k^p(\mathbf{R}^n) = L_k^p(\mathbf{R}^n)$$

in the sense that $f \in \mathscr{L}_k^p(\mathbf{R}^n)$ if and only if $f \in L_k^p(\mathbf{R}^n)$, and the two norms, given respectively in (39) and (16), are equivalent.

This identity between the spaces $\mathscr{L}_k^p$ and L_k^p fails when $p = 1$ or $p = \infty$. See the discussion in §6.6 below.

3.4 Proof of Theorem 3. The proof of the theorem is based on the following lemma.

LEMMA 3. *Suppose $1 < p < \infty$, and $\alpha \ge 1$. Then $f \in \mathscr{L}_\alpha^p(\mathbf{R}^n)$ if and only if $f \in \mathscr{L}_{\alpha-1}^p(\mathbf{R}^n)$ and for each j, $\dfrac{\partial f}{\partial x_j} \in \mathscr{L}_{\alpha-1}^p(\mathbf{R}^n)$. Moreover, the two norms,*

$$\|f\|_{p,\alpha} \ and \ \|f\|_{p,\alpha-1} + \sum_{j=1}^{n} \left\| \frac{\partial f}{\partial x_j} \right\|_{p,\alpha-1} \quad are \ equivalent.$$

Assume first that $f \in \mathscr{L}_\alpha^p(\mathbf{R}^n)$. Then $f = \mathscr{J}_\alpha(g)$ with $g \in L^p$. We claim that

(42) $\dfrac{\partial f}{\partial x_j} = \mathscr{J}_{\alpha-1}(g^{(j)})$,   where $g^{(j)} = -R_j(\mu_1 * g)$,   if $f = \mathscr{J}_\alpha(g)$.

This is immediately verifiable when g (and then f) are in $\mathscr{S}$. In fact, in that case

$$\left(\frac{\partial f}{\partial x_j} \right)^{\wedge}(x) = -2\pi i x_j \hat{f}(x)$$

$$= -2\pi i x_j (1 + 4\pi^2 |x|^2)^{-\alpha/2} \hat{g}(x)$$

$$= (1 + 4\pi^2 |x|^2)^{-(\alpha-1)/2} \hat{g}^{j}(x),$$

where $\hat{g}^j(x) = -\dfrac{ix_j}{|x|} \dfrac{(2\pi |x|)}{(1 + 4\pi^2 |x|^2)^{1/2}} \hat{g}(x)$. This proves (42), when $g \in \mathscr{S}$. In the general case, if $g \in L^p(\mathbf{R}^n)$, there exists a sequence $g_m \in \mathscr{S}$, so that $g_m \to g$ in L^p norm. The mapping $g \to \mu_1 * g$, and consequently the mapping $g \to R_j(\mu_1 * g)$ is bounded in the L^p norm. The first is bounded since μ_1 is a finite measure, according to Lemma 2; the second is bounded since when $1 < p < \infty$, the Riesz transforms R_j are bounded (see Chapter II, and Chapter III, §1). This shows that the sequence $\left\{ \dfrac{\partial f_m}{\partial x_j} \right\}$ converges in the $\mathscr{L}_{\alpha-1}^p$ norm and that (42) holds. Thus $\dfrac{\partial f}{\partial x_j} \in \mathscr{L}_{\alpha-1}^p$ and $\sum_{j=1}^{n} \left\| \dfrac{\partial f}{\partial x_j} \right\|_{p,\alpha-1} = \sum_{j=1}^{n} \|g^{(j)}\|_p \le A_p \|g\|_p = A_p \|f\|_{p,\alpha}$. Combining this with the trivial estimate that $\|f\|_{p,\alpha-1} \le \|f\|_{p,\alpha}$, (see (40)), we get

(43) $$\|f\|_{p,\alpha-1} + \sum_{j=1}^{n} \left\| \frac{\partial f}{\partial x_j} \right\|_{p,\alpha-1} \le A_p^1 \|f\|_{p,\alpha}.$$

To prove the converse we first observe that if f, and the $\dfrac{\partial f}{\partial x_j}$ are all in $\mathcal{L}^p_{\alpha-1}$, then

(44) $$f = \mathcal{J}_{\alpha-1}(g), \quad \text{and} \quad \frac{\partial f}{\partial x_j} = \mathcal{J}_{\alpha-1}\left(\frac{\partial g}{\partial x_j}\right)$$

where the $\dfrac{\partial g}{\partial x_j}$ exist in the weak sense and g and $\dfrac{\partial g}{\partial x_j}$ belong to L^p.

In fact, suppose $\dfrac{\partial f}{\partial x_j} = \mathcal{J}_{\alpha-1}(g^{(j)})$. We let φ' and $\varphi \in \mathcal{S}$. Then

$$\int_{R^n} f\varphi'\, dx = \int_{R^n} \mathcal{J}_{\alpha-1}(g)\varphi'\, dx = \int_{R^n} g\mathcal{J}_{\alpha-1}(\varphi')\, dx.$$

Similarly $\displaystyle\int_{R^n} \frac{\partial f}{\partial x_j}\varphi\, dx = \int_{R^n} g^{(j)}\mathcal{J}_{\alpha-1}(\varphi)\, dx$. But

$$\int_{R^n} f\frac{\partial \varphi}{\partial x_j}\, dx = -\int_{R^n} \frac{\partial f}{\partial x_j}\varphi\, dx;$$

this holds whenever $\varphi \in \mathcal{D}$; and a simple limiting argument extends it to φ which are in $\mathcal{S}$. Therefore, with $\varphi' = \dfrac{\partial \varphi}{\partial x_j}$,

$$\int_{R^n} g\frac{\partial}{\partial x_j}(\mathcal{J}_{\alpha-1}\varphi)\, dx = -\int_{R^n} g^{(j)}\mathcal{J}_{\alpha-1}(\varphi)\, dx$$

since $\dfrac{\partial}{\partial x_j}\mathcal{J}_{\alpha-1}(\varphi) = \mathcal{J}_{\alpha-1}\left(\dfrac{\partial \varphi}{\partial x}\right)$, for each φ in $\mathcal{S}$, as can be verified by taking the Fourier transform. However, the mapping $\varphi \to \mathcal{J}_{\alpha-1}(\varphi)$ is onto all of $\mathcal{S}$ as we have already observed, and so we get that whenever $\psi \in \mathcal{S}$ (in particular if $\psi \in \mathcal{D}$)

$$\int_{R^n} g\frac{\partial \psi}{\partial x_j}\, dx = -\int_{R^n} g^{(j)}\psi\, dx,$$

which proves (44).

Since $g \in L^p_1$, we can approximate it according to the proposition in §2.1; this gives us a sequence $\{g_m\}$ in $\mathcal{D}$, (hence in $\mathcal{S}$), so that $g_m \to g$ and $\dfrac{\partial g_m}{\partial x_j} \to \dfrac{\partial g}{\partial x_j}$ in L^p norm. We can write $g_m = \mathcal{J}_1(h_m)$, $h_m \in \mathcal{S}$. According to Lemma 2, part (ii), with $\alpha = 1$,

$$h_m = \nu_1 * g_m + \lambda_1 * \left(\sum_{j=1}^n R_j\left(\frac{\partial}{\partial x_j}g_m\right)\right)$$

and so $\|h_m\|_p \le A_p \left[\|g_m\|_p + \sum_{j=1}^{n} \left\| \dfrac{\partial g_m}{\partial x_j} \right\|_p \right]$, for $1 < p < \infty$, since the R_j are bounded in that range.

However $f_m = \mathscr{I}_\alpha(h_m)$, because $f_m = \mathscr{I}_{\alpha-1}(g_m)$, so $\|f_m\|_{p,\alpha} = \|h_m\|_p$. Thus

$$\|f_m\|_{p,\alpha} \le A_p \left[\|g_m\|_p + \sum_{j=1}^{n} \left\| \frac{\partial g_m}{\partial x_j} \right\|_p \right].$$

The same inequality holds with f_m replaced by $f_m - f_{m'}$, and g_m replaced by $g_m - g_{m'}$. This shows that the sequence f_m also converges in the $\mathscr{L}_\alpha^p$ norm; in the limit we get $f \in \mathscr{L}_\alpha^p$ and

$$\|f\|_{p,\alpha} \le A_p \left[\|g\|_p + \sum_{j=1}^{n} \left\| \frac{\partial g}{\partial x_j} \right\|_p \right] = A_p \left[\|f\|_{p,\alpha-1} + \sum_{j=1}^{n} \left\| \frac{\partial f}{\partial x_j} \right\|_{p,\alpha-1} \right].$$

This together with (43) concludes the proof of Lemma 3.

The proof of Theorem 3 is now immediate. The identity between L_k^p and $\mathscr{L}_\alpha^p$ is complete, and obvious, when $\alpha = k = 0$. However, it is clear that if $k \ge 1$, then $f \in L_k^p(\mathbf{R}^n)$ if and only if f and $\dfrac{\partial f}{\partial x_j} \in L_{k-1}^p(\mathbf{R}^n)$, $j = 1, \ldots, n$. The two norms $\|f\|_{L_k^p(\mathbf{R}^n)}$ and

$$\|f\|_{L_{k-1}^p(\mathbf{R}^n)} + \sum_{j=1}^{n} \left\| \frac{\partial f}{\partial x_j} \right\|_{L_{k-1}^p(\mathbf{R}^n)}$$

are also obviously equivalent. Thus Lemma 3 extends the identity of L_k^p and $\mathscr{L}_k^p$ from $k = 0$ to $k = 1, 2, \ldots$.

3.5 Modulus of continuity. Suppose that $f \in L^p(\mathbf{R}^n)$. We introduce again the L^p modulus of continuity: $\omega_p(t) = \|f(x + t) - f(x)\|_p$, where the $L^p(\mathbf{R}^n)$ norm is taken with respect to the x variables. We know that $\omega_p(t) \to 0$ as $|t| \to 0$ when $1 \le p < \infty$ (see Chapter III, §2.2).

We ask ourselves the following natural question. Can the property that $f \in \mathscr{L}_\alpha^p(\mathbf{R}^n)$ be characterized in terms of the order of smallness of $\omega_p(t)$, as $|t| \to 0$? If so this would give a simple characterization of the elements of $\mathscr{L}_\alpha^p$ in terms of their smoothness.

Unfortunately this hope cannot be realized, except in certain special circumstances. The particular situations when the spaces $\mathscr{L}_\alpha^p$ can be characterized in terms of moduli of continuity are simple and worth recording. These circumstances occur only when α is integral (then p may be arbitrary), or when $p = 2$ (then α may be arbitrary). We set down the details only in the cases of small α, which cases are already entirely typical.

PROPOSITION 3. *Suppose $1 < p < \infty$. Then $f \in \mathscr{L}_1^p(\mathbf{R}^n)$ if and only if $f \in L^p(\mathbf{R}^n)$, and $\omega_p(t) = O(|t|)$, as $|t| \to 0$.*

Suppose $f \in \mathscr{D}$. If $t = |t| t'$, with $|t'| = 1$, then $f(x + t) - f(x) = \int_0^{|t|} (\nabla f, t')(x + st') \, ds$. Therefore by Minkowski's inequality

$$(45) \qquad \|f(x + t) - f(x)\|_p \leq |t| \sum_{j=1}^n \left\| \frac{\partial f}{\partial x_j} \right\|_p .$$

The approximation technique of §2.1 (Proposition 1) allows us to extend (45) to any $f \in L_1^p(\mathbf{R}^n)$. By Theorem 3 this holds whenever $f \in \mathscr{L}_1^p(\mathbf{R}^n)$, $1 < p < \infty$, and so $f \in \mathscr{L}_1^p(\mathbf{R}^n)$ implies $\omega_p(t) = O(|t|)$.

Conversely if $\omega_p(t) = O(|t|)$, as $|t| \to 0$, then if e_j is the unit vector along the x_j axis, the sequence

$$\left\{ \frac{f(x + e_j/m) - f(x)}{1/m} \right\}$$

is uniformly bounded in the $L^p(\mathbf{R}^n)$ norm. By the weak compactness of the unit sphere for L^p ($1 < p$), we can find a subsequence $\{m_k\}$, and an $f^{(j)} \in L^p(\mathbf{R})$, so that $\dfrac{f(x + e_j/m_k) - f(x)}{1/m_k} \to f^{(j)}$ weakly. In particular

$$\int_{\mathbf{R}^n} \left[\frac{f(x + e_j/m_k) - f(x)}{1/m_k} \right] \varphi(x) \, dx = \int_{\mathbf{R}^n} f(x) \left[\frac{\varphi(x - e_j/m_k) - \varphi(x)}{1/m_k} \right] \, dx$$

$$\to \int_{\mathbf{R}^n} f^{(j)} \varphi \, dx = - \int_{\mathbf{R}^n} f \frac{\partial \varphi}{\partial x_j} \, dx .$$

This shows that $\dfrac{\partial f}{\partial x_j} \in L^p(\mathbf{R}^n)$, and so $f \in L_1^p(\mathbf{R}^n) = \mathscr{L}_1^p(\mathbf{R}^n)$.

PROPOSITION 4. *Suppose $0 < \alpha < 1$. Then $f \in \mathscr{L}_\alpha^2(\mathbf{R}^n)$ if and only if $f \in L^2(\mathbf{R}^n)$ and $\displaystyle\int_{\mathbf{R}^n} \frac{(\omega_2(t))^2}{|t|^{n+2\alpha}} \, dt < \infty$.*

It is to be noted that because $\omega_p(t) \leq 2 \|f\|_p$, only that part of the integral $\displaystyle\int_{\mathbf{R}^n} \frac{(\omega_2(t))^2}{|t|^{n+2\alpha}} \, dt$ near the origin is critical.

In view of Plancherel's theorem, the fact that $f = \mathscr{J}_\alpha(g)$, $g \in L^2(\mathbf{R}^n)$, is equivalent with the statement that

$$(46) \qquad \int_{\mathbf{R}^n} |\hat{f}(x)|^2 (1 + 4\pi^2 |x|^2)^\alpha dx < \infty.$$

Now again by Plancherel's theorem

$$(\omega_2(t))^2 = \|f(x + t) - f(x)\|_2^2 = \int_{\mathbf{R}^n} |\hat{f}(x)|^2 |e^{-2\pi i x \cdot t} - 1|^2 dx.$$

Therefore

$$\int_{\mathbf{R}^n} \frac{(\omega_2(t))^2}{|t|^{n+2\alpha}} \, dt = \int_{\mathbf{R}^n} |\hat{f}(x)|^2 \mathscr{I}(x) \, dx,$$

with

(47)
$$\mathscr{I}(x) = \int_{\mathbf{R}^n} \frac{|e^{-2\pi i x \cdot t} - 1|^2}{|t|^{n+2\alpha}} \, dt.$$

An evaluation of the integral $\mathscr{I}(x)$ is easy. First $\mathscr{I}(x) = \mathscr{I}(\rho x)$, where ρ is any rotation about the origin. Therefore $\mathscr{I}(x) = \mathscr{I}_0(|x|)$. Next, by homogeneity $\mathscr{I}(x) = |x|^{2\alpha} \mathscr{I}(\eta)$, where η is any fixed unit vector. Clearly the constant $\mathscr{I}(\eta)$ satisfies the properties that $0 < \mathscr{I}(\eta) < \infty$; finiteness of $\mathscr{I}(\eta)$ follows because $|e^{-2\pi i \eta \cdot t} - 1| \leq 2$, and $|e^{-2\pi i \eta \cdot t} - 1| \leq c \, |t|$, so the integral giving $\mathscr{I}(\eta)$ converges. Therefore the conditions $f \in L^2(\mathbf{R}^n)$ and $\int_{\mathbf{R}^n} \frac{(\omega_2(t))^2}{|t|^{n+2\alpha}} \, dt < \infty$ are equivalent with the statements that $\int_{\mathbf{R}^n} |\hat{f}(x)|^2 \, dx < \infty$ and $\int_{\mathbf{R}^n} |x|^{2\alpha} |\hat{f}(x)|^2 \, dx < \infty$. These two together are clearly equivalent with (46) and so the proposition is proved.

3.5.1   It is interesting to observe that Proposition 3, for $p = 2$, $\alpha = 1$, is not (at least in the way things are stated) a limiting case of Proposition 4, as $\alpha \to 1$. There is however a statement of the case $\alpha = 1$, in the spirit of Proposition 4, whose form anticipates some of the expressions that will interest us later.

We consider a modified modulus of continuity, $\tilde{\omega}_p(t)$, given by $\tilde{\omega}_p(t) = \|f(x + t) + f(x - t) - 2f(x)\|_p$. The point is that $\tilde{\omega}_p(t)$ is never really larger than $\omega_p(t)$, because $\tilde{\omega}_p(t) \leq \omega_p(t) + \omega_p(-t)$. On the other hand it may sometimes be effectively smaller than $\omega_p(t)$.*

PROPOSITION 5.   *Suppose* $0 < \alpha < 2$. *Then* $f \in \mathscr{L}_\alpha^2(\mathbf{R}^n)$ *if and only if* $f \in L^2(\mathbf{R}^n)$ *and* $\int_{\mathbf{R}^n} \frac{(\tilde{\omega}_2(t))^2 \, dt}{|t|^{n+2\alpha}} < \infty.$

The proof of this proposition is nearly identical with that of the previous one. The only difference is that the integral (47) (which converged only when $0 < \alpha < 1$) is replaced by

$$\int_{\mathbf{R}^n} \frac{|e^{-2\pi i x \cdot t} + e^{2\pi i x \cdot t} - 2|^2}{|t|^{n+2\alpha}} \, dt,$$

---

* For example, suppose $f \in \mathscr{S}$. Then $\omega_p(t) = O(|t|)$; but if $\dfrac{\omega_p(t)}{|t|} \to 0$, as $|t| \to 0$, then $f \equiv 0$. However, $\tilde{\omega}_p(t) = O(|t|^2)$, as $|t| \to 0$, always. For a deeper insight see §4.3.1 below.

which converges when $0 < \alpha < 2$. (See also the proposition in §5.2 of Chapter VIII.)

3.5.2 While in general the space $\mathscr{L}_\alpha^p(\mathbf{R}^n)$ cannot be characterized in terms of the modulus of continuity, there are nevertheless some interesting relations, as we shall see in §5 below. Suppose that $0 < \alpha < 1$. Then if $f \in \mathscr{L}_\alpha^p$ we shall have

$$\int_{\mathbf{R}^n} \frac{(\omega_p(t))^p\, dt}{|t|^{n+\alpha p}} < \infty \quad \text{if} \quad p \geq 2,$$

and

$$\int_{\mathbf{R}^n} \frac{(\omega_p(t))^2\, dt}{|t|^{n+2\alpha}} < \infty \quad \text{if} \quad p \leq 2.$$

Conversely, suppose $f \in L^p$, then $f \in \mathscr{L}_\alpha^p$ if

$$\int_{\mathbf{R}^n} \frac{(\omega_p(t))^p\, dt}{|t|^{n+\alpha p}} < \infty \quad \text{when} \quad p \leq 2,$$

or if

$$\int_{\mathbf{R}^n} \frac{(\omega_p(t))^2\, dt}{|t|^{n+2\alpha}} < \infty \quad \text{when} \quad p \geq 2.$$

This indicates that it would be interesting to study function spaces defined in terms of their moduli of continuity. We begin by considering the simplest and most familiar example of such a function space, the space of "Lipschitz" (or "Hölder") continuous functions.*

4. The spaces Λ_α of Lipschitz continuous functions

4.1 We start with the case $0 < \alpha < 1$. We define Λ_α as follows.

$$\Lambda_\alpha = \{f : f \in L^\infty(\mathbf{R}^n), \quad \text{and} \quad \omega_\infty(t) = \|f(x+t) - f(x)\|_\infty \leq A\,|t|^\alpha\}$$

The Λ_α norm is then given by

$$(48) \qquad \|f\|_{\Lambda_\alpha} = \|f\|_\infty + \sup_{|t|>0} \frac{\|f(x+t) - f(x)\|_\infty}{|t|^\alpha}.$$

The first thing to observe is that the functions in Λ_α may be taken to be continuous, and so the relation $|f(x+t) - f(x)| \leq A\,|t|^\alpha$ holds for every x. More precisely,

PROPOSITION 6. *Every $f \in \Lambda_\alpha$ may be modified on a set of measure zero so that it becomes continuous.*

* The terminologies of functions satisfying a "Hölder condition" or a "Lipschitz condition" are equally common. Our personal preference is for the latter.

The proof can be carried out by using the device of regularization of §2.1. Any smooth regularization will do, and we shall use here that of the Poisson integral (see Chapter III, §2). Thus consider

$$u(x, y) = \int_{\mathbf{R}^n} P_y(t) f(x - t)\, dt, \qquad P_y(t) = \frac{c_n y}{(|t|^2 + y^2)^{(n+1)/2}}.$$

Then

$$u(x, y) - f(x) = \int_{\mathbf{R}^n} P_y(t)[f(x - t) - f(x)]\, dt,$$

and so

$$\|u(x, y) - f(x)\|_\infty \le \int_{\mathbf{R}^n} P_y(t)\omega_\infty(-t)\, dt \le A c_n y \int_{\mathbf{R}^n} \frac{|t|^\alpha}{(|t|^2 + y^2)^{(n+1)/2}}\, dt$$

$$= A' y^\alpha$$

(if $\alpha < 1$). In particular $\|u(x, y_1) - u(x, y_2)\|_\infty \to 0$, as y_1 and $y_2 \to 0$, and since $u(x, y)$ is continuous in x, then $u(x, y)$ converges uniformly as $y \to 0$. Therefore $f(x)$ may be taken to be continuous.

4.2 A characterization.   Contrary to the situation just considered, the particular properties of the Poisson integral will play a more decisive role in the reasoning that follows. We begin by giving a characterization of $f \in \Lambda_\alpha$ in terms of their Poisson integrals $u(x, y)$.

PROPOSITION 7. *Suppose $f \in L^\infty(\mathbf{R}^n)$, and $0 < \alpha < 1$. Then $f \in \Lambda_\alpha(\mathbf{R}^n)$ if and only if*

$$(49) \qquad\qquad \left\| \frac{\partial u(x, y)}{\partial y} \right\|_\infty \le A y^{-1+\alpha}.$$

ADDENDUM: *If A_1 is the smallest constant A for which (49) holds, then $\|f\|_\infty + A_1$ and $\|f\|_{\Lambda_\alpha}$ give equivalent norms.*

We make use of the following readily verified facts about the Poisson kernel:

$$(50) \qquad \int_{\mathbf{R}^n} \left| \frac{\partial P_y(x)}{\partial y} \right| dx \le c/y; \qquad \int_{\mathbf{R}^n} \frac{\partial P_y(x)}{\partial y}\, dx = 0, \qquad y > 0.$$

The first holds because of the obvious estimates for $\dfrac{\partial P_y}{\partial y}$,

$$\left| \frac{\partial P_y}{\partial y} \right| \le c' y^{-n-1}, \qquad \left| \frac{\partial P_y}{\partial y} \right| \le c' |x|^{-n-1}.$$

The second holds because $\int_{\mathbf{R}^n} P_y(x)\,dx \equiv 1$. Thus

$$\frac{\partial u}{\partial y}(x, y) = \int \frac{\partial P_y}{\partial y}(t) f(x - t)\,dt = \int \frac{\partial P_y}{\partial y}(t)[f(x - t) - f(x)]\,dt.$$

Hence

$$\left\|\frac{\partial u}{\partial y}\right\|_\infty \leq \frac{cc_n}{y}\|f\|_{\Lambda_\alpha}\int_{\mathbf{R}^n}\left|\frac{\partial P_y}{\partial y}\right||t|^\alpha\,dt = c'\|f\|_{\Lambda_\alpha}y^{-1+\alpha}.$$

The converse, although not much more difficult, is far more enlightening, as it reveals an essential feature of the spaces in question. This insight is contained in the lemma below and the comments that follow.

LEMMA 4.   *Suppose $f \in L^\infty(\mathbf{R}^n)$ and $0 < \alpha < 1$. Then the single condition* (49) *is equivalent with the n conditions*

(51) $$\left\|\frac{\partial u(x, y)}{\partial x_j}\right\|_\infty \leq A'y^{-1+\alpha}, \qquad j = 1, \ldots, n.$$

ADDENDUM: *The smallest A in* (49) *is comparable to the smallest A' in* (51).

We know that the relation between $\dfrac{\partial u}{\partial y}$, and $\dfrac{\partial u}{\partial x_1}, \dfrac{\partial u}{\partial x_2}, \cdots \dfrac{\partial u}{\partial x_n}$ is implemented by the Riesz transforms; see Chapter III, §2.3. However the Riesz transforms do not preserve the class of bounded functions (see Chapter II, §6.1, (b)), and so there would be no identity between condition (49) and (51), when $\alpha = 1$. The meaning of the lemma is that, nevertheless, the Riesz transforms are effectively bounded on the Λ_α. (The elementary proof for this stands in sharp contrast to the difficult proof for the case of the boundedness for L^p, $1 < p < \infty$; see also Chapter II, §6.9.)

To prove the lemma we use the following estimates

(52) $$\left\|\frac{\partial P_y}{\partial y}\right\|_1 \leq cy^{-1}, \qquad \left\|\frac{\partial P_y}{\partial x_j}\right\|_1 \leq cy^{-1}, \qquad (y > 0).$$

The first already appears in (50), and the second is proved the same way.

Because $P_y = P_{y_1} * P_{y_2}$, $y = y_1 + y_2$, $y_j > 0$, we get $u(x, y) = P_{y_1} * u(x, y_2)$, and therefore with $y_1 = y_2 = y/2$,

$$\frac{\partial^2 u}{\partial y\,\partial x_j} = \left(\frac{\partial P_{y/2}}{\partial x_j}\right) * \left(\frac{\partial u}{\partial y}\right)_{y/2}.$$

Thus by (52) the assumption $\left\|\dfrac{\partial u}{\partial y}\right\| \le Ay^{-1+\alpha}$ implies the fact that

(53)
$$\left\|\frac{\partial^2 u}{\partial y\, \partial x_j}\right\|_\infty \le A_1 y^{-2+\alpha}.$$

However,

$$\left\|\frac{\partial}{\partial x_j} u(x, y)\right\|_\infty = \left\|\frac{\partial P_y}{\partial x_j} * f\right\|_\infty \le \left\|\frac{\partial P_y}{\partial x_j}\right\|_1 \|f\|_\infty \le c y^{-1} \|f\|_\infty$$

by (52). So

$$\frac{\partial}{\partial x_j} u(x, y) \to 0, \quad \text{as} \quad y \to \infty$$

and therefore

$$\frac{\partial}{\partial x_j} u(x, y) = - \int_y^\infty \frac{\partial^2 u(x, y')}{\partial y\, \partial x_j}\, dy'.$$

(53) the gives us that

$$\left\|\frac{\partial u}{\partial x_j}\right\|_\infty \le A_2 y^{-1+\alpha}, \quad \text{if} \quad \alpha < 1.$$

Conversely suppose that (49) is satisfied. Reasoning as before we get that $\left\|\dfrac{\partial^2 u}{\partial x_j^2}\right\|_\infty \le A_3 y^{-2+\alpha}$, $j = 1, 2, \ldots, n$. However since u is harmonic, that is because

$$\frac{\partial^2 u}{\partial y^2} = - \sum_{j=1}^n \frac{\partial^2 u}{\partial x_j^2},$$

we therefore have $\left\|\dfrac{\partial^2 u}{\partial y^2}\right\|_\infty \le A_4 y^{-2+\alpha}$, and a similar integration argument then shows that $\left\|\dfrac{\partial y}{\partial y}\right\|_\infty \le A_5 y^{-1+\alpha}$.

We can now prove the converse part of Proposition 7. Suppose $\left\|\dfrac{\partial}{\partial y} u(x, y)\right\|_\infty \le Ay^{-1+\alpha}$. Then the lemma also shows that $\left\|\dfrac{\partial}{\partial x_j} u(x, y)\right\|_\infty \le A'y^{-1+\alpha}$. We write

$$f(x + t) - f(x) = \{u(x + t, y) - u(x, y)\}$$
$$+ \{f(x + t) - u(x + t, y)\} - \{f(x) - u(x, y)\}.$$

Here y does not necessarily depend on t but it is best to choose $y = |t|$. Now $|u(x + t, y) - u(x, y)| \le \int_L |\nabla u(x + s, y)|\, ds$ where L is the line segment (of length $|t|$) joining x with $x + t$. Thus

$$|u(x + t, y) - u(x, y)| \le |t| \sum_{j=1}^n \|u_{x_j}(x, y)\|_\infty \le A_5 |t|\, |t|^{-1+\alpha} = A_5 |t|^\alpha.$$

Also

$$f(x + t) - u(x + t, y) = -\int_0^y \frac{\partial}{\partial y'} u(x + t, y') \, dy',$$

and so

$$|f(x + t) - u(x + t, y)| \leq y \int_0^y \left\| \frac{\partial u}{\partial y'} \right\|_\infty dy' \leq A_6 y^\alpha = A_6 |t|^\alpha.$$

With a similar estimate for $f(x) - u(x, y)$ the proof of the proposition is concluded.

4.2.1   The reader should have no difficulty in verifying that the reasoning of Lemma 4 also proves the following lemma.

LEMMA 5.   *Suppose* $f \in L^\infty(\mathbf{R}^n)$, *and* $0 < \alpha$. *Let* k *and* l *be two integers, both greater than* α. *Then the two conditions*

$$\left\| \frac{\partial^k u(x, y)}{\partial y^k} \right\|_\infty \leq A_k y^{-k+\alpha}, \quad and \quad \left\| \frac{\partial^l u(x, y)}{\partial y^l} \right\|_\infty \leq A_l y^{-l+\alpha}$$

are equivalent. Moreover, the smallest A_k *and* A_l *holding in the above inequalities are comparable.*

The utility of this lemma will be apparent soon.

4.3 Λ_α, all $\alpha > 0$.   We can now define the space $\Lambda_\alpha(\mathbf{R}^n)$ for any $\alpha > 0$. Suppose that k is the smallest integer greater than α. We set

(54)
$$\Lambda_\alpha = \left\{ f \in L^\infty(\mathbf{R}^n) : \left\| \frac{\partial^k}{\partial y^k} u(x, y) \right\|_\infty \leq A y^{-k+\alpha} \right\}$$

If A_k denotes the smallest A appearing in the inequality in (54), then we can define the Λ_α norm by

(55)
$$\|f\|_{\Lambda_\alpha} = \|f\|_\infty + A_k.$$

According to Proposition 7, when $0 < \alpha < 1$, this definition is equivalent with the previous one and the resulting norms are also equivalent. Lemma 5 also shows us that we could have replaced the $\dfrac{\partial^k u(x, y)}{\partial y^k}$ by the corresponding estimate for $\dfrac{\partial^l u(x, y)}{\partial y^l}$ where l is *any* integer greater than α.

A remark about the condition in (54) is in order. The estimate

$$\left\| \frac{\partial^k}{\partial y^k} u(x, y) \right\|_\infty \leq A y^{-k+\alpha}$$

is of interest only for y near zero, since the inequality $\left\| \dfrac{\partial^k}{\partial y^k} u(x, y) \right\|_\infty \leq Ay^{-k}$ (which is stronger away from zero) follows already from the fact that $f \in L^\infty$, (as the argument of Lemma 4 shows). This observation allows us to assert the inclusion $\Lambda_\alpha \subset \Lambda_{\alpha'}$, if $\alpha > \alpha'$.

The definitions (54) and (55) for Λ_α, when $\alpha > 0$ have a rather artificial appearance when compared with the definition given in the case $0 < \alpha < 1$ in §4.1. The reader should not be troubled about this point because the definitions just given are temporary expedients. The natural characterizations are formulated in the two propositions below.

We consider first the situation $0 < \alpha < 2$. We shall need to consider second differences, as in Proposition 5 in §3.5.1.

PROPOSITION 8. *Suppose $0 < \alpha < 2$. Then $f \in \Lambda_\alpha$ if and only if $f \in L^\infty(\mathbf{R}^n)$ and $\|f(x + t) + f(x - t) - 2f(x)\|_\infty \leq A\,|t|^\alpha$. The expression*

$$\|f\|_\infty + \sup_{|t| > 0} \frac{\|f(x + t) + f(x - t) - 2f(x)\|_\infty}{|t|^\alpha}$$

is equivalent with the Λ_α norm.

We need the following observations about the differentiated Poisson kernel:

(a) $$\int_{\mathbf{R}^n} \frac{\partial^2 P_y(t)}{\partial y^2}\, dt = 0;$$

(b) $$\frac{\partial^2 P_y(t)}{\partial y^2} = \frac{\partial^2 P_y(-t)}{\partial y^2};$$

(c) $$\left| \frac{\partial^2 P_y(t)}{\partial y^2} \right| \leq cy^{-n-2};$$

(d) $$\left| \frac{\partial^2 P_y(t)}{\partial y^2} \right| \leq c\,|t|^{-n-2}.$$

The details of the calculation giving (c) and (d) are best left to the reader but it may be helpful to point out that $\dfrac{\partial^2 P_y(t)}{\partial y^2}$ is jointly homogeneous of degree $-n - 2$.

With these remarks we see that

$$\frac{\partial^2}{\partial y^2} u(x, y) = \frac{1}{2} \int_{\mathbf{R}^n} \frac{\partial^2}{\partial y^2} P_y(t)[f(x + t) + f(x - t) - 2f(x)]\, dt,$$

and so

$$\left\| \frac{\partial^2 u(x, y)}{\partial y^2} \right\|_\infty \leq \frac{Ac}{2} \left\{ y^{-n-2} \int_{|t| \leq y} |t|^\alpha\, dt + \int_{|t| \geq y} |t|^{-n-2+\alpha}\, dt \right\}.$$

Therefore

$$\left\|\frac{\partial^2 u}{\partial y^2}\right\|_\infty \le A'y^{-2+\alpha} \quad \text{if} \quad \alpha < 2.$$

To prove the converse, write $(\Delta_t^2 F)(x) = F(x+t) + F(x-t) - 2F(x)$, and observe that if F has two continuous derivatives, then

$$\Delta_t^2 F(x) = \int_0^{|t|} \left\{ \int_{-s}^s \frac{d^2}{d\tau^2}(F(x+t'\tau))\,d\tau \right\} ds, \quad \text{where} \quad t' = t/|t|.$$

It follows immediately that

(56)
$$\|\Delta_t^2 F\|_\infty \le |t|^2 \left\{ \sum_{i,j} \left\|\frac{\partial^2 F}{\partial x_i \,\partial x_j}\right\|_\infty \right\}.$$

By the definition (54) it is clear that if $f \in \Lambda_\alpha \Rightarrow f \in \Lambda_{\alpha'}$, where $\alpha' < \alpha$. If we choose an $\alpha' < 1$, then by the results in Propositions 6 and 7 we get

(57) $\|u(x,y) - f(x)\|_\infty \to 0$, and $y\,\|u_y(x,y)\|_\infty \to 0$ as $y \to 0$.

Now the identity

(58) $f(x) = u(x,0) = \displaystyle\int_0^y y' \frac{\partial^2}{(\partial y')^2} u(x,y')\,dy' - y \frac{\partial u}{\partial y}(x,y) + u(x,y)$

is obtained by noticing that the derivative with respect to y of the extreme right-hand side vanishes, and by the use of the end-point conditions (57). However, the arguments of Lemma 4 and 5 show that the inequality $\left\|\dfrac{\partial^2 u(x,y)}{\partial y^2}\right\|_\infty \le Ay^{-2+\alpha}$ implies the estimates

$$\left\|\frac{\partial^2 u}{\partial x_i \,\partial x_j}\right\|_\infty \le A'y^{-2+\alpha}, \qquad \left\|\frac{\partial^3 u}{\partial y\,\partial x_i\,\partial x_j}\right\|_\infty \le A'y^{-3+\alpha}.$$

Thus by (56) and (58)

$$\|\Delta_t^2 f\|_\infty \le A'' \left\{ \int_0^y y'(y')^{-2+\alpha}\,dy' + (y)^{-2+\alpha}\cdot|t|^2 \right\}.$$

Taking $y = |t|$ gives the desired result

$$\|\Delta_t^2 f\|_\infty \le A''|t|^\alpha, \quad \text{if} \quad 0 < \alpha.$$

PROPOSITION 9. *Suppose* $\alpha > 1$. *Then* $f \in \Lambda_\alpha$ *if and only if* $f \in L^\infty$ *and*
$\dfrac{\partial f}{\partial x_j} \in \Lambda_{\alpha-1}, j = 1, \ldots, n$. *The norms* $\|f\|_{\Lambda_\alpha}$ *and* $\|f\|_\infty + \displaystyle\sum_{j=1}^n \left\|\dfrac{\partial f}{\partial x_j}\right\|_{\Lambda_{\alpha-1}}$
are equivalent.

Let us suppose for simplicity that $1 < \alpha \leq 2$; the other cases are argued similarly.

Observe first that $\dfrac{\partial f}{\partial x_j} \in L^\infty$. We have $\left\| \dfrac{\partial^3 u}{\partial y^3} \right\|_\infty \leq A y^{-3+\alpha}$, which implies, as we know, $\left\| \dfrac{\partial^3 u}{\partial y^2 \, \partial x_j} \right\|_\infty \leq A y^{-3+\alpha}$. We restrict to $0 < y \leq 1$, then we see that $\left\| \dfrac{\partial^3 y}{\partial y^2 \, \partial x_j} \right\|_\infty \leq A y^{-1-\beta}$, where $\beta < 1$. An integration in y then gives

$$\left\| \frac{\partial^2 u}{\partial y \, \partial x_j} \right\|_\infty \leq A \, y^{-\beta} + A \left\| \left[\frac{\partial^2 u}{\partial y \, \partial x_j} \right]_{y=1} \right\|_\infty.$$

Another integration then shows that $\left\{ \dfrac{\partial}{\partial x_j} u(x, y) \right\}$ is Cauchy in the L^∞ norm (as $y \to 0$) and so its limits can be taken to be $\dfrac{\partial f}{\partial x_j}$. The argument also gives the bound

$$\left\| \frac{\partial f}{\partial x_j} \right\|_\infty \leq C \, \|f\|_{\Lambda_\alpha}.$$

Since the (weak) derivative of f is $\dfrac{\partial f}{\partial x_j}$, the Poisson integral of the latter is $\dfrac{\partial u}{\partial x_j}$. But $\left\| \dfrac{\partial^3 u}{\partial y^2 \, \partial x_j} \right\|_\infty \leq A y^{-3+\alpha}$. Therefore $\dfrac{\partial f}{\partial x_j} \in \Lambda_{\alpha-1}$. The converse implication is proved the same way.

The last proposition reduces the study of the spaces Λ_α to those α such that $0 < \alpha \leq 1$.

4.3.1 *An example.* Concerning the Λ_α, $0 < \alpha \leq 1$, the following additional remarks are in order. First, when $0 < \alpha < 1$, Proposition 8 shows that if $f \in L^\infty$ the two conditions $\|f(x + t) - f(x)\|_\infty \leq A |t|^\alpha$ and $\|f(x + t) + f(x - t) - 2f(x)\|_\infty \leq A' |t|^\alpha$ are equivalent. However this is not the case when $\alpha = 1$.

EXAMPLE. *There exist $f \in L^\infty(\mathbf{R}^n)$ so that*

$$\|f(x + t) + f(x - t) - 2f(x)\|_\infty \leq A |t|, \qquad |t| > 0,$$

but so that $\|f(x + t) - f(x)\|_\infty \leq A' |t|$ fails for all A'.

One can construct such f by lacunary series, and more particularly as Weierstrass-Hardy non-differentiable functions. To do this we consider the function of one variable x, given by $f(x) = \sum\limits_{k=1}^\infty a^{-k} e^{2\pi i a^k x}$. Here $a > 1$;

for simplicity we take a to be an integer and this makes f periodic.* Now

$$f(x + t) + f(x - t) - 2f(x) = 2 \sum a^{-k} [\cos 2\pi a^k t - 1] e^{2\pi i a^k x}.$$

Therefore

$$\|f(x + t) + f(x - t) - 2f(x)\|_\infty \leq 2 \sum_{a^k |t| \leq 1} a^{-k} A(a^k t)^2 + 4 \sum_{a^k |t| \geq 1} a^{-k} \leq A' |t|.$$

We have used merely the facts that $|\cos 2\pi a^k t - 1| \leq A(a^k t)^2$, and ≤ 2.

If however we had $\|f(x + t) - f(x)\|_\infty \leq A' |t|$, then by Bessel's inequality for L^2 periodic functions we would get

$$(A' |t|)^2 \geq \int_0^1 |f(x + t) - f(x)|^2 \, dx$$
$$= \sum a^{-2k} |e^{2\pi i a^k t} - 1|^2 \geq \sum_{a^k |t| \leq 1} |e^{2\pi i a^k t} - 1|^2.$$

In the range $a^k |t| \leq 1$ we have $|e^{2\pi i a^k t} - 1|^2 \geq c(a^k t)^2$, and so we would arrive at the contradiction

$$(A' |t|)^2 \geq c |t|^2 \sum_{a^k |t| \leq 1} 1.$$

4.4 $\mathscr{J}_\beta : \Lambda_\alpha \leftrightarrow \Lambda_{\alpha+\beta}$. We shall now connect the Bessel potentials $\mathscr{J}_\beta$ with the Lipschitz spaces, Λ_α.

THEOREM 4. *Suppose $\alpha > 0$, $\beta \geq 0$. Then $\mathscr{J}_\beta$ maps Λ_α isomorphically onto $\Lambda_{\alpha+\beta}$.*

We should explain that by "isomorphism" in this case we do not mean that the norms $\|f\|_{\Lambda_\alpha}$ and $\|\mathscr{J}_\beta f\|_{\Lambda_{\alpha+\beta}}$ are identical, but only that they are *equivalent*.

We have already noted in §3.3 that the mapping $\mathscr{J}_\beta$ is one-one. To prove that the image of Λ_α under $\mathscr{J}_\beta$ lies in $\Lambda_{\alpha+\beta}$, and that the mapping is continuous, we argue as follows. Let u equal the Poisson integral of f, and U be the Poisson integral of $\mathscr{J}_\beta(f) = G_\beta * f$. Then $u = P_y * f$, and $U = P_y * G_\beta * f$. Thus $U(x, y) = G_\beta(x, y) * f(x)$, where $G_\beta(x, y)$ is the Poisson integral of $G_\beta(x)$. The following property of $G_\beta(x, y)$ will be proved in §5.4 below.

Suppose l is an integer and $l > \beta$. Then

(59)
$$\left\| \frac{\partial^l G_\beta(x, y)}{\partial y^l} \right\|_1 \leq A y^{-l+\beta}, \qquad y > 0.$$

However, we know that $P_{y_1+y_2} = P_{y_1} * P_{y_2}$, $y_1 > 0$, $y_2 > 0$; consequently,

$$U(x, y_1 + y_2) = P_{y_1+y_2} * G_\beta * f = P_{y_1} * G_\beta * P_{y_2} * f = G_\beta(x, y_1) * u(x, y_2).$$

---

* The result also holds if a is non-integral.

Let k be the smallest integer larger than α, and differentiate in the above l times with respect to y_1 and k times with respect to y_2. The result is

$$\frac{\partial^{k+l} U(x, y)}{\partial y^{k+l}} = \frac{\partial^l}{\partial y_1^l} G_\beta(x, y_1) * \frac{\partial^k}{\partial y_2^k} u(x, y_2), \qquad y = y_1 + y_2$$

Thus in view of (59), with $y_1 = y_2 = \dfrac{y}{2}$, we obtain

$$\left\| \frac{\partial^{k+l} U(x, y)}{\partial y^{k+l}} \right\|_1 \le A \left(\frac{y}{2} \right)^{-l+\beta} \cdot A' \left(\frac{y}{2} \right)^{-k+\alpha}.$$

Now $f \in \Lambda_\alpha$ implies that $\left\| \dfrac{\partial^k}{\partial y^k} u(x, y) \right\|_\infty \le A y^{-k+\alpha}$; (see the definition (59)).

Moreover, clearly, $\mathscr{J}_\beta(f) \in L^\infty$, since $f \in L^\infty$. Therefore $\mathscr{J}_\beta f \in \Lambda_{\alpha+\beta}$ and the proof also shows that $\| \mathscr{J}_\beta f \|_{\Lambda_{\alpha+\beta}} \le C \| f \|_{\Lambda_\alpha}$.

We claim next that the image of Λ_α under $\mathscr{J}_2$ is all of $\Lambda_{\alpha+2}$. To see this let $f \in \Lambda_{\alpha+2}$. Then $f \in \Lambda_\alpha$; also $\Delta f \in \Lambda_\alpha$, the latter fact is a consequence of Proposition 9 in §4.3. Therefore $(I - \Delta)f \in \Lambda_\alpha$. However $\mathscr{J}_2[(I - \Delta)f] = f$. To prove this identity it suffices to verify that

$$\int_{\mathbf{R}^n} (\mathscr{J}_2(1 - \Delta)f)\varphi \, dx = \int_{\mathbf{R}^n} f \, \varphi \, dx$$

whenever $\varphi \in \mathscr{S}$. But

$$\int_{\mathbf{R}^n} (\mathscr{J}_2(1 - \Delta)f)\phi \, dx = \int_{\mathbf{R}^n} (I - \Delta)f \mathscr{J}_2(\phi) \, dx$$

$$= \int f(1 - \Delta) \mathscr{J}_2 \phi \, dx = \int f\phi \, dx,$$

since obviously $(1 - \Delta)\mathscr{J}_2\varphi = \varphi$ as the Fourier transform shows. Because $\mathscr{J}_2$ is onto, $\mathscr{J}_{2-\beta}$ is one-one, and $\mathscr{J}_2 = \mathscr{J}_{2-\beta}\mathscr{J}_\beta$, for $0 < \beta < 2$, then $\mathscr{J}_\beta$ must be onto, for that range of β. Finally by superimposing such $\mathscr{J}_\beta$ we arrive at the conclusion that $\mathscr{J}_\beta$ is onto for *any* positive β and the theorem is then proved if we appeal to the closed graph theorem.*

5. The spaces $\Lambda_\alpha^{p,q}$

5.1 In analogy with our definition of Λ_α, and motivated by Proposition 4 in §3.5 we define the spaces $\Lambda_\alpha^{p,q}$, where $1 \le p, q \le \infty$. We begin with the case $0 < \alpha < 1$. Then $\Lambda_\alpha^{p,q}(\mathbf{R}^n)$ consists of all function f in $L^p(\mathbf{R}^n)$

---

* We defer to the closed-graph theorem only so that we may give a quick proof of the continuity of the mappings inverse to $\mathscr{J}_\beta$. But as the reader may guess, with a little extra effort the matter could have been treated directly.

for which the norm

(60)
$$\|f\|_p + \left(\int_{\mathbf{R}^n} \frac{(\|f(x+t) - f(x)\|_p)^q \, dt}{|t|^{n+\alpha q}} \right)^{1/q}$$

is finite. When $q = \infty$, the expression (60) is interpreted in the normal limiting way, namely

(60')
$$\|f\|_p + \sup_{|t|>0} \frac{\|f(x+t) - f(x)\|_p}{|t|^\alpha}.$$

We see therefore that $\Lambda_\alpha^{\infty,\infty} = \Lambda_\alpha$, and that Proposition 4 states, in effect, that $\Lambda_\alpha^{2,2} = \mathscr{L}_\alpha^2$, $0 < \alpha < 1$. (The identity $\Lambda_\alpha^{2,2} = \mathscr{L}_\alpha^2$ will later be seen to be valid for all α.)

The basic properties of the spaces, Λ_α, given in Propositions 7, 8, and 9, Lemmas 4 and 5, and Theorem 4 hold with the obvious modifications for the space $\Lambda_\alpha^{p,q}$. We formulate this generally, but prove it in detail only for the direct part of the analogue of Proposition 7.

PROPOSITION 7'. *Suppose* $f \in L^p(\mathbf{R}^n)$, *and* $0 < \alpha < 1$. *Then* $f \in \Lambda_\alpha^{p,q}$ *if and only if*

(61)
$$\left(\int_0^\infty \left(y^{\alpha-1} \left\| \frac{\partial}{\partial y} u(x,y) \right\|_p \right)^q \frac{dy}{y} \right)^{1/q} < \infty$$

The $\Lambda_\alpha^{p,q}$ *norm is equivalent with the norm*

$$\|f\|_p + \left(\int_0^\infty \left(y^{\alpha-1} \left\| \frac{\partial u}{\partial y} \right\|_p \right)^q \frac{dy}{y} \right)^{1/q}$$

We have

$$\frac{\partial}{\partial x} u(x,y) = \int_{\mathbf{R}^n} \frac{\partial P_y(t)}{\partial t} [f(x-t) - f(x)] \, dt$$

and therefore by the elementary estimate $\left| \dfrac{\partial P_y(t)}{\partial y} \right| \leq c' y^{-n-1}$, $\left| \dfrac{\partial P_y(t)}{\partial y} \right| \leq$ $c' |t|^{-n-1}$, already used, we see that

$$\left\| \frac{\partial}{\partial y} u(x,y) \right\|_p \leq c' y^{-n-1} \int_{|t| \leq y} \|f(x+t) - f(x)\|_p \, dt$$
$$+ c' \int_{|t|>y} \|f(x+t) - f(x)\|_p \frac{dt}{|t|^{n+1}}.$$

Next set $t = r\xi \in \mathbf{R}^n$, with $r = |t|$, and $|\xi| = 1$. Then with

$$\|f(x+t) - f(x)\|_p = \omega_p(t) = \omega_p(r\xi),$$

we write $\Omega(r) = \int_{S^{n-1}} \omega_p(r\xi)\, d\xi$. The inequality above becomes

$$\left\|\frac{\partial u}{\partial y}\right\|_p \le c' y^{-n-1} \int_0^y \Omega(r) r^{n-1}\, dr + c' \int_y^\infty \Omega(r) r^{-2}\, dr.$$

(because $dt = d\xi r^{n-1}\, dr$)

Therefore by the Hardy inequalities (see Appendix A)

$$\left(\int_0^\infty \left(y^{\alpha-1}\left\|\frac{\partial u}{\partial y}\right\|_p\right)^q \frac{dy}{y}\right)^{1/q} \le c\left(\int_0^\infty [\Omega(r) r^{-\alpha}]^q \frac{dr}{r}\right)^{1/q}.$$

However, $\Omega(r)^q \le c \int_{S^{n-1}} (\omega(r\xi))^q\, d\xi$, by Hölder's inequality.* Substituting this in the above leads to the bound

$$c''\left(\int_{S^{n-1}} \int_0^\infty (\omega(r\xi))^q r^{-\alpha q} \frac{dr}{r}\, d\xi\right)^{1/q} = c''\left(\int_{\mathbf{R}^n} \frac{(\|f(x+t) - f(x)\|_p^q}{|t|^{n+\alpha q}}\, dt\right)^{1/q}.$$

In the same way we can prove

LEMMA 4'. *Suppose $f \in L^p(\mathbf{R}^n)$, $0 < \alpha < 1$. Then the single condition (61) is equivalent with the n conditions*

(62) $$\left(\int_0^\infty \left(y^{\alpha-1}\left\|\frac{\partial u}{\partial x_j}\right\|_p\right)^q \frac{dy}{y}\right)^{1/q} < \infty, \qquad j = 1, 2, \ldots, n.$$

The norm which results if we replace the quantity in (61) *by the sum of the n quantities in* (62) *is equivalent with the original $\Lambda_\alpha^{p,q}$ norm.*

Before going further it will be well to record the more general assertion which is in back of Lemma 4'. It is the inequality

(62') $$\left(\int_0^\infty (y^{\alpha-k} \|D^k u\|_p)^q \frac{dy}{y}\right)^{1/q} \le A\left(\int_0^\infty \left(y^{\alpha-k}\left\|\frac{\partial^k u}{\partial y^k}\right\|_p\right)^q \frac{dy}{y}\right)^{1/q}.$$

Here k is a positive integer, $0 < \alpha < k$, and D^k is any differential monomial in $x_1, x_2, \ldots, x_n, y$, of total order k. The proof follows by the same arguments.

Proceeding as before we define next the spaces $\Lambda_\alpha^{p,q}(\mathbf{R}^n)$ for any $\alpha > 0$. Let k be the smallest integer greater than α. We set

(63) $$\Lambda_\alpha^{p,q}(\mathbf{R}^n) = \left\{f \in L^p(\mathbf{R}^n) : \left(\int_0^\infty \left(y^{\alpha-k}\left\|\frac{\partial^k u}{\partial y^k}\right\|_p\right)^q \frac{dy}{y}\right)^{1/q} < \infty\right\}$$

Then $\Lambda_\alpha^{p,q}$ norm is defined by

$$\|f\|_{\Lambda_\alpha^{p,q}} = \|f\|_p + \left(\int_0^\infty \left(\left\|y^{\alpha-k}\frac{\partial^k u}{\partial y^k}\right\|_p\right)^q \frac{dy}{y}\right)^{1/q}.$$

* Use $\Omega(r) \le \sup_\xi \omega(r, \xi)$, when $q = \infty$.

We should remark, as we did in the special case of Λ_α, than an equiv-alent definition and equivalent norm would have been obtained had we replaced the integer k by any other integer l, with $l > \alpha$. (This is already implicit in (62').) In complete analogy with Propositions 8 and 9 and Theorem 4 we state the following.

PROPOSITION 8'. *Suppose* $0 < \alpha < 2$. *Then* $f \in \Lambda_\alpha^{q,p}$ *if and only if* $f \in L^p(\mathbf{R}^n)$ *and*

$$\left(\int_{\mathbf{R}^n} \frac{(\|f(x+t) + f(x-t) - 2f(x)\|_p)^q}{|t|^{n+\alpha q}} \, dt \right)^{1/q} < \infty.$$

The expression

$$\|f\|_p + \left(\int_{\mathbf{R}^n} \frac{(\|f(x+t) + f(x-t) - 2f(x)\|_p)^q}{|t|^{n+\alpha q}} \, dt \right)^{1/q}$$

is equivalent with the $\Lambda_\alpha^{p,q}$ *norm.*[*]

PROPOSITION 9'. *Suppose* $\alpha > 1$. *Then* $f \in \Lambda_\alpha^{p,q}$ *if and only if* $f \in L^p(\mathbf{R}^n)$ *and* $\dfrac{\partial f}{\partial x_j} \in \Lambda_{\alpha-1}^{p,q}$. *The norm* $\|f\|_{\Lambda_{\alpha-1}^{p,q}}$ *and* $\|f\|_p + \displaystyle\sum_{j=1}^n \left\| \frac{\partial}{\partial x_j} f \right\|_{\Lambda_{\alpha-1}^{p,q}}$ *are equivalent.*

THEOREM 4'. *Suppose* $\alpha > 0$, $\beta \geq 0$. *Then* $\mathscr{J}_\beta$ *maps* $\Lambda_\alpha^{p,q}$ *isomorphically onto* $\Lambda_{\alpha+\beta}^{p,q}$.

5.2 A further look at $\Lambda_\alpha^{p,q}$. After these rather mechanical prelimi-naries we intend to make some more interesting observations about the space $\Lambda_\alpha^{p,q}$. The first question is, what are the roles of the indices α, p, and q? The answer is roughly as follows. First the index p indicates the basic norm that is used; next α gives the order of smoothness involved, and the index q represents a second-order (and rather subtle) correction to this order of smoothness. A precise result is as follows.

PROPOSITION 10. *The inclusion* $\Lambda_{\alpha_1}^{p,q_1}(\mathbf{R}^n) \subset \Lambda_{\alpha_2}^{p,q_2}(\mathbf{R}^n)$ *holds if either* (a) *if* $\alpha_1 > \alpha_2$ *(then* q_1 *and* q_2 *need not be related), or* (b) *if* $\alpha_1 = \alpha_2$ *and* $q_1 \leq q_2$.

* For $q = \infty$, we interpret

$$\left(\int_{\mathbf{R}^n} \frac{(\|f(x+t) + f(x-t) - 2f(x)\|_p)^q}{t^{n+\alpha q}} \, dt \right)^{1/q}$$

as

$$\sup_{|t|>0} \frac{\|f(x+t) + f(x-t) - 2f(x)\|_p}{|t|^\alpha}$$

The proof is based on the following lemma, which in reality is nothing but a variant of the usual maximum principle for harmonic functions.

LEMMA 6.   *Suppose* $f \in L^p(\mathbf{R}^n)$; *then for any integer k the function* $\left\| \dfrac{\partial^k}{\partial y^k} u(x, y) \right\|_p$ *is a non-increasing function of y for $0 < y < \infty$.*

Consider first the case $k = 0$. Since $P_{y_1} * P_{y_2} = P_{y_1+y_2}$ we have

(64) $u(x, y_1 + y_2) = P_{y_1} * u(x, y_2),$

and so $\|u(x, y_1 + y_2)\|_p \le \|P_{y_1}\|_1 \|u(x, y_2)\|_p$. Because $\|P_{y_1}\|_1 = 1$ we obtain $\|u(x, y_1 + y_2)\|_p \le \|u(x, y_2)\|_p$ and the assertion is proved in this case. To prove the general case of the lemma, differentiate the identity (64) k times with respect to y_2 and argue similarly.

Let us now prove part (b) of the Proposition (part (a) is argued similarly, and anyway that conclusion is even less delicate). Assume $q_1 < \infty$, and

(65) $\left(\int_0^\infty \left(y^{\alpha-k} \left\| \dfrac{\partial^k u}{\partial y^k} \right\|_p \right)^{q_1} \dfrac{dy}{y} \right)^{1/q_1} = A$

Then

$$\int_{y_0/2}^{y_0} y^{(\alpha-k)q_1} \left\| \dfrac{\partial^k u}{\partial y^k} \right\|_p^{q_1} \dfrac{dy}{y} \le A^{q_1}.$$

However, by the lemma $\left\| \dfrac{\partial^k u}{\partial y^k} \right\|_p$ takes its minimum value at the end point $(y = y_0)$ of the above integral. So we get

$$\left\| \dfrac{\partial^k}{\partial y^k} u(x, y_0) \right\|_p^{q_1} \int_{y_0/2}^{y_0} y^{(\alpha-k)q_1} \dfrac{dy}{y} \le A^{q_1},$$

that is

(66) $\left\| \dfrac{\partial^k u}{\partial y^k} \right\|_p \le c A y^{-k+\alpha}$

In other words $f \in \Lambda_\alpha^{p,q_1}$ implies also that $f \in \Lambda_\alpha^{p,\infty}$. Combining (66) with (65) shows easily that

$$\left(\int_0^\infty \left(y^{\alpha-k} \left\| \dfrac{\partial^k u}{\partial y} \right\|_p \right)^{q_2} \dfrac{dy}{y} \right)^{1/q_2} < \infty, \qquad q_2 \ge q$$

and so $f \in \Lambda_\alpha^{p,q_2}$.

For other inclusion relations of this type see also §6.7 below.

5.3   Comparison of $\mathscr{L}_\alpha^p$ with $\Lambda_\alpha^{p,q}$ We come now to one of our main goals whose interest justifies much of the preparatory work described in

§4 and §5. The comparison between the potential spaces $\mathscr{L}_\alpha^p$ and the Lipschitz spaces $\Lambda_\alpha^{p,q}$ achieves the deepest insight in this chapter and incidentally the only one which uses the Littlewood-Paley theory of Chapter IV. The result is as follows

THEOREM 5. *Suppose* $1 < p < \infty$ *and* $\alpha > 0$. *Then*

$$(A) \quad \mathscr{L}_\alpha^p \subset \Lambda_\alpha^{p,p} \quad if \quad p \geq 2$$

$$(B) \quad \mathscr{L}_\alpha^p \subset \Lambda_\alpha^{p,2} \quad if \quad p \leq 2$$

$$(C) \quad \Lambda^{p,p} \subset \mathscr{L}_\alpha^p \quad if \quad p \leq 2$$

$$(D) \quad \Lambda_\alpha^{p,p} \subset \mathscr{L}_\alpha^p \quad if \quad p \geq 2.$$

The fact that sharper inclusion relations of this type are not possible is contained in §6.8 and §6.9 below.

Because of the isomorphisms given by the operators $\mathscr{J}_\beta$ (see Theorem 4, and (41) in §3.3), it suffices to prove the inclusion relations for any particular value of α. It will be convenient for us to take $\alpha = 1$. In view of Theorem 3 of §3.3 the space $\mathscr{L}_1^p$ is equivalent with L_1^p, (when $1 < p < \infty$) and we have therefore reduced considerations to the proof of the inclusion relations for $\alpha = 1$, and with $\mathscr{L}_1^p$ replaced by L_1^p.

The norms in $\Lambda_1^{p,q}$ are expressed in terms of quantities which involve the second derivatives of u, with u the Poisson integral of f. It is for this reason we consider the following variants of the Littlewood-Paley functions:

$$(67) \quad \begin{cases} \mathscr{G}_p(x) = \left(\int (y\,|\nabla^2 u(x, y)|)^p \dfrac{dy}{y} \right)^{1/p}, & if \quad p < \infty \\[2mm] \mathscr{G}_\infty(x) = \sup_{y>0} y\,|\nabla^2 u(x, y)| \end{cases}$$

Here

$$|\nabla^2 u(x, y)|) = \sum_{k=0}^{n} \sum_{j=0}^{n} \left| \frac{\partial^2}{\partial x_j\,\partial x_k} u(x, y) \right|^2, \quad with \quad x_0 = y.$$

Assume that $\dfrac{\partial f}{\partial x_j} \in L^p(\mathbf{R}^n)$, $j = 1, \ldots, n$. Then since u is the Poisson integral of f, the Poisson integral of $\dfrac{\partial f}{\partial x_j}$ is $\dfrac{\partial u}{\partial x_j}$, as we have already pointed out in §4.3. Recalling the definition of the g-function, (see Chapter IV, §1.1), we see that

$$\left[g\left(\frac{\partial f}{\partial x_j} \right)(x) \right]^2 = \sum_{k=0}^{n} \int_0^\infty y \left| \frac{\partial^2}{\partial x_j\,\partial x_k} u(x, y) \right|^2 dy, \qquad x_0 = y.$$

However, $\dfrac{\partial^2 u}{\partial y^2} = -\displaystyle\sum_{j=1}^{n} \dfrac{\partial^2 u}{\partial x_j^2}$, and therefore

$$\mathscr{G}_2(x) \le c \sum_{j=1}^{n} g\!\left(\frac{\partial f}{\partial x_j}\right)(x).$$

By §4.9 of Chapter III

$$\sup_{y>0} \left| y\, \frac{\partial^2}{\partial y^2}\, u(x, y) \right| \le A \sum_{j=1}^{n} M\!\left(\frac{\partial f}{\partial x_j}\right)(x)$$

Thus

(68) $$\| \mathscr{G}_2(x) \|_p \le A_p \sum_{j=1}^{n} \left\| \frac{\partial f}{\partial x_j} \right\|_p$$

and

(69) $$\| \mathscr{G}_\infty(x) \|_p \le A_p \sum_{j=1}^{n} \left\| \frac{\partial f}{\partial x_j} \right\|_p.$$

The inequality for $\mathscr{G}_2$ holds because of Theorem 1 in Chapter VI, and that for $\mathscr{G}_\infty$ is a consequence of the maximal theorem in Chapter I. Now it is clear that $\mathscr{G}_p^p(x) \le \mathscr{G}_2^2(x)\mathscr{G}_\infty^{p-2}(x)$, if $p \ge 2$. Hence

$$\mathscr{G}_p(x) \le \mathscr{G}_2^{2/p}(x)\,\mathscr{G}_\infty^{(p-2)/p}(x) = \mathscr{G}_2^\theta(x)\mathscr{G}_\infty^{1-\theta}(x),$$

(with $\theta = 2/p$).

Therefore by Hölder's inequality (68), and (69),

$$\|\mathscr{G}_p(x)\|_p \le \| \mathscr{G}_2(x) \|_p^\theta \| \mathscr{G}_\infty(x) \|_p^{1-\theta} \le A_p \sum_{j=1}^{\infty} \left\| \frac{\partial f}{\partial x_j} \right\|_p$$

We have then that in particular

$$\int_0^\infty \left(y \left\| \frac{\partial^2 u}{\partial y} \right\|_p \right)^p \frac{dy}{y} < \infty, \quad \text{if} \ \ 2 \le p < \infty.$$

This shows that if $2 \le p < \infty$, then $L_1^p(\mathbf{R}^n) \subset \Lambda_1^{p,p}(\mathbf{R}^n)$, and conclusion (A) is proved.

To prove (B) apply Minkowski's inequality for integrals in the form that if $F(x, y) \ge 0$, and $r \ge 1$

(70) $$\left(\int_0^\infty \left\{ \int_{\mathbf{R}^n} F(x,\, y)\, dx \right\}^r y\, dy \right)^{1/r} \le \int_{\mathbf{R}^n} \left(\int_0^\infty F^r(x,\, y)y\, dy \right)^{1/r} dx,$$

to the effect that the norm of an integral is not greater than the integral of the norms. Take $r = 2/p$ (here $p \le 2$), and $F(x, y) = |\nabla u(x, y)|^p$. Then (70) can be rewritten as

$$\int_0^\infty y \, \| \, |\nabla^2 u| \, \|_p^2 \, dy \le \left(\int_{\mathbf{R}^n} (\mathscr{G}_2(x))^p \, dx \right)^{2/p},$$

and the latter is finite by (68), if $f \in L_1^p$. Therefore $f \in \mathscr{L}_1^p \Rightarrow f \in \Lambda_1^{p,2}$, if $p \leq 2$ and (B) is proved.

The arguments above also show that $\|f\|_{\Lambda_1^{p,q}} \leq A_p \|f\|_{L_1^p}$, with $q = p$, if $2 \leq p < \infty$, and $q = 2$, if $1 < p \leq 2$.

We shall prove the converse inclusions, (C) and (D) by establishing the *a priori* inequalities

(71) $$\|f\|_{L_1^p} \leq A_p \|f\|_{\Lambda_1^{p,q}},$$

with $q = p$ if $1 < p \leq 2$, and $q = 2$, if $2 \leq p < \infty$, under the assumption that f belongs to L_1^p.

This turns out to be merely an inversion of the arguments just given. In fact when $r \leq 1$, then Minkowski's inequality for integrals shows that (70) holds with a reversal of the sign of inequality. Thus we get

$$\left(\int_{\mathbf{R}^n} (\mathscr{G}_p(x))^p \, dx \right)^{2/p} \leq \int_0^\infty y \, \|\nabla^2 u\|_p^2 \, dy, \quad \text{if} \quad p \geq 2.$$

But since $A_p' \left\| \dfrac{\partial f}{\partial x_j} \right\|_p \leq \|\mathscr{G}_2(x)\|_p$, according to the converse of Theorem 1 of Chapter IV, we obtain (71) for $2 \leq p < \infty$.

Similarly if $1 < p \leq 2$, then $\mathscr{G}_2(x) \leq \mathscr{G}_p^\theta(x) \mathscr{G}_\infty^{1-\theta}(x)$, with $\theta = p/2$, and therefore by Hölder's inequality

$$\|\mathscr{G}_2(x)\|_p \leq \|\mathscr{G}_p(x)\|_p^\theta \, \|\mathscr{G}_\infty(x)\|_p^{1-\theta}.$$

Again by the Littlewood-Paley Theorem, $\|\mathscr{G}_2\|_p$ exceeds $A_p' \left\| \dfrac{\partial f}{\partial x_j} \right\|_p$ and then by (69)

$$\left(A'' \sum \left\| \frac{\partial f}{\partial x_j} \right\|_p \right)^\theta \leq \|\mathscr{G}_p(x)\|_p^\theta$$

However,

$$\|\mathscr{G}_p(x)\|_p = \left(\int_0^\infty y^p \, \|\nabla^2 u\|_p^p \, \frac{dy}{y} \right)^{1/p}$$

$$\leq c \left(\int_0^\infty y^p \left\| \frac{\partial^2 u}{\partial y^2} \right\|_p^p \frac{dy}{y} \right)^{1/p} \leq c \, \|f\|_{\Lambda_1^{p,p}}$$

This is because of (62') and the definition of the $\Lambda^p{}_p$ norm. Therefore (71) is also proved for $1 < p \leq 2$.

Finally, to lift the restriction $f \in L_1^p$, we consider $u(x, \varepsilon)$ instead of f, with $\varepsilon > 0$. Then clearly $u(x, \varepsilon) \in L_1^p(\mathbf{R}^n)$, if $f \in \Lambda_1^{p,q}$ (since then $f \in L^p(\mathbf{R}^n)$). Therefore by (71)

$$\|u(x, \varepsilon)\|_{L_1^p} \leq A_p \|u(x, \varepsilon)\|_{\Lambda_1^{p,q}} \leq A_p \|f\|_{\Lambda_1^{p,q}}$$

So the family $u(x, \varepsilon)$ converges in L^p norm to f, and its L_1^p norm remains uniformly bounded. From this, we see that for each j,

$$\int_{\mathbf{R}^n} \frac{\partial}{\partial x_j} u(x, \varepsilon)\varphi \, dx \to -\int_{\mathbf{R}^n} f(x) \frac{\partial \varphi}{\partial x_j} \, dx,$$

whenever $\varphi \in \mathscr{D}$, and that the linear functional $\varphi \to \int_{\mathbf{R}^n} f \frac{\partial \varphi}{\partial x_j} \, dx$ is bounded in the norm dual to that of $L^p(\mathbf{R}^n)$. Therefore by the Riesz representation theorem there exists a g_j so that

$$\int_{\mathbf{R}^n} f \frac{\partial \varphi}{\partial x_j} = -\int_{\mathbf{R}^n} g_j \varphi \, dx,$$

with $g_j \in L^p$. This shows that $f \in L_1^p$ and the theorem is completely proved.

5.4   A point left open.   We return to the proof of (59) we had postponed until now. If we look back to the definition of $\Lambda_\beta^{p,q}$ given in §5.1 we see that what we need to show has the following interpretation:

(72) $$G_\beta(x) \in \Lambda_\beta^{1,\infty}, \quad \text{if} \quad \beta > 0.$$

Let us first consider the case $0 < \beta < 1$. Since $G_\beta \in L^1(\mathbf{R}^n)$, we must see, according to Proposition 7', that

$$\int_{\mathbf{R}^n} |G_\beta(x + t) - G_\beta(x)| \, dx \leq A \, |t|^\beta.$$

We write

$$\int_{\mathbf{R}^n} |G_\beta(x + t) - G_\beta(x)| \, dt = \int_{|x| \leq 2|t|} |\cdot| \, dt + \int_{|x| > 2|t|} |\cdot| \, dt.$$

The first integral can be estimated by

$$\int_{|x| \leq 2|t|} [|G_\beta(x + t)| + |G_\beta(x)|] \, dx \leq 2 \int_{|x| \leq 3|t|} |G_\beta(x)| \, dx$$

Because $G_\beta(x) \leq c \, |x|^{-n+\beta}$ (see (29) and (30) in §3.1) we see that

$$2 \int_{|x| \leq 3|t|} |G_\beta(x)| \, dx \leq A \, |t|^\beta.$$

Next by differentiating the formula (26) in §3.1 we are led quickly to the bound

$$\left| \frac{\partial G_\beta}{\partial x_j} \right| = c \, |x_j| \int_0^\infty e^{-\pi |x|^2/\delta} e^{-\delta/4\pi} \delta^{(\beta-n-2)/2} \frac{d\delta}{\delta}$$

$$\leq c \, |x_j| \int_0^\infty e^{-\pi |x|^2/\delta} \delta^{(\beta-n-2)/2} \frac{d\delta}{\delta} = c' \, |x_j| \, |x|^{-n+\beta-2}$$

$$\leq c' \, |x|^{-n+\beta-1}$$

Therefore $|G_\beta(x+t) - G_\beta(x)| \leq c'' |t| |x|^{-n+\beta-1}$, if $|x| \geq 2 |t|$ and hence $\int_{|x| \geq 2|t|} |G_\beta(x+t) - G_\beta(x)| \, dx \leq A |t|^\beta$, if $0 < \beta < 1$. Thus (59) and also (72) are proved when $0 < \beta < 1$.

To pass to the general case we observe that wherever r is a positive integer, $G_{\beta r} = G_\beta * G_\beta * \cdots * G_\beta$, ($r$ factors), and $P_y = P_{y_1} * P_{y_2} \cdots * P_{y_r}$, if $y = y_1 + y_2 \cdots + y_r$, and $y_k > 0$. Consequently

$$G_{\beta r}(\cdot, y) = G_\beta(\cdot, y_1) * G_\beta(\cdot, y_2) \cdots * G_\beta(\cdot, y_r)$$

Now differentiate this relation once with respect to each of the variables $y_1, y_2, \ldots, y_r$, and then set $y_1 = y_2 \cdots y_2 = y/r$. The result is

$$\left\| \frac{\partial^r G_{\beta r}}{\partial y^r}(x, y) \right\| \leq A y^{-\beta r}$$

Since $0 < \beta < 1$, and is otherwise arbitrary we get the required estimate (for βr in place of β) in (59); this also implies (72).

6. *Further results*

6.1 f belongs to $L_1^p(\mathbf{R}^n)$ if and only if $f \in L^p(\mathbf{R}^n)$ and (i) f can be modified on a set of measure zero so that it is absolutely continuous in the sense of Tonelli; (ii) $\dfrac{\partial f}{\partial x_j} \in L^p(\mathbf{R}^n)$, $j = 1, \ldots, n$ (the derivatives exist almost everywhere).

6.2 f belongs to $L_k^\infty(\mathbf{R}^n)$, $k \geq 1$ if and only if f can be modified on a set of measure zero so that either of the following two equivalent conditions are satisfied.

(a) f has continuous partial derivatives of total order $\leq k - 1$. Moreover, whenever $g = \dfrac{\partial^\alpha f}{\partial x^\alpha}$, $|\alpha| \leq k - 1$, then

$$\sup_x |g(x)| < \infty \quad \text{and} \quad \sup_{x, x'} \frac{|g(x) - g(x')|}{|x - x'|} < \infty.$$

(b) There exists a sequence $\{\varphi_n\}$, $\varphi_n \in \mathscr{D}$, so that $\varphi_n \to f$ uniformly on every compact set and

$$\sup_{|\alpha| \leq k} \sup_n \left\| \frac{\partial^\alpha \varphi_n}{\partial x^\alpha} \right\|_\infty < \infty.$$

(Hint: See the proof of Proposition 3 in §3.5, and Proposition 1 in §2.1 respectively.)

6.3 Suppose $1 < p < \infty$, and $1/p = k/n$. Then there exists an $f \in L_k^p(\mathbf{R}^n)$ which is essentially unbounded in the neighborhood of every point.

Hint: Consider, for example, the case $n = 2$, $k = 1$, (then $p = 2$). Let $\varphi(x) = |x|^{-1}(\log 1/|x|)^{-1}$, if $|x| \leq 1/2$, $\varphi = 0$, otherwise. Set $f_0 = I_1(\varphi)$. Then $\dfrac{\partial f_0}{\partial x_j} = R_j(\varphi) \in L^2$. However, f_0 is not bounded near the origin. One may also construct a similar f_0 more directly by taking $f_0(x) = \log \log 1/|x|$, for small x, and f_0 positive, smooth, and with compact support, away from the origin. Finally set $f(x) = \sum\limits_{k=1}^{\infty} 2^{-k} f_0(x - r_k)$, where $\{r_k\}$ is dense set in $\mathbf{R}^n$.

6.4 The following is the generalization of inequality (23) in §2.5. Suppose $1 < k \leq n$.

Then with $1/q = 1 - k/n$, $f \in \mathscr{D}$,

$$\|f\|_q \leq \left(\prod_{i_1, \cdots i_k} \left\| \frac{\partial^k f}{\partial x_{i_1} \cdots \partial x_{i_k}} \right\|_1 \right)^{1/\binom{n}{k}}$$

where the product ranges over the $\dbinom{n}{k} = \dfrac{n!}{k!\,(n-k)!}$ ways of picking distinct $i_1, i_2, \ldots, i_k$ from $1, 2, \ldots, n$.

Hint: Consider, for example, $k = n - 1$. Write $I_j(x_j) = \displaystyle\int_{\mathbf{R}^k} \left| \dfrac{\partial^{n-1} f}{\partial \hat{x}_j^{n-1}} \right| d\hat{x}_j$, where the symbol $\hat{x}_j$ indicates that the variable has been omitted. Clearly $|f(x)| \leq I_j(x_j)$, and so $|f(x)|^n \leq \prod\limits_{j=1}^{n} I_j(x_j)$. Integrate. (If we start with the identity $f(x) = \frac{1}{2} \displaystyle\int_{-\infty}^{\infty} \operatorname{sign}(x - t) f'(t)\, dt$, instead of $f(x) = \displaystyle\int_{-\infty}^{x} f'(t)\, dt$, then the above inequality is improved by a factor of 2^{-k}.)

6.5 An alternative formula to (26) for the Bessel kernel is

$$G_\alpha(x) = c_\alpha e^{-|x|} \int_0^\infty e^{-|x|t} \left(t + \frac{t^2}{2} \right)^{(n-\alpha-1)/2} dt, \quad \text{for} \quad 0 < \alpha < n + 1$$

$$c_\alpha^{-1} = (2\pi)^{(n-1)/2} 2^{\alpha/2} \, \Gamma\!\left(\frac{\alpha}{2}\right) \Gamma\!\left(\frac{n-\alpha+1}{2}\right).$$

See Aronszajn and Smith [1].

6.6 The following describe the possible inclusion relations between $L_k^p(\mathbf{R}^n)$ and $\mathscr{L}_k^p(\mathbf{R}^n)$, in the extreme cases $p = 1$, and $p = \infty$.

(a) When $n = 1$, then $L_k^p(\mathbf{R}^1) = \mathscr{L}_k^p(\mathbf{R}^1)$, when k is even and $p = 1$, or ∞.

(b) When $n > 1$, then $L_k^p(\mathbf{R}^n) \subset \mathscr{L}_k^p(\mathbf{R}^n)$, when k is even, and $p = 1$ or ∞; the reverse inclusion fails for both $p = 1$ and $p = \infty$.

(c) For all n, if k is odd then neither $L_k^p(\mathbf{R}^n) \subset \mathscr{L}_k^p(\mathbf{R}^n)$ nor $\mathscr{L}_k^p(\mathbf{R}^n) \subset L_k^p(\mathbf{R}^n)$.

Hints: For (a) use the fact that $f \in L^p(\mathbf{R}^1)$ and $\dfrac{d^2 f}{dx^2} \in L^p(\mathbf{R}^1)$ implies $\dfrac{df}{dx} \in L^p(\mathbf{R}^1)$.

To show that $\mathscr{L}_k^p(\mathbf{R}^n) \not\subset L_k^p(\mathbf{R}^n)$, use the unboundedness of the higher Riesz transforms for L^1 and L^∞ (see §6.1 in Chapter II). To see, e.g. that

$L_1^\infty(\mathbf{R}^n) \not\subset \mathscr{L}_1^\infty(\mathbf{R}^n)$, use the function $G_{n+1}(x)$. From formula (26) it follows easily that G_{n+1} and $\dfrac{\partial G_{n+1}}{\partial x_j} \in L^\infty$, thus $G_{n+1} \in L_1^\infty$. However, $G_{n+1} \notin \mathscr{L}_1^\infty(\mathbf{R}^n)$, since $G_n(x) \approx \log \dfrac{1}{|x|}$, as $|x| \to 0$, and so $G_n \notin L^\infty$. The fact that $G_n \approx \log \dfrac{1}{|x|}$ as $|x| \to 0$ also follows from (26), in the same way as (29) is proved. Special functions useful in this connection are studied in Wainger [1].

6.7

(a) $\Lambda_{\alpha_1}^{p_1,q}(\mathbf{R}^n) \subset \Lambda_{\alpha_2}^{p_2,q}(\mathbf{R}^n)$ if $\alpha_1 \geq \alpha_2$ and $\alpha_1 - \dfrac{n}{p_1} = \alpha_2 - \dfrac{n}{p_2}$.

(b) If $f \in \Lambda_{\alpha_j}^{p_j,q_j}(\mathbf{R}^n)$ where $j = 0, 1$, then $f \in \Lambda_\alpha^{p,q}(\mathbf{R}^n)$, where

$$\alpha = \alpha_0(1 - \theta) + \alpha_1\theta,$$

$$\frac{1}{p} = \frac{1}{p_0}(1 - \theta) + \frac{\theta}{p_1}, \qquad \frac{1}{q} = \frac{1}{q_0}(1 - \theta) + \frac{\theta}{q_1},$$

for each $0 < \theta < 1$. See Hardy-Littlewood [2], Taibleson [2].

6.8 Suppose that $f_{\alpha,\sigma}(x) = e^{-\pi x^2} \cdot \displaystyle\sum_{k=1}^\infty a^{-k\alpha} k^{-\sigma} e^{2\pi i a^k x}$, $x \in \mathbf{R}^1$, where a is an integer > 1.

(i) $f_{\alpha,\sigma} \in \mathscr{L}_\alpha^p(\mathbf{R}^1) \Leftrightarrow \sigma > \dfrac{1}{2}$, for $1 \leq p < \infty$.

(ii) $f_{\alpha,\sigma} \in \Lambda_\alpha^{p,q}(\mathbf{R}^1) \Leftrightarrow \sigma > \dfrac{1}{q}$, (for $1 \leq p \leq \infty$).

Thus $\mathscr{L}_\alpha^p(\mathbf{R}^1) \not\subset \Lambda_\alpha^{p,q}(\mathbf{R}^1)$ if $q < 2$, and $\Lambda_\alpha^{p,q}(\mathbf{R}^1) \not\subset \mathscr{L}_\alpha^p(\mathbf{R}^1)$ if $q > 2$.

6.9 Let $g_{\alpha,\delta,p}(x) = |x|^{\alpha - n/p}(\log 1/|x|)^{-\delta}$ for $|x| < \dfrac{1}{2}$, and assume that $g_{\alpha,\delta,p}$ is smooth away from the origin and has compact support. Assume $\alpha < n/p$.

(i) $g_{\alpha,\delta,p} \in \mathscr{L}_\alpha^p(\mathbf{R}^n) \Leftrightarrow \delta p > 1$

(ii) $g_{\alpha,\delta,p} \in \Lambda_\alpha^{p,q}(\mathbf{R}^n) \Leftrightarrow \delta q > 1$

Thus $\mathscr{L}_\alpha^p(\mathbf{R}^n) \not\subset \Lambda_\alpha^{p,q}(\mathbf{R}^n)$ if $q < p$, and $\Lambda_\alpha^{p,q}(\mathbf{R}^n) \not\subset \mathscr{L}_\alpha^p(\mathbf{R}^n)$ if $q > p$. For examples closely related to §6.8 and §6.9 see Taibleson [2].

6.10 Suppose $0 < \alpha < 2$. Then $f \in \mathscr{L}_\alpha^p(\mathbf{R}^n) \Leftrightarrow f \in L^p$ and

(i) $\lim\limits_{\varepsilon \to 0} I_\varepsilon \equiv \lim\limits_{\varepsilon \to 0} \displaystyle\int_{|t| \geq \varepsilon} \dfrac{[f(x + t) - f(x)]}{|t|^{n+\alpha}} \, dt$ converges in L^p norm, if $1 \leq p < \infty$.

(ii) I_ε remains bounded in L^∞ norm, when $p = \infty$. See Stein [7]; also Wheeden [2].

Hint: Verify that if $f \in \mathscr{D}$, $\displaystyle\lim_{\varepsilon \to 0} \int_{|t| \geq \varepsilon} \frac{f(x + t) - f(x)}{|t|^{n+\alpha}} \, dt = c_\alpha(-\Delta)^{\alpha/2} f$. Conversely, suppose $f = \mathscr{I}_\alpha(g)$, with $g \in L^p$. Then $f = I_\alpha(\gamma)$, $\gamma \in L^p$. (See §3.2). Also

$$\int_{|t| \geq \varepsilon} \frac{f(x + t) - f(x)}{|t|^{n+\alpha}} \, dt = \int K_\varepsilon(t) \gamma(x + t) \, dt,$$

where $K_\varepsilon(t) = \varepsilon^{-n} K(t/\varepsilon)$; it can be shown that $|K(x)| \leq A \, |x|^{-n+\alpha}$ and $|K(x)| \leq A \, |x|^{-n+\alpha-2}$, thus $K \in L^1(\mathbf{R}^n)$.

6.11 (a) The space $\mathscr{L}_\alpha^p(\mathbf{R}^n)$ is an algebra under pointwise multiplication if and only if every element of $\mathscr{L}_\alpha^p(\mathbf{R}^n)$ is continuous. This holds if and only if $p > n/\alpha$.

(b) Let χ_K be the characteristic function of an arbitrary convex set $K \subset \mathbf{R}^n$. Then the mapping $f \to \chi_K \cdot f$ is continuous in $\mathscr{L}_\alpha^p(\mathbf{R}^n)$ when $0 \leq \alpha < 1/p$. For these and related results see Strichartz [1]. For $p = 2$, see also Hirschmann [2].

6.12 Suppose $F = I_\alpha(f)$, and $0 < \alpha < 1$.

$$(\mathscr{D}_\alpha F)(x) = \left(\int_{\mathbf{R}^n} \frac{|F(x + t) - F(x)|^2}{|t|^{n+2\alpha}} \, dt \right)^{1/2}$$

Then

$$A_\alpha g_1(f)(x) \leq \mathscr{D}_\alpha(F)(x) \leq B_\alpha g_\lambda^*(f)(x)$$

where $\lambda < 1 + 2\alpha/n$, and the functions g_1 and g_* have been defined in Chapter IV, (§1.2 and §2.2 respectively). A_α and B_α are appropriate constants.

Hint: Write $U(x, y)$ and $u(x, y)$ for the Poisson integrals of F and f respectively. Since

$$\frac{\partial^2 U}{\partial y^2} = \int \frac{\partial^2 P_y}{\partial y^2} (t) \, [F(x + t) - F(x)] \, dt,$$

simple estimates show that $\displaystyle\int_0^\infty y^{3+2\alpha} \left| \frac{\partial^2 U}{\partial y^2} \right|^2 dy \leq c_1(\mathscr{D}_\alpha(F))^2$. Next

$$\frac{\partial u}{\partial y}(x, y) = \frac{1}{\Gamma(1 - \alpha)} \int_0^\infty \frac{\partial^2 U}{\partial y^2}(x, y + s) s^\alpha \, ds$$

$$= \frac{1}{\Gamma(1 - \alpha)} \int_y^\infty \frac{\partial^2 U}{\partial y^2}(x, s)(s - y)^\alpha \, ds$$

Therefore $\displaystyle\int_0^\infty y \left| \frac{\partial u}{\partial y} \right|^2 dy \leq c_2 \int_0^\infty y^{3+2\alpha} \left| \frac{\partial^2 U}{\partial y^2} \right|^2 dy$, and so $A_\alpha g_1(f)(x) \leq \mathscr{D}_\alpha(F)(x)$. Conversely, we have

$$|F(x + t) - F(x)| \leq \int_{L_1} |\nabla U| \, ds + \int_{L_2} |\nabla U| \, ds + \int_{L_3} |\nabla U| \, ds$$

where L_1, L_2, and L_3 are respectively the line segments joining $(x, 0)$ with (x, y); $(x + t, 0)$ with $(x + t, y)$; and $(x + t, y)$ with (x, y). However $U(x, y) = \frac{1}{\Gamma(\alpha)} \int_0^\infty u(x, y + s)s^{-1+\alpha} \, ds$, therefore

$$|\nabla U(x, y)| \leq \frac{1}{\Gamma(\alpha)} \int_0^\infty |\nabla u(x, y + s)| \, s^{-1+\alpha} \, ds.$$

If we substitute this estimate in the above, set $y = |t|$, and carry out the indicated integrations, we get after some further reduction the result $\mathscr{D}_\alpha(F)(x) \leq B_\alpha g_\lambda^*(f)(x)$, with $\lambda < 1 + \frac{2\alpha}{n}$. See also the bibliographical references in §6.13 below.

6.13 (a) Suppose that $0 < \alpha < 1$, and $\frac{2n}{n + \alpha} < p < \infty$ (the latter holds in particular if $2 \leq p < \infty$). Then $f \in \mathscr{L}_\alpha^p(\mathbf{R}^n)$ if and only if $f \in L^p(\mathbf{R}^n)$ and $\mathscr{D}_\alpha(f) \in L^p(\mathbf{R}^n)$. Also $\|f\|_{p,\alpha}$ is comparable with $\|f\|_p + \|\mathscr{D}_\alpha(f)\|_p$.

(b) The similar result holds in the larger range $0 < \alpha < 2$ if $\mathscr{D}_\alpha(f)(x)$ is replaced by

$$\left(\int_{\mathbf{R}^n} \frac{|f(x + t) + f(x - t) - 2f(x)|^2}{|t|^{n+2\alpha}} \, dt \right)^{1/2}$$

The results of §6.12 and §6.13 (a) and (b) are stated in Stein [7]. For the earlier (one-dimensional) theory see Marcinkiewicz [2], Zygmund [1], and Hirschmann [1]. For a recent stronger result, dealing with the critical case $p = \frac{2n}{n + \alpha}$, see Fefferman [1].

(c) A variant of (a) above holds for all p in $1 < p < \infty$ if $\mathscr{D}_\alpha(f)$ is replaced by

$$\left(\int_0^\infty \left\{ \int_B |f(x + rt) - f(x)| \, dt \right\}^2 \frac{dr}{r^{1+2\alpha}} \right)^{1/2}$$

where B is the unit ball. See Strichartz [1].

6.14 Let $\beta > \alpha$. Then T is a bounded linear transformation from $\Lambda_\alpha(\mathbf{R}^n)$ to $\Lambda_\beta(\mathbf{R}^n)$ which commutes with translations if and only if T is of the form $Tf = K * f$, with $K \in \Lambda_{\beta-\alpha}^{1,\infty}(\mathbf{R}^n)$. See Zygmund [6] for the case $n = 1$, and Taibleson [2] for the general case.

6.15 (a) Suppose T is of the type discussed in §6.14 above. Then T maps $L^p(\mathbf{R}^n)$ boundedly into $L^q(\mathbf{R}^n)$ if $1 < p, q < \infty$ and $1/q = 1/p - \frac{(\beta - \alpha)}{n}$.

(b) However, there exists a T commuting with translations and mapping $\Lambda_\alpha(\mathbf{R}^n)$ to $\Lambda_\alpha(\mathbf{R}^n)$ boundedly, but which is not bounded on $L^p(\mathbf{R}^n)$, for $p \neq 2$. See Stein and Zygmund [2]; earlier results in this direction are in Hardy and Littlewood [3].

6.16 The last set of results deal with the space of functions of *bounded mean oscillation*; it illustrates the fact that this class arises often as a substitute for the space L^∞ in the usual limiting cases where results break down for L^∞.

(a) Suppose f is defined on $\mathbf{R}^n$. Then it is said to be of bounded mean oscillation (on $\mathbf{R}^n$), (abbreviated as BMO), if there exists a constant M, so that $\dfrac{1}{m(Q)} \displaystyle\int_Q |f(x) - a_Q| \, dx \le M$, for every cube Q in $\mathbf{R}^n$; a_Q is the mean-value of f over Q. Notice that every bounded function is BMO; however, the function $\log |x|$ can be seen to be BMO, so the converse does not hold. That this example is to a certain extent typical can be seen by the fact that it is possible to make the estimate

$$m\{x \in Q : |f(x) - a_Q| > \alpha\} \le e^{-c\alpha/M} m(Q), \quad \text{every} \quad \alpha > 0.$$

In particular if f is BMO, then $\displaystyle\int_Q e^{a|f|} \, dx < \infty$, for every cube Q, for appropriate positive a. See John and Nirenberg [1].

(b) Let T be one of the singular integrals transforms dealt with by Theorem 1, its corollary, or Theorems 2 and 3 of Chapter II. Suppose f is bounded. Then Tf is BMO. See Stein [8].

6.17 (a) Suppose f is in BMO. Then $\mathcal{I}_\alpha f \in \Lambda_\alpha(\mathbf{R}^n)$ for $\alpha > 0$.

(b) Suppose f is of weak type p; that is $m\{x : |f(x)| > \lambda\} \le A\lambda^{-p}, 0 < \lambda < \infty$, with $1 < p < \infty$. Then $\mathcal{I}_\alpha(f) \in$ BMO, if $\alpha = n/p$. (Compare with Theorem 1 in this chapter.)

See Stein and Zygmund [2].

6.18 Suppose $f \in \Lambda_1^{\infty,2}$. That is, suppose $f \in L^\infty(\mathbf{R}^n)$ and

$$\int_{\mathbf{R}^n} \frac{\|f(x+t) + f(x-t) - 2f(x)\|_\infty^2}{|t|^{n+2}} \, dt < \infty.$$

Then $\dfrac{\partial f}{\partial x_j} \in$ BMO, $j = 1, \ldots, n$.

Hint: Using the reasoning of §4 and §5, the assumptions can be shown to imply the existence of a function $\delta(s)$ on $0 < s < \infty$, so that

(i) $\delta(s)$ is non-decreasing on $0 < s < \infty$

(ii) $\displaystyle\int_0^1 \frac{\delta^2(s)}{s} \, ds < \infty$

(iii) $\|f(x+t) + f(x-t) - 2f(x)\|_\infty \le |t| \, \delta(|t|)$.

With this done one can then adapt the reasoning in John and Nirenberg [1] given for the case $\delta(s) = \left(\log \dfrac{1}{s}\right)^{-1/2-\varepsilon}$, $\varepsilon > 0$ ($s < 1/2$). An earlier related result is in M. Weiss and Zygmund [1].

Notes

Section 1. For the Riesz potentials see M. Riesz [2], and an earlier one-dimensional version in Weyl [1]. The L^p inequalities of Theorem 1 are due to Hardy and Littlewood [2], when $n = 1$, and to Sobolov [1] for general n. The fact that the fractional integration mapping is of weak-type $(1, n/(n - \alpha)$ appears first in Zygmund [4]. For a general treatment in terms of Lorentz spaces see O'Neil [1]; the simple proof given in the text is taken from Muckenhoupt and Stein [1].

Section 2. Theorem 2 goes back to Sobolov [1]. The case $p = 1$, was however not dealt with until later by Gagliardo [2] and Nirenberg [1].

Section 3. The Bessel potentials and the corresponding spaces $\mathscr{L}^p_\alpha$ were introduced by Aronszajn and Smith [1], and Calderón [4]. Lemma 2 connecting the Bessel and Riesz potentials is stated in Stein [7]. The identification of $\mathscr{L}^p_k$ with L^p_k, $1 < p < \infty$, is proved in Calderón [4], and the characterization of $\mathscr{L}^2_\alpha$ given by Proposition 4 is taken from Aronszajn and Smith [1].

Sections 4 and 5. For the case $n = 1$ most of the results given here were formulated and proved in one form or another (and sometimes only implicitly) by Hardy and Littlewood [2], and [3], Zygmund [6], and Hirschmann [1]. For the first explicit treatment of the spaces $\Lambda^{p,q}_\alpha$ (in n-dimensions) see Besov [1]; this was preceded however by a significant paper of Gagliardo [1]. The presentation given in these two sections leans heavily on the systematic treatment of Taibleson [2]; Theorems 4' and 5 in particular are due to him. The reader may also consult the earlier survey paper of Nikolskïï [1].

CHAPTER VI

Extensions and Restrictions

If we want to apply the results of harmonic analysis of $\mathbf{R}^n$ to a variety of other problems we are often faced with the following situation. Let S be a subset of $\mathbf{R}^n$, (the nature of S will be specified later), and consider one of the Banach spaces of functions on $\mathbf{R}^n$ we have already studied. Two problems then arise. The *restriction* problem: What is the space of functions that arise by restriction to S of the functions in the given Banach space? There is also the closely related *extension* problem: Given an appropriate space of functions defined on S, how can these functions be extended to $\mathbf{R}^n$?

The techniques and results differ depending on the nature of the set S, although there is some overlapping. We shall single out three cases which seem to be of genuine interest.

(i) The set S is an arbitrary closed set F. The appropriate function spaces are those composed of functions which have continuous partial derivatives up to a certain order, together with bounds on their moduli of continuity. The type of extension considered goes back to Whitney and we follow his construction except for details.

(ii) The set S is a domain (open subset of $\mathbf{R}^n$) whose boundary satisfies a certain minimal smoothness condition. If the domain had a smooth (say C^∞) boundary the extension result would be much easier and a simpler construction would do the job. The main point of the given extension is that one needs to assume what amounts roughly to only one order of differentiability of the boundary, and obtain extensions for all orders of differentiability. The presentation we shall give (in §3) will be based on ideas different from the one initially introduced by Calderón in this context. The gist of his method is outlined in §4.8 below.

(iii) The set S is a linear sub-variety $\mathbf{R}^m$ of $\mathbf{R}^n$. Looked at from the point of view of the restriction problem, there is in general a loss of smoothness in going from appropriate functions on $\mathbf{R}^n$ to functions on $\mathbf{R}^m$. Since $\mathbf{R}^m$ has Lebesgue measure zero in $\mathbf{R}^n$, there is also the problem of giving the functions in $\mathbf{R}^n$ their *natural* definition on $\mathbf{R}^m$, so that the restriction may be well defined. This kind of difficulty did not arise in case (i) because there continuous functions are dealt with exclusively. For the present problem the functions considered may be discontinuous at every point, but they do have certain average continuity. A striking aspect turns out to be the fact that the function spaces appropriate for $\mathbf{R}^m$ (for the restrictions), may in character be quite different from those appropriate for $\mathbf{R}^n$.

We begin this chapter by giving the details of the decomposition of an arbitrary open set in $\mathbf{R}^n$ into a suitable "disjoint" union of cubes. The usefulness of this decomposition has already been indicated in Chapter I. Here we apply it again, and the partition of unity based on it, as the main tool in the extension of the type (i). It arises again, if only implicitly, in the extension of type (ii).

1. *Decomposition of open sets into cubes*

In what follows, F will denote an arbitrary non-empty closed set in $\mathbf{R}^n$, Ω its complement. By a *cube* we mean a closed cube in $\mathbf{R}^n$, with sides parallel to the axes, and two such cubes will be said to be *disjoint* if their interiors are disjoint. For such a cube Q, diam (Q) denotes its diameter, and dist (Q, F) its distance from F.

1.1 Theorem 1. *Let F be given. Then there exists a collection of cubes $\mathscr{F}$, $\mathscr{F} = \{Q_1, Q_2, \ldots Q_k, \ldots\}$ so that*

(1) $\bigcup_k Q_k = \Omega = (^cF)$,

(2) *The Q_k are mutually disjoint,*

(3) $c_1 \text{ diam } (Q_k) \leq \text{dist } (Q_k, F) \leq c_2 \text{ diam } (Q_k)$.

The constants c_1 and c_2 are independent of F. In fact we may take $c_1 = 1$ and $c_2 = 4$.

1.2 Consider the lattice of points in $\mathbf{R}^n$ whose coordinates are integral. This lattice determines a *mesh* $\mathscr{M}_0$, which is a collection of cubes: namely all cubes of unit length, whose vertices are points of the above lattice. The mesh $\mathscr{M}_0$ leads to a two-way infinite chain of such meshes $\{\mathscr{M}_k\}_{-\infty}^{\infty}$, with $\mathscr{M}_k = 2^{-k}\mathscr{M}_0$.

Thus each cube in the mesh $\mathscr{M}_k$ gives rise to 2^n cubes in the mesh $\mathscr{M}_{k+1}$ by bisecting the sides. The cubes in the mesh $\mathscr{M}_k$ each have sides of length 2^{-k} and are thus of diameter $\sqrt{n}\, 2^{-k}$.

In addition to the meshes $\mathscr{M}_k$ we consider the layers Ω_k, defined by $\Omega_k = \{x : c2^{-k} < \text{dist } (x, F) \leq c2^{-k+1}\}$; c is a positive constant we shall fix momentarily. Obviously $\Omega = \bigcup_{k=-\infty}^{\infty} \Omega_k$.

We now make an initial choice of cubes, and denote the resulting collection by $\mathscr{F}_0$. Our choice is made as follows. We consider the cubes of the mesh $\mathscr{M}_k$, (each such cube is of size approximately 2^{-k}), and include a cube of this mesh in $\mathscr{F}_0$ if it intersects Ω_k, (the points of the latter are all approximately at a distance 2^{-k} from F). That is we take

$$\mathscr{F}_0 = \bigcup_k \{Q \in \mathscr{M}_k : Q \cap \Omega_k \neq 0\}.$$

We then have

$$\bigcup_{Q \in \mathscr{F}_0} Q = \Omega.$$

For appropriate choice of c,

(3) $\operatorname{diam}(Q) \leq \operatorname{dist}(Q, F) \leq 4 \operatorname{diam}(Q), \qquad Q \in \mathscr{F}_0.$

Let us prove (3) first. Suppose $Q \in \mathscr{M}_k$; then the diameter of $Q = \sqrt{n}\, 2^{-k}$. Since $Q \in \mathscr{F}_0$, there exists $x \in Q \cap \Omega_k$. Thus $\operatorname{dist}(Q, F) \leq \operatorname{dist}(x, F) \leq c 2^{-k+1}$, and $\operatorname{dist}(Q, F) \geq \operatorname{dist}(x, F) - \operatorname{diam}(Q) > c 2^{-k} - \sqrt{n}\, 2^{-k}$. If we choose $c = 2\sqrt{n}$ we get (3).

Then by (3) the cubes $Q \in \mathscr{F}_0$ are disjoint from F and clearly cover Ω. Therefore (1) is also proved. Notice that the collection $\mathscr{F}_0$ has all our required properties, except that the cubes in it are not necessarily disjoint. To finish the proof of the theorem we need to refine our choice leading to $\mathscr{F}_0$, eliminating those cubes which were really unnecessary.

We require the following simple observation. Suppose Q_1 and Q_2 are two cubes (taken respectively from the mesh $\mathscr{M}_{k_1}$ and $\mathscr{M}_{k_2}$). Then if Q_1 and Q_2 are not disjoint, one of the two must be contained in the other. (In particular $Q_1 \subset Q_2$, if $k_1 \geq k_2$.)

Start now with any cube $Q \in \mathscr{F}_0$, and consider *the maximal* cube in $\mathscr{F}_0$ which contains it. In view of the inequality (3) for any cube Q' in $\mathscr{F}_0$, which contains Q in $\mathscr{F}_0$ we have $\operatorname{diam}(Q') \leq 4 \operatorname{diam}(Q)$. Moreover any two cubes Q' and Q'' which contain Q have obviously a non-trivial intersection. Thus by the observation made above each cube $Q \in \mathscr{F}_0$ has a *unique* maximal cube in $\mathscr{F}_0$ which contains it. By the same token these maximal cubes are also disjoint. We let $\mathscr{F}$ denote the collection of maximal cubes of $\mathscr{F}_0$. Then obviously

(1) $\bigcup_{Q \in \mathscr{F}} Q = \Omega,$

(2) The cubes of $\mathscr{F}$ are disjoint,

(3) $\operatorname{diam}(Q) \leq \operatorname{dist}(Q, F) \leq 4 \operatorname{diam}(Q), \qquad Q \in \mathscr{F}.$

Theorem 1 is therefore proved.

1.3 A partition of unity. We shall now make a few observations about the family $\mathscr{F}$ of cubes whose existence is guaranteed by Theorem 1. Let us say that two distinct cubes of $\mathscr{F}$, Q_1 and Q_2, *touch* if their boundaries have a common point. (We remind the reader that two distinct cubes of $\mathscr{F}$ always have disjoint interiors.)

PROPOSITION 1. *Suppose Q_1 and Q_2 touch. Then*

$$(1/4) \operatorname{diam} (Q_2) \leq \operatorname{diam} (Q_1) \leq 4 \operatorname{diam} (Q_2).$$

We know that $\operatorname{dist} (Q_1, F) \leq 4 \operatorname{diam} (Q_1)$. Then $\operatorname{dist} (Q_2, F) \leq 4 \operatorname{diam} (Q_1) + \operatorname{diam} (Q_1) = 5 \operatorname{diam} (Q_1)$, since Q_1 and Q_2 touch. But $\operatorname{diam} (Q_2) \leq \operatorname{dist} (Q_2, F)$, therefore $\operatorname{diam} (Q_1) \leq 5 \operatorname{diam} (Q_2)$. However $\operatorname{diam} (Q_2) = 2^k \operatorname{diam} (Q_1)$ for some integral k, thus $\operatorname{diam} (Q_1) \leq 4 \operatorname{diam} (Q_2)$, and the symmetrical implication proves the proposition.

We now set $N = (12)^n$. The exact size of N needed in what follows is of no importance; what matters is that it can be chosen to depend only on the dimension n, and in particular to be independent of the closed set F.

PROPOSITION 2. *Suppose $Q \in \mathscr{F}$. Then there are at most N cubes in $\mathscr{F}$ which touch Q.*

If the cube Q belongs to the mesh $\mathscr{M}_k$, then as is easily seen, there are 3^n cubes (including Q) which belong to the mesh $\mathscr{M}_k$ and touch Q. Next, each cube in the mesh $\mathscr{M}_k$ can contain at most 4^n cubes of $\mathscr{F}$, of diameter $\geq (1/4)$ diameter of Q. If we combine this with Proposition 1 we get the proof of Proposition 2.

Let now Q_k denote any cube in $\mathscr{F}$. Write x^k as the center of this cube and l_k the common length of its sides. Then of course $\operatorname{diam} (Q_k) = \sqrt{n}\, l_k$. For any ε, $0 < \varepsilon < 1/4$, which is arbitrary but will be kept fixed in what follows, denote by Q_k^* the cube which has the same center as Q_k but is expanded by the factor $1 + \varepsilon$; that is, $Q_k^* = (1 + \varepsilon)[Q_k - x^k] + x^k$. Clearly $Q_k \subset Q_k^*$, and the cubes Q_k^* no longer have disjoint interiors. However the following holds:

PROPOSITION 3. *Each point of Ω is contained in at most N of the cubes Q_k^*.*

Let Q and Q_k be two cubes of $\mathscr{F}$. We claim that Q_k^* intersects Q only if Q_k touches Q. In fact consider the union of Q_k with all the cubes in $\mathscr{F}$ which touch Q_k; since the diameters of these cubes are all $\geq (1/4)$ diameter of Q_k, it is clear that this union contains Q_k^*. Therefore Q intersects Q_k^* only if Q touches Q_k. However any point $x \in \Omega$ belongs to some cube Q, and therefore by Proposition 2 there are at most N cubes Q_k^* which contain x.

The proof also shows that every point of Ω is contained in a small neighborhood intersecting at most N cubes Q_k^*.

Now let Q_0 denote the cube of unit length centered at the origin. Fix a C^∞ function φ with the properties: $0 \leq \varphi \leq 1$; $\varphi(x) = 1$, $x \in Q_0$; and $\varphi(x) = 0$, $x \notin (1 + \varepsilon)Q_0$.

Let φ_k denote the function φ adjusted to the cube Q_k; that is

$$\varphi_k(x) = \varphi\left(\frac{x - x^k}{l_k}\right).$$

Recall that x^k is the center of Q_k and l_k is the common length of its sides. Notice that therefore $\varphi_k(x) = 1$ if $x \in Q_k$, and $\varphi_k(x) = 0$ if $x \notin Q_k^*$. It is to be observed that

(4)
$$\left|\left(\frac{\partial}{\partial x}\right)^\alpha \varphi_k(x)\right| \leq A_\alpha (\operatorname{diam} Q_k)^{-|\alpha|}.$$

We now define $\varphi_k^*(x)$ for $x \in {}^cF$ by

$$\varphi_k^*(x) = \frac{\varphi_k(x)}{\Phi(x)}, \quad \text{where} \quad \Phi(x) = \sum_k \varphi_k(x).$$

The obvious identity

(5)
$$\sum_k \varphi_k^*(x) \equiv 1, \qquad x \in {}^cF$$

then defines our required partition of unity.

2. Extension theorems of Whitney type

2.1 The regularized distance. The ideas of the extension theorem of Whitney are implicitly contained in the partition of unity (5) just developed, and are further suggested by the construction of the regularized distance function which we shall now describe.

Let F be an arbitrary closed set in $\mathbf{R}^n$, and following the notation of Chapter I, let $\delta(x)$ denote the distance of x from F. While this function is smooth on F (it vanishes there) it is in general not more differentiable on ${}^cF$ than the obvious Lipschitz-condition-inequality $|\delta(x) - \delta(y)| \leq |x - y|$ would indicate.

For several applications it is desirable to replace $\delta(x)$ by a regularized distance which is smooth for $x \in {}^cF$, as x stays away from F. In addition this regularized distance is to have essentially the same profile as $\delta(x)$.

Its existence is guaranteed by the following theorem.

THEOREM 2. *There exists a function* $\Delta(x) = \Delta(x, F)$ *defined in* ^cF *such that*

(a) $c_1 \delta(x) \leq \Delta(x) \leq c_2 \delta(x)$, $x \in {}^cF$

(b) $\Delta(x)$ *is* C^∞ *in* ^cF *and*

$$\left| \frac{\partial^\alpha}{\partial x^\alpha} \Delta(x) \right| \leq B_\alpha (\delta(x))^{1-|\alpha|}.$$

B_α, c_1 *and* c_2 *are independent of* F.

The construction of $\Delta(x)$ is given in one stroke. In fact we can set

(6) $$\Delta(x) = \sum_k \text{diam} (Q_k) \varphi_k(x).$$

Observe that if $x \in Q_k$, then $\delta(x) = \text{dist} (x, F) \leq \text{dist} (Q_k, F) + \text{diam} (Q_k) \leq 5 \text{ diam} (Q_k)$ by inequality (3). Also if $x \in Q_k^*$, then $\delta(x) \geq \text{dist} (Q_k, F) - 1/4 \text{ diam} (Q_k) \geq (3/4) \text{ diam} (Q_k)$, because of (3). To summarize:

(7) *If* $x \in Q_k$ *then* $\delta(x) \leq 5 \text{ diam} (Q_k)$. *If* $x \in Q_k^*$, *then* $\delta(x) \geq (3/4) \text{ diam} (Q_k)$.

However if $x \in Q_k$, then $\varphi_k(x) = 1$, so $\Delta(x) \geq \text{diam} (Q_k) \geq \dfrac{\delta(x)}{5}$. On the other hand, any given x lies in at most N of the Q_k^*, and thus $\Delta(x) \leq \sum_{x \in Q_k^*} \text{diam} (Q_k) \leq (4/3) N \delta(x)$.

We have therefore proved conclusion (a) with $c_1 = \frac{1}{5}$ and $c_2 = (4/3)N$.

To prove conclusion (b) we argue similarly but invoke inequality (4), and the observation (analogous to (7)) that if $x \in Q_k^*$, then $\delta(x) \leq 6 \text{ diam } Q_k$. This gives the desired result with $B_\alpha = A_\alpha N 6^{|\alpha|-1}$.

We shall not follow up this construction now, and defer its application until §3. We wish here to remark that the bounds given by (b) on the derivatives of $\Delta(x)$, although they blow up as x approaches F, are in general the best possible under the circumstances. This can already be seen in the case of $\mathbf{R}^1$ if we take for F the complement of the open set

$$\bigcup_{j=-\infty}^{\infty} (2^{-j}, 2^{-j+1}).$$

On each interval the regularized distance function must rise from zero to at least $c_1 2^{-j-1}$, in passing over a distance of length 2^{-j-1} and so the first derivative must attain a size at least as large as c_1; by the same token the first derivative must attain a size smaller than $-c_1$, and so the second derivative must somewhere in that interval be at least as large as $c_1 2^{j+1}$, etc.

2.2   The first extension operator, $\mathscr{E}_0$.   Let F be a closed set in $\mathbf{R}^n$. Our purpose here will be to describe an operator $\mathscr{E}_0$ which extends functions defined on F to functions defined on $\mathbf{R}^n$. Its main properties are expressible in terms of function spaces involving one order or less of differentiability. As such $\mathscr{E}_0$ is the simplest of a hierarchy of extension operators required for differentiability of higher order.

The definition of $\mathscr{E}_0$ is as follows. Consider the set F and the family of cubes $\{Q_k\}$ given in Theorem 1. For each cube Q_k fix a point p_k in F with the property that dist $(Q_k, F) = $ dist (Q_k, p_k).

Such a point p_k of course exists since F is closed. While it is not unique any fixed choice of a point of minimum distance from F will do. In fact any choice of a point $p_k \in F$ with the property that the distance of p_k to Q_k is comparable to the distance of Q_k from F would do just as well, but it is somewhat simpler if we specify p_k as above.

Let now f be given on F. Consider the function $\mathscr{E}_0(f)$ defined by $\mathscr{E}_0(f)(x) = f(x)$, $x \in F$, and

(8) $$\mathscr{E}_0(f)(x) = \sum_k f(p_k)\varphi_k^*(x), \qquad x \in {}^c F,$$

where $\{\varphi_k^*(x)\}$ is the partition of unity described at the end of §1.3.

It is to be observed that if $x \in {}^c F$, then it belongs to at most N cubes Q_k^*, and since the φ_k^* are supported in Q_k^*, the sum in (8) is really a finite sum and thus $\mathscr{E}_0(f)(x)$ is well-defined. The first properties of $\mathscr{E}_0$ will now be given.

PROPOSITION. *Suppose f is a given function on F. Then $\mathscr{E}_0(f)$ is an extension of f to $\mathbf{R}^n$. Assume, in addition, that f is continuous on F. Then $\mathscr{E}_0(f)$ is continuous on $\mathbf{R}^n$, and in fact is C^∞ in ${}^c F$.*

That $\mathscr{E}_0(f)$ is an extension of f is by definition. To prove the continuity of $\mathscr{E}_0(f)$, and to make later estimates it is convenient to use the following notational convention: Suppose A and B are two positive quantities; then we write $A \approx B$ to mean that A and B are *comparable*. In the context of this chapter this means that there exist two constants c_1 and c_2 so that $c_1 A \leq B \leq c_2 A$; it will be understood that these constants, c_1 and c_2, may depend on the dimension n, but are otherwise independent of the set F, the cubes Q_k, the function f, etc.

With this notation we observe first that

(9) $$\text{if} \quad x \in Q_k^*, \quad \text{then} \quad |x - p_k| \approx \text{diam} (Q_k).$$

Also

(10) $$\text{dist} (Q_k^*, F) \approx \text{diam} (Q_k), \qquad \text{(see (7))}.$$

Now if $y \in F$, $x \in Q_k^*$, then $|y - p_k| \leq |y - x| + |p_k - x|$. But clearly

$|y - x| \geq$ dist (Q_k^*, F), and therefore by (9) and (10):

(11) $\qquad$ if $\quad y \in F \quad$ and $\quad x \in Q_k^*, \quad$ then $\quad |y - p_k| \leq c\,|y - x|.$

We are now ready to prove the continuity of $\mathscr{E}_0(f)$. We have already observed that each point of $x \in {}^cF$ belongs to a neighborhood which intersects at most N of the cubes Q_k^*. Since each of the functions φ_k^* are C^∞ in ${}^cF$, this shows that $\mathscr{E}_0(f)(x)$ is C^∞ in ${}^cF$, and hence certainly continuous in ${}^cF$.

Now let y be a fixed point of F. We want to prove the continuity of $\mathscr{E}_0(f)(x)$ at $x = y$. Consider therefore $\mathscr{E}_0(f)(y) - \mathscr{E}_0(f)(x) = f(y) - \mathscr{E}_0(f)(x)$, with $x \to y$. For those x which belong to F this difference is $f(y) - f(x)$ and matters are reduced to the given continuity of f on F. Suppose therefore that $x \to y$, with $x \in {}^cF$. Then

$$f(y) - \mathscr{E}_0(f)(x) = f(y) - \sum f(p_k)\varphi_k^*(x) = \sum (f(y) - f(p_k))\varphi_k^*(x),$$

because $\sum \varphi_k^*(x) \equiv 1$, $x \in {}^cF$.

We now use the observation (11) together with the fact that $\varphi_k^*(x)$ is supported in Q_k^* to see that

$$|f(y) - \mathscr{E}_0(f)(x)| \leq \sup_{y' \in F} |f(y) - f(y')| \to 0,$$

$$\text{as} \quad x \to y, \quad \text{with} \quad |y - y'| \leq c\,|y - x|.$$

2.2.1 Theorem 3. It is desirable to go further and express the continuity properties of the linear operator $f \to \mathscr{E}_0(f)$ in terms of Banach spaces. The most appropriate function spaces are those given in terms of the modulus of continuity, and in particular the Lipschitz spaces. For this purpose let $0 < \gamma \leq 1$, and define

Lip $(\gamma, \mathbf{R}^n) = \{f : |f(x)| \leq M, |f(x) - f(y)| \leq M\,|x - y|^\gamma, x, y \in \mathbf{R}^n\}.$

Lip $(\gamma, \mathbf{R}^n)$ becomes a Banach space if we take for norm the smallest M in the above definition.*

It is to be noted that when $0 < \gamma < 1$, then Lip $(\gamma, \mathbf{R}^n)$ is equivalent with the space $\Lambda_\gamma = \Lambda_\gamma^{\infty,\infty}$ studied in Chapter V, §4. However it is important to point out that in the present context we have a different transition as $\gamma \to 1$. Namely Lip $(1, \mathbf{R}^n)$ is isomorphic to $L_1^\infty(\mathbf{R}^n)$ the space of bounded function on $\mathbf{R}^n$ whose first derivatives are bounded, and *not* to $\Lambda_1 (= \Lambda_1^{\infty,\infty})$; see §4.3.1 and §6.2 of Chapter V.

If F is any closed set we define Lip (γ, F) similarly as consisting of those f defined on F for which

(12) $\quad |f(x)| \leq M \quad$ and $\quad |f(x) - f(y)| \leq M\,|x - y|^\gamma, \qquad x, y \in F.$

Again Lip (γ, F) is a Banach space, with the smallest M as norm.

* When $\gamma > 1$, the space defined above consists of constants only.

THEOREM 3. *The linear extension operator $\mathscr{E}_0$ maps* Lip (γ, F) *continuously into* Lip $(\gamma, \mathbf{R}^n)$, *if* $0 < \gamma \leq 1$. *The norm of this mapping has a bound independent of the closed set F.*

2.2.2 In order to prove the theorem we begin by recording the inequality

$$(13) \qquad \left| \frac{\partial^\alpha}{\partial x^\alpha} \varphi_k^*(x) \right| \leq A_\alpha' (\text{diam } Q_k)^{-|\alpha|}.$$

It can be derived as an easy consequence of the analogous inequality (4) for φ_k in §1.3; we leave the straightforward details to the reader.

Now let us assume that f satisfies the inequality (12) with $M = 1$. We have already observed that whatever f is, $\mathscr{E}_0(f)$ is C^∞ in ^cF. Here we shall need the quantitative estimate

$$(14) \qquad \left| \frac{\partial}{\partial x_i} \mathscr{E}_0(f)(x) \right| \leq c(\delta(x))^{\gamma-1} \qquad i = 1, \dots n, \quad x \in {}^cF;$$

and $\delta(x)$ denotes the distance of x from F.

In fact

$$\frac{\partial}{\partial x_i} \mathscr{E}_0(f)(x) = \sum_k f(p_k) \frac{\partial \varphi_k^*(x)}{\partial x_i} = \sum_k (f(p_k) - f(y)) \frac{\partial \varphi_k^*(x)}{\partial x_i},$$

in view of the fact that $\sum \dfrac{\partial \varphi_k^*}{\partial x_i}(x) \equiv 0$, by (5). For any $x \in {}^cF$ choose y to be a point in F closest to x, that is $|x - y| = \delta(x)$.

Consider next those cubes Q_k^* so that $x \in Q_k^*$. There are at most N of these and we always have $|y - p_k| \leq c |x - y| = c\delta(x)$ for these cubes, as was observed in (11) (see §2.2). Therefore

$$\left| \frac{\partial}{\partial x_i} \mathscr{E}_0(f)(x) \right| \leq A_1 \sum_{x \in Q_k^*} |f(p_k) - f(y)| (\text{diam } Q_k)^{-1}.$$

Clearly, however, if $x \in Q_k^*$ then $\delta(x) \approx \text{diam } (Q_k)$. Thus

$$\left| \frac{\partial}{\partial x_i} \mathscr{E}_0(f)(x) \right| \leq A' \left(\sum_{x \in Q_k^*} |p_k - y|^\gamma \right) \delta^{-1}(x) \leq c' \, \delta(x)^{\gamma-1},$$

which proves (14).

The estimate (14) is the appropriate one for points away from F. For points near F we observe that if $y \in F$, $x \in {}^cF$, then

$$\mathscr{E}_0(f)(y) - \mathscr{E}_0(f)(x) = f(y) - \mathscr{E}_0 f(x) = \sum_k (f(y) - f(p_k)) \varphi_k^*(x);$$

therefore by (11)

(15) $|\mathscr{E}_0(f)(y) - \mathscr{E}_0(f)(x)| \leq \sup_{|y-p_k| \leq c|y-x|} |f(y) - f(p_k)| \leq c\,|y - x|^{\gamma},$

$$\text{if} \quad y \in F, \quad x \in {}^c F.$$

Suppose now that both y and x are in ${}^c F$. Let L be the line segment joining them and we consider the two cases: (i) the distance of L from F exceeds the length of L $(= |x - y|)$, (ii) the distance of L from F is not larger than the length of L. In the first case we have simply

$$|\mathscr{E}_0(f)(y) - \mathscr{E}_0(f)(x)| \leq |y - x| \sup_{x' \in L} |\nabla \mathscr{E}_0(f)(x')|$$

$$\leq c\,|y - x| \sup_{x' \in L} (\delta(x'))^{\gamma-1},$$

by (14) since in this case $\delta(x') > |y - x|$, $x' \in L$. We obtain as a result $|\mathscr{E}_0(f)(y) - \mathscr{E}_0(f)(x)| \leq c\,|y - x|^{\gamma}$. In the second case we can find a point $x' \in L$, and a point $y' \in F$ so that $|x' - y'| \leq |y - x|$. Therefore $|y' - x| \leq 2\,|y - x|$ and $|y' - y| \leq 2\,|y - x|$. If we apply (15) to $\mathscr{E}_0(f)(y') - \mathscr{E}_0(f)(x)$ and $\mathscr{E}_0(f)(y') - \mathscr{E}_0(f)(y)$ we get again

$$|\mathscr{E}_0(f)(y) - \mathscr{E}_0(f)(x)| \leq c'\,|y - x|^{\gamma}.$$

Finally if x and $y \in F$ we have trivially $|\mathscr{E}_0(f)(y) - \mathscr{E}_0(f)(x)| \leq |y - x|^{\gamma}$.

Observe also that if the absolute value of f is bounded by 1, then the same is true for the absolute value of $\mathscr{E}_0(f)$. Theorem 3 is therefore completely proved if we note that all our bounds are independent of the closed set F.

2.2.3 A corollary. The proof of Theorem 3 leads to a simple generalization of itself. Let $\omega(\delta)$, $0 < \delta < \infty$, be a *modulus of continuity*, that is, a positive increasing function of δ; assume it is *regular* in the sense that:

(1) $\dfrac{\omega(\delta)}{\delta}$ is increasing as $\delta \to 0$, and (2) $\omega(2\delta) \leq c\omega(\delta)$. (The first condition among other things excludes $\omega(\delta) = \delta^{\gamma}$ for $\gamma > 1$. The second condition makes the statement of the result neater.) Define Lip $(\omega, F) = \{f : |f(x)| \leq M, |f(x) - f(y)| \leq M\omega(|x - y|), \ x, y \in F\}$ with norm the smallest M. Then:

COROLLARY. $\mathscr{E}_0$ *is a continuous mapping of* Lip (ω, F) *into* Lip $(\omega, \mathbf{R}^n)$.

The proof is merely a repetition of that of Theorem 3. Notice that the condition $\omega(2\delta) \leq c\omega(\delta)$ and its non-decreasing character implies that for every positive c_1 there exists a positive c_2 so that $\omega(c_1\delta) \leq c_2\omega(\delta)$, $0 < \delta < \infty$.

2.3   The extension operators $\mathscr{E}_k$.   In generalizing the results of §2.2 to higher derivatives the first requirement is the corresponding definition of Lip (γ, F) when $\gamma > 1$. For this purpose let k be a non-negative integer and assume that $k < \gamma \leq k + 1$.

We shall say that a function f, defined on F belongs to Lip (γ, F) if there exists functions $f^{(j)}$, $0 \leq |j| \leq k$ defined on F, with $f^{(0)} = f$, and so that if

(16) $$f^{(j)}(x) = \sum_{|j+l| \leq k} \frac{f^{(j+l)}(y)}{l!} (x - y)^l + R_j(x, y)$$

then

(17) $|f^{(j)}(x)| \leq M$ and $|R_j(x, y)| \leq M |x - y|^{\gamma - |j|}$,   all $x, y \in F$,

$$|j| \leq k.$$

Several explanations are in order concerning this definition. j and l denote multi-indices $j = (j_1, j_2, \ldots, j_n)$, $l = (l_1, l_2, \ldots, l_n)$ with $j! = j_1! j_2! \cdots j_n!$, and $|j| = j_1 + j_2 \cdots + j_n$; $x^l = x_1^{l_1} x_2^{l_2} \cdots x_n^{l_n}$.

It is to be noted that the function $f = f^{(0)}$ does not necessarily determine the $f^{(j)}$, $(0 < |j| \leq k)$, uniquely; (consider, for example, the case of an f defined on F, where F is a finite set). Thus in order to avoid this ambiguity, when speaking of an element of Lip (γ, F) we shall mean in fact the collection $\{f^{(j)}(x)\}_{|j| \leq k}$. The norm of an element in Lip (γ, F) will then be taken to be the smallest M for which the inequality (17) holds. In the definition just adopted we make an exception if $F = \mathbf{R}^n$. By Lip $(\gamma, \mathbf{R}^n)$ we shall then mean the linear space of the $f = f^{(0)}$ only; for which, of course there exists $f^{(j)}$ satisfying (16) and (17).

Again the norm is taken to be the smallest M satisfying (17). This convention, which is adopted purely for a notational ease is consistent with the general definition of Lip (γ, F), since it can easily be seen that the $f^{(j)}$, $1 \leq |j|$, are uniquely determined by f, if $F = \mathbf{R}^n$.

More particularly if $f \in$ Lip $(\gamma, \mathbf{R}^n)$ according to the definition just given, then f is continuous and bounded and has continuous bounded partial derivatives of order not greater than k; furthermore $\frac{\partial^j f}{\partial x^j} = f^{(j)}$, $|j| \leq k$, and the functions $f^{(j)}$, for $|j| = k$ belong to the space Lip $(\gamma - k, \mathbf{R}^n)$ considered in §2.2.1. The converse is also true and easily established. Therefore if γ is not integral Lip $(\gamma, \mathbf{R}^n)$ is equivalent with Λ_γ; see Chapter V, §4, and Proposition 9, page 147, in particular. When γ is integral, $\gamma = k + 1$, then Lip $(k + 1, \mathbf{R}^n)$ is equivalent with $L_{k+1}^\infty(\mathbf{R}^n)$; see §6.2 in Chapter V.

Let now $\{f^{(j)}\}_{|j| \leq k}$ be a collection of functions defined on F. The linear mapping $\mathscr{E}_k$ will assign to any such collection a function $\mathscr{E}_k(f^{(j)})$

defined on $\mathbf{R}^n$, which will visibly be an extension of $f^{(0)} = f$ to $\mathbf{R}^n$. For simplicity of notation we shall *denote this extension by f also*. Our definition of $\mathscr{E}_k$ is as follows:

$$(18) \quad \begin{cases} \mathscr{E}_k(f^{(j)}) = f^{(0)}(x), & x \in F \\ \mathscr{E}_k(f^{(j)}) = \sum_i' P(x, p_i)\varphi_i^*(x), & x \in {}^c F. \end{cases}$$

$P(x, y)$ denotes the polynomial in x giving the Taylor expansion of f about the point $y \in F$, that is

$$P(x, y) = \sum_{|l| \leq k} \frac{f^{(l)}(y)(x - y)^l}{l!}, \quad x \in \mathbf{R}^n, \quad y \in F.$$

p_i is as in §2.2, a point in F of minimum difference from the cube Q_i. Finally the symbol $\sum'$ indicates that the summation is taken only over those cubes Q_i near F; more precisely, those whose distance from F is not greater than one.

THEOREM 4. *Suppose k is a non-negative integer, $k < \gamma \leq k + 1$, and F a closed set in $\mathbf{R}^n$.*

Then the mapping $\mathscr{E}_k$ is a continuous mapping of Lip (γ, F) *to* Lip $(\gamma, \mathbf{R}^n)$ *which gives an extension of $f^{(0)}$ to all of $\mathbf{R}^n$. The norm of this mapping has a bound independent of F.*

2.3.1 In addition to the notation $P(x, y)$ just introduced it is convenient to write $P_j(x, y)$ for $\sum_{|j+l| \leq k} \frac{f^{(j+l)}(y)}{l!} (x - y)^l$, $|j| \leq k$. Then of course $f^{(j)}(x) = P_j(x, y) + R_j(x, y)$, $x, y \in F$. We have $P(x, y) = P_0(x, y)$, and so consistent with this we set $R(x, y) = R_0(x, y)$.

LEMMA. *Suppose $b, a \in F$, $x \in \mathbf{R}^n$, then*

$$P(x, b) - P(x, a) = \sum_{|l| \leq k} R_l(b, a) \frac{(x - b)^l}{l!},$$

and more generally

$$P_j(x, b) - P_j(x, a) = \sum_{|j+l| \leq k} R_{j+l}(b, a) \frac{(x - b)^l}{l!}.$$

To prove the lemma we expand the polynomial in x, $P(x, b) - P(x, a)$, in its Taylor expansion about the point b. The coefficient of $\frac{(x - b)^l}{l!}$ is then $\frac{\partial^l}{\partial x^l}(P(x, b) - P(x, a)]_{x=b}$. However $\frac{\partial^l}{\partial x^l}(P(x, y)) = P_l(x, y)$, and $P_l(b, b) = f^{(l)}(b)$ which proves the lemma for $j = 0$. The case for $j \neq 0$ can of course be considered as a special instance of the case already proved.

Let us now turn our attention to the sum $\sum'$ and see how it differs from $\sum$. The observations we have already made, see (7), (10), and property (3) show that

(19) if $x \in Q_k^*$,   then $\delta(x) = \text{dist}(x, F) \approx \text{dist}(Q_k, F) \approx \text{dist}(Q_k^*, F)$.

Therefore if $\delta(x) \leq c_1$, for some appropriate positive c_1, which is sufficiently small, then the sum $\sum'$ is the full sum taken over all cubes. If $\delta(x) \geq c_2$, for another positive constant which is sufficiently large, then there are no terms in the sum $\sum'$, and thus f vanishes for $\delta(x) \geq c_2$. Finally if $c_1 \leq \delta(x) \leq c_2$, then only a bounded number of terms occur, and in view of the bounds of the derivatives of $\varphi_i^*(x)$ given by (13) and the

bounds on $f^{(j)}$ given by (17), we see that there $\left| \dfrac{\partial^\alpha}{\partial x^\alpha}(f(x)) \right| \leq A'_\alpha M$, all α.

Because of this we shall be able to limit our considerations to x close to F, namely $\delta(x) \leq c_1$. We shall also assume the normalization $M = 1$.

2.3.2   We now claim that the following hold:

(a) $|f(x) - P(x, a)| \leq A |x - a|^\gamma$,   for $x \in \mathbf{R}^n$, $a \in F$

(a′) $|f^{(j)}(x) - P_j(x, a)| \leq A |x - a|^{\gamma - |j|}$,   for $x \in \mathbf{R}^n$, $a \in F$, $|j| \leq k$

(b) $|f^{(j)}(x)| \leq A$,   for $|j| \leq k$,

(b′) $|f^{(j)}(x)| \leq A(\delta(x))^{\gamma - k - 1}$,   for $x \in {}^c F$, $|j| = k + 1$.

To prove (a) notice first that it holds for $A = 1$, when $x \in F$, as a result of our assumptions. Suppose that $x \in {}^c F$, and $\delta(x) \leq c_1$ (otherwise matters become obvious in view of the remarks made earlier). Then $f(x) - P(x, a) = \sum_i \{P(x, p_i) - P(x, a)\}\varphi_i^*(x)$. We invoke now the lemma and get in view of our assumptions

$$|f(x) - P(x, a)| \leq \sum_{|t| \leq k} \sum |p_i - a|^{\gamma - |t|} |x - a|^{|t|}$$

where the inner (un-indexed) sum is taken over those cubes Q_i, (there are at most N), so that $x \in Q_i^*$. By (11), $|p_i - a| \leq c_1 |x - a|$, and therefore (a) is proved.

In proving (a′) we may restrict ourselves again to points $x \in {}^c F$ with $\delta(x) \leq c_1$. Here $f^{(j)}(x) = \left(\dfrac{\partial^j f}{\partial x^j}\right)(x)$. Thus

$$f^{(j)}(x) = \sum_i \left(\frac{\partial}{\partial x}\right)^j (P(x, p_i))\varphi_i^*(x) + \text{other terms.}$$

If we disregard the "other terms" and notice that $\left(\dfrac{\partial}{\partial x}\right)^j P(x, p_i) = P_j(x, p_i)$ then we get (a') just as (a). The other terms are themselves sums of expressions like

(20)
$$\sum_i P_{j-l}(x, p_i)\left(\frac{\partial}{\partial x}\right)^l \varphi_i^*(x)$$

where $0 < |l|$, and $l_i \le j_i$, $i = 1, \ldots, n$. Since $\sum_i \left(\dfrac{\partial}{\partial x}\right)^l \varphi_i^*(x) = 0$, then these sums are in turn equal to

(21)
$$\sum_i \{P_{j-l}(x, p_i) - P_{j-l}(x, a)\}\left(\frac{\partial}{\partial x}\right)^l \varphi_i^*(x).$$

The same argument as before together with the estimate (13) for $\left(\dfrac{\partial}{\partial x}\right)^l \varphi_i^*$, and the inequalities (19) then prove (a').

Inequality (b) (again for $\delta(x) \le c_1$) is an immediate consequence of (a') if for the point a we take a point in F of bounded distance from x. (Incidentally it was at the stage of the proof of inequality (b) that it really mattered that we defined f in terms of $\sum_i' P(x, p_i)\varphi_i^*(x)$ instead of the full sum over all cubes.)

Finally we come to the proof of (b'). If we carry out the differentiation for $\delta(x) \le c_1$ then we get that $f^{(j)}(x)$ equals a sum of expressions of the form (20). Since $|j| = k + 1$, it must follow that $|l| > 0$, otherwise $P_j(x, p_i) \equiv 0$. Thus each sum can be rewritten as one of the form (21), where we choose a to be a point in F of minimum distance from x. In view of the lemma, and inequalities (11), (13) and (19) we get that each sum (21) is dominated by a finite sum of terms of the form $A\,|p_i - a|^{\gamma - |j| + |l|}$ $(\delta(x))^{l-1} \le A'\delta(x)^{\gamma-k-1}$, and (b') is also proved.

2.3.3 Having disposed the inequalities (a), (a'), (b), and (b') we can now finish the proof of Theorem 4. The case $k = 0$ is of course Theorem 3 (§2.2.1). We shall consider in detail the case $k = 1$, which is already entirely typical; we have then $1 < \gamma \le 2$.

The inequality (a) shows that the function f has first partial derivatives at every point of F, and these are the $f^{(j)}(x)$, with $|j| = 1$. However f is C^∞ in ^cF, and thus $\left(\dfrac{\partial}{\partial x}\right)^j f = f^{(j)}$ exists for each point in ^cF. Inequality (a') shows that the resulting $f^{(j)}$, $|j| = 1$ are continuous through $\mathbf{R}^n$. Now let g denote one of these first partial derivatives, then (a') and (b') respectively imply that

$$|g(x) - g(a)| \le A\,|x - a|^{\gamma-1}, \qquad x \in \mathbf{R}^n, a \in F$$

and

$$\left| \frac{\partial}{\partial x_i} g(x) \right| \leq A(\delta(x))^{\gamma - 2}, \qquad i = 1, \ldots, n, \ x \in {}^c F.$$

These two inequalities are of the same form as (15) and (14) in the proof of the theorem for $k = 0$, (but with γ instead of $\gamma - 1$). Following the argument given there it follows that each $g(x) \in \mathrm{Lip}\ (\gamma - 1, \mathbf{R}^n)$ and this is the desired result when $k = 1$. The proof for $k \geq 2$ can then be carried out by induction, the inductive step being very similar to the case $k = 1$ just given.

For the variant of the theorem analogous to the corollary in §2.2.3 see §4.6 below; see also §4.7 where another version is given.

3. Extension theorem for a domain with minimally smooth boundary

Let D be an open set in $\mathbf{R}^n$. Our purpose will be to describe an operator $\mathfrak{E}$ which extends functions defined in D to $\mathbf{R}^n$. The operator that will be given will be universal in the sense that it will simultaneously extend all orders of differentiability. This is to be contrasted with the hierarchy of operators $\mathscr{E}_k$, of increasing complexity in k, that we needed to perform the job for an arbitrary closed set F. The construction of $\mathfrak{E}$ will be possible if the boundary of D satisfies some minimal smoothness property, which is approximately equivalent to saying that it is of class Lip 1. It will be seen momentarily that this condition cannot really be relaxed. What is striking therefore is that one order of differentiability of the boundary is roughly speaking just the right requirement to allow extension of all orders of differentiability.

3.1 Statement of the theorem. The appropriate function spaces to be used here are the Sobolov spaces $L_k^p(D)$, defined in analogy with the special case $D = \mathbf{R}^n$ as follows. Let $C_0^\infty(D)$ denote the class of C^∞ functions with compact support, lying in D. Then a locally integrable function f defined in D has a (weak) derivative $\dfrac{\partial^\alpha f}{\partial x^\alpha} = g$ which is locally integrable, if

$$\int_D f \frac{\partial^\alpha}{\partial x^\alpha} \varphi \, dx = (-1)^{|\alpha|} \int_D g\varphi \, dx, \quad \text{for all} \quad \varphi \in C_0^\infty(D).$$

If $g \in L^p(D)$, then we say $\dfrac{\partial^\alpha f}{\partial x^\alpha} \in L^p(D)$. Now if k is an integer then

$$L_k^p(D) = \left\{ f \in L^p(D) : \frac{\partial^\alpha f}{\partial x^\alpha} \in L^p(D), \quad \text{all} \ |\alpha| \leq k \right\}.$$

The norm, on the resulting equivalence classes is given by

$$\|f\|_{L_k^p(D)} = \sum_{|\alpha| \le k} \left\| \frac{\partial^\alpha f}{\partial x^\alpha} \right\|_{L^p(D)}.$$

Our objective will be to prove the following theorem.

THEOREM 5. *Let D be a domain whose boundary satisfies the minimal smoothness condition given by* (i), (ii), *and* (iii) *in §3.3. Then there exists a linear operator $\mathfrak{E}$ mapping functions on D to functions on $\mathbf{R}^n$ with the properties*

(a) $\mathfrak{E}(f)|_D = f$, *that is, $\mathfrak{E}$ is an extension operator.*

(b) $\mathfrak{E}$ *maps $L_k^p(D)$ continuously into $L_k^p(\mathbf{R}^n)$ for all p, $1 \le p \le \infty$, and all non-negative integral k.*

Notice that for these domains the theorem also solves the restriction problem for $L_k^p(\mathbf{R}^n)$. In fact if D is any domain in $\mathbf{R}^n$ it is obvious that the restriction to D of any element of $L_k^p(\mathbf{R}^n)$ belongs to $L_k^p(D)$.

3.2 A basic special case.   The main element of the proof of the theorem is contained in a substantial special case which we formulate and discuss separately.

For this purpose it is convenient for the sake of notation to change our setting from $\mathbf{R}^n$ to $\mathbf{R}^{n+1}$. We consider the points in $\mathbf{R}^{n+1}$ as pairs (x, y), where $x \in \mathbf{R}^n$, and $y \in \mathbf{R}^1$. The domains D (now open sets of $\mathbf{R}^{n+1}$) we shall consider are the *special Lipschitz domains* defined as follows.

Let $\varphi : \mathbf{R}^n \to \mathbf{R}^1$ be a function which satisfies the Lipschitz condition

(22) $|\varphi(x) - \varphi(x')| \le M |x - x'|$,   all $x, x \in \mathbf{R}^n$.

In terms of this function we can define the special Lipschitz domain it determines to be the set of points lying above the hypersurface $y = \varphi(x)$ in $\mathbf{R}^{n+1}$, i.e.,

(23) $D = \{(x, y) \in \mathbf{R}^{n+1} : y > \varphi(x)\}$.

The smallest M for which (22) holds will be called the *bound* of the special Lipschitz domain D.

The special case we have in mind can be formulated as follows.

THEOREM 5′. *Let D be a special Lipschitz domain in $\mathbf{R}^{n+1}$. Then there exists a linear extension operator $\mathfrak{E}$ taking appropriate functions on D to functions on $\mathbf{R}^{n+1}$ with the property that $\mathfrak{E}$ maps $L_k^p(D)$ continuously into $L_k^p(\mathbf{R}^{n+1})$, $1 \le p \le \infty$, k integral. Moreover the norms of these mappings have bounds which depend only on the number n, the order of differentiability k, and the bound of the special Lipschitz domain.*

We shall first point out that the Lipschitz condition (22) for the boundary of the domain is in the nature of the best possible. Suppose we consider in $\mathbf{R}^2$ the domain where $\varphi(x) = |x|^\gamma$, $\gamma < 1$, that is $D = \{(x, y) : y > |x|^\gamma\}$. Here φ satisfies a Lipschitz condition of order γ, and violates condition (22) near the origin only. Let us set $f(x, y) = y^{-\beta}$ in D near the origin, $f \in C^\infty$ away from the origin and suppose f has bounded support. Notice that $f \in L_1^{2+\varepsilon}(D)$ for some $\varepsilon > 0$ as soon as $\dfrac{1}{\gamma} + 2(\beta - 1) >$ -1, which can always be achieved with *negative* β, no matter how close γ is to 1. However if the extension theorem were valid for this type D then the extended f would have to be in $L_1^{2+\varepsilon}(\mathbf{R}^2)$, and so by Sobolov's theorem (Theorem 2 in §2.2 of Chapter V) would have to be continuous, which is a contradiction.

3.2.1 Outline. Let us consider the domain D, and the points (x, y) which lie outside its closure. Our problem is to define $\mathfrak{E}(f)(x, y)$, (here $y < \varphi(x)$), where f is given in D. For fixed x we shall define $\mathfrak{E}(f)(x, y)$ for $\varphi(x) > y$ in terms of a suitable average of the values of f on the segment where $\varphi(x) < y$. Two things are needed so that we can implement this idea. First, an appropriate weighting function in terms of which the averages will be defined. Secondly, a device to get around the difficulty that the difference $\varphi(x) - y$ allows for at most one order of differentiation (in x). Matters are taken care of by the following two lemmas.

LEMMA 1. *There exists a continuous function ψ defined on $[1, \infty)$ which is rapidly decreasing at ∞, that is $\psi(\lambda) = O(\lambda^{-N})$, as $\lambda \to \infty$ for every N, and which satisfies in addition the properties*

$$\int_1^\infty \psi(\lambda)\, d\lambda = 1, \qquad \int_1^\infty \lambda^k \psi(\lambda)\, d\lambda = 0, \quad for \quad k = 1, 2, \ldots.$$

LEMMA 2. *Let $F = \bar{D}$. Suppose $\Delta(x, y)$ is the regularized distance from F, as given in Theorem 2, in §2.1. Then there exists a constant c, (which depends only on the Lipschitz bound of D), so that if $(x, y) \in {}^cF$, then $c\Delta(x, y) \geq \varphi(x) - y$.*

For simplicity of notation we shall write $\delta^* = 2c\Delta$, so then $\delta^*(x, y) \geq 2(\varphi(x) - y)$.

We now write down what will turn out to be the extension operator for D. If $(x, y) \in {}^c\bar{D}$, we will set

$$(24) \qquad \mathfrak{E}(f)(x, y) = \int_1^\infty f(x, y + \lambda\, \delta^*(x, y))\psi(\lambda)\, d\lambda,$$

where the integral will be defined in an appropriate limiting sense.

The plan of how we are to proceed is as follows. First in §3.2.2 we give the proof of the two lemmas. Next we show that the operator $\mathfrak{E}$ defined by (24) accomplishes the goal of Theorem 5′, namely the extension when D is a special Lipschitz domain. Finally the $\mathfrak{E}$ corresponding to the more general domain considered will be constructed in terms of the operators corresponding to the special domains.

3.2.2 Proof of Lemmas 1 and 2. An elementary function can be given which satisfies the conclusion of Lemma 1, namely

$$\psi(\lambda) = \frac{e}{\pi\lambda} \cdot \text{Im}\, (e^{-\omega(\lambda-1)^{1/4}})$$

where $\omega = e^{-i\pi/4}$.

In fact we consider the single-valued analytic function $e^{-\omega(z-1)^{1/4}}$ in the complex plane which is slit along the real axis from 1 to $+\infty$. In this connection we take the contour γ which goes from $+\infty$ to 1 above the slit, makes an infinitesimal half-loop about 1 and returns to $+\infty$ below the slit. Then since $e^{-\omega(z-1)^{1/4}}$ decreases rapidly as $z \to \infty$ we get by Cauchy's theorem that

$$-\frac{1}{2\pi i}\int_\gamma \frac{1}{z} e^{-\omega(z-1)^{1/4}}\, dz = e^{-\omega(z-1)^{1/4}}\Big]_{z=0} = e^{-1}$$

while

$$\frac{1}{2\pi i}\int_\gamma \frac{z^k}{z} e^{-\omega(z-1)^{1/4}}\, dz = 0, \quad \text{if}\quad k = 1, 2, \ldots.$$

To prove Lemma 2 we use a simple geometrical interpretation of the Lipschitz condition for the boundary of D. Let Γ_- be the (lower) cone with vertex at the origin given by $\Gamma_- = \{(x, y): M\,|x| < |y|, y < 0\}$. For any point $p \in \mathbf{R}^{n+1}$, denote by $\Gamma_-(p)$ the cone Γ_- translated so that its vertex is p.

Now it is immediate that the Lipschitz condition (22) implies that if p is any point on the boundary of D, i.e. $p = (x_1, y_1)$, with $y_1 = \varphi(x_1)$, then $\Gamma_-(p) \subset {}^c\bar{D} = {}^cF$. Next, let (x, y) denote any point in ${}^c\bar{D}$, and let $p = (x, \varphi(x))$ be the point on the boundary of D lying above it. Then of course $(x, y) \in \Gamma_-(p)$, and no point of $\bar{D}$ is closer to (x, y) than the boundary of $\Gamma_-(p)$. Clearly (x, y) lies along the central axis of the circular cone $\Gamma_-(p)$, and a simple geometrical argument shows that this minimum distance must be at least $\dfrac{\varphi(x) - y}{\sqrt{1 + M^{-2}}}$. Therefore

$$\delta(x, y) \geq (1 + M^{-2})^{-1/2}(\varphi(x) - y),$$

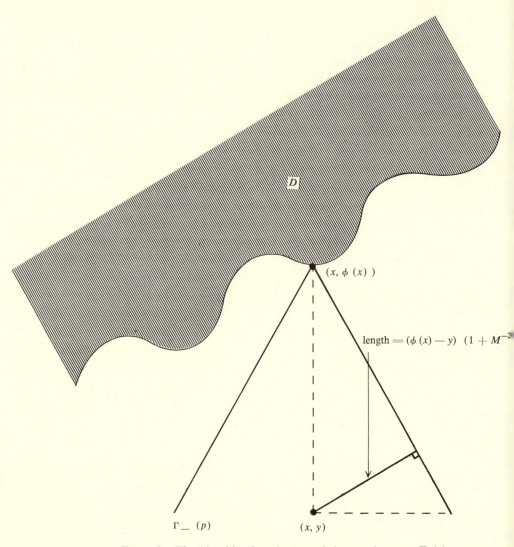

Figure 3. The Lipschitz domain D, and the exterior cone $\Gamma_-(p)$.

and hence $c\Delta(x, y) \geq \varphi(x) - y$, with $c = 5(1 + M^{-2})^{1/2}$, because by Theorem 2 we have $c_1\delta \leq \Delta$, and the proof of theorem shows we may take $c_1 = 1/5$; (see §2.1).

3.2.3 Armed with Lemmas 1 and 2 and the resulting definition (24) for $\mathfrak{E}$ we come now to the proof of Theorem 5′. We assume that $f \in L_k^p(D)$;

we suppose also that f is C^∞ in D and it, together with all its partial derivatives are continuous and bounded in $\bar{D}$. The last set of conditions are of course not satisfied for all $f \in L_k^p(D)$. But it will be our intention to show that for such f we have the inequality

$$(25) \qquad \|\mathfrak{E}(f)\|_{L_k^p(\mathbf{R}^{n+1})} \leq A_{k,n}(M) \|f\|_{L_k^p(D)}.$$

With this *a priori* inequality we shall be able to treat the general f in $L_k^p(D)$ by a passage to the limit.

Now define $\mathfrak{E}(f)(x,y) = f(x,y)$, if $(x,y) \in \bar{D}$ and $\mathfrak{E}(f)(x,y) = \int_1^\infty \psi(\lambda) f(x, y + \lambda \delta^*(x,y)) \, d\lambda$, for $(x,y) \in {}^c\bar{D}$. Notice that in view of the fact that $\delta^*(x,y) \geq 2(\varphi(x) - y)$, we get that

$$y + \lambda\delta^*(x,y) \geq y + 2(\varphi(x) - y) = \varphi(x) + \varphi(x) - y > \varphi(x), \text{ if } \lambda \geq 1.$$

This, and the assumed boundedness of f, shows that the integral giving $\mathfrak{E}$ is well defined.

We are now faced by the following situation: let D be as before and $D_- = \{(x,y) : \varphi(x) > y\}$, that is those points lying strictly below our Lipschitz hypersurface. Then of course $\bar{D} \cup \bar{D}_- = \mathbf{R}^n$, but $\bar{D}$ and $\bar{D}_-$ intersect. The properties of f that we are momentarily taking for granted assure that $\mathfrak{E}(f)$ is continuous with all its partial derivatives in $\bar{D}$. Next observe that for such f, $\mathfrak{E}(f)$ is C^∞ in D_- and all its partial derivatives are continuous (and bounded in $\bar{D}_-$). The argument is entirely typical for say, $\dfrac{\partial^2\mathfrak{E}(f)}{\partial x_j^2}$. Carrying out the differentiation gives

$$(26) \qquad \begin{aligned} \frac{\partial^2\mathfrak{E}(f)}{\partial x_j^2} &= \int_1^\infty f_{jj}(\cdot)\psi(\lambda)\,d\lambda + 2\int_1^\infty f_{jy}(\cdot)\lambda\,\delta_j^*\psi(\lambda)\,d\lambda \\ &+ \int_1^\infty f_{yy}(\cdot)(\lambda\,\delta_j^*)^2\psi(\lambda)\,d\lambda + \int_1^\infty f_y(\cdot)\lambda\,\delta_{jj}^*\psi(\lambda)\,d\lambda. \end{aligned}$$

We have used the following abbreviation: the notation f_{jj} means $\dfrac{\partial^2 f}{\partial x_j^2}$; similarly $\dfrac{\partial f}{\partial y}$ is designated by f_y. The $(\cdot)$ stands for $(x, y + \lambda\delta^*)$.

In view of the assumed differentiability of f in D the above shows $\dfrac{\partial^2}{\partial x_j^2}\mathfrak{E}(f)(x,y)$ is well defined for $(x,y) \in D_-$. (In fact it is just as obvious that $\mathfrak{E}f(x,y)$ is C^∞ in D_-.) Next let (x,y) tend to a boundary point of D_-, namely $(x^0, y^0) \in \bar{D}_- \cap \bar{D}$. Then $\delta^*(x,y) \to 0$ and since δ_j^* remains bounded (see conclusion (b) of Theorem 2) while

$$\int_1^\infty \psi(\lambda)\,d\lambda = 1, \qquad \int_1^\infty \lambda\psi(\lambda)\,d\lambda = 0, \qquad \int_1^\infty \lambda^2\psi(\lambda)\,d\lambda = 0,$$

the first three terms on the right side converge to a total which equals

$$\lim_{\substack{(x,y)\in D \\ (x,y)\to(x^0,y^0)}} \frac{\partial^2 f}{\partial x_j^2}(x, y).$$

The difficulty with the last term is that it involves more than one derivative of the regularized distance δ^*, which is then no longer bounded. However we can write

$$f_y(x, y + \lambda\delta^*) = f_y(x, y + \delta^*) + (\lambda - 1)\delta^* f_{yy}(x, y + \delta^*)$$
$$+ O([(\lambda - 1)\delta^*]^2).$$

Substituting in the last term gives two further integrals which vanish identically together with a remainder, which is

$$O\left((\delta^*)^2\, \delta^*_{x_j x_j} \int_1^\infty (\lambda - 1)^2 \lambda \psi(\lambda)\, d\lambda\right).$$

Now ψ is rapidly decreasing, therefore by Theorem 2 we have that the whole quantity is $O(\delta) \to 0$, as $(x, y) \to (x^0, y^0)$.

To summarize, since f is continuous (and bounded) with all its partial derivatives in $\bar{D}$, then the same can be said about $\mathfrak{E}f$ in $\bar{D}_-$, and f and $\mathfrak{E}f$ agree on $\bar{D} \cap \bar{D}_-$, together with all their partial derivatives. This shows also that $\mathfrak{E}f \in C^\infty(\mathbf{R}^{n+1})$, which in effect means that the two pieces of $\mathfrak{E}f$, one coming from D, the other from D_-, have in reality been joined properly. Let us prove this by first showing that $\mathfrak{E}f$ is of class C^1. (The continuity of $\mathfrak{E}f$ has already been demonstrated.)

It will be necessary to show that

$$\mathfrak{E}(f)(u) - \mathfrak{E}(f)(v) = (u - v) \cdot (\nabla\mathfrak{E}f)(u) + o(|u - v|),$$

for any point $v \in \mathbf{R}^{n+1}$, as $u \to v$. There is nothing further to prove if v is in either D or D_-. Assume therefore that v is on the (common) boundary of these two domains. Suppose $u \in \bar{D}_-$; the argument if $u \in \bar{D}$ is entirely similar. We claim that v and u can be joined by a broken line segment which except for v and u is entirely in D_-, and its total length is not greater than $c\,|v - u|$. In fact there exists a $w \in D_-$, so that the segments joining v to w, and w to u have the required property. To find this w, notice that either $u \in \Gamma_-(v)$, (then we can pick $w = u$) or else the cones $\Gamma_-(v)$ and $\Gamma_-(u)$ must intersect. The nearest point of the intersection will do for w. Now $\mathfrak{E}(f)(u) - \mathfrak{E}(f)(w) = (u - w)(\nabla\mathfrak{E}f)(u) + o(|u - w|)$, and $\mathfrak{E}(f)(w) - \mathfrak{E}(f)(v) = (w - v)(\nabla\mathfrak{E}f)(w) + o(|w - v|)$. Adding the two and using the fact that $(\nabla\mathfrak{E}f)(w) - (\nabla\mathfrak{E}f)(u) = o(1)$, as $|u - v| \to 0$ in $\bar{D}$ gives the required result. Similarly it is seen that $\mathfrak{E}f \in C^k(\mathbf{R}^{n+1})$ for every k.

The next step is to prove the inequality (25), which we will do by proving a corresponding inequality for each fixed x, and then integrating the result in x.

Consider first the case $k = 0$. Let us fix on x^0 and assume (for notational convenience) that $\varphi(x^0) = 0$. Then

$$|\mathfrak{E}f(x^0, y)| \leq A \int_1^\infty |f(x^0, y + \lambda \delta^*)| \frac{d\lambda}{\lambda^2}, \qquad y < 0.$$

We are of course using the fact that $|\psi(\lambda)| \leq A/\lambda^2$. Since $\delta^* \geq 2(\varphi(x) - y)$, we have $\delta^* \geq 2|y|$ in this case. Also of course in general $\varphi(x) - y \geq$ distance (x, y) from $\bar{D}$, therefore in this case $\delta^* \leq a|y|$. Put $s = y + \lambda\delta^*$, with y fixed, then $ds = d\lambda$ and the above inequality becomes

$$|\mathfrak{E}f(x, y)| \leq A \delta^* \int_{|y|}^\infty f(x^0, s)(s - y)^{-2} ds, \qquad y < 0.$$

Therefore

(27) $$|\mathfrak{E}(f)(x^0, y)| \leq Aa |y| \int_{|y|}^\infty |f(x^0, s)| \frac{ds}{s^2}, \qquad y < 0.$$

Hardy's inequality (see Appendix A, p. 272) then shows that

$$\left(\int_{-\infty}^0 |\mathfrak{E}(f)(x^0, y)|^p \, dy \right)^{1/p} \leq A' \left(\int_0^\infty |f(x^0, y)|^p \, dy \right)^{1/p}.$$

If we drop the condition that $\varphi(x^0) = 0$, which we may after a suitable translation in y, we get

$$\left(\int_{-\infty}^\infty |\mathfrak{E}(f)(x^0, y)|^p \, dy \right)^{1/p} \leq A \left(\int_{\varphi(x^0)}^\infty |f(x^0, y)|^p \, dy \right)^{1/p}.$$

Raising both sides to the pth power and integrating for $x^0 \in \mathbf{R}^n$ then gives the inequality (25), for $k = 0$. The proof for $k > 0$ is similar.

Consider for example the case $k = 2$. Here the consideration of $\frac{\partial^2 \mathfrak{E}}{\partial x_j^2}$ is again typical. The first three terms in the right side of (26) are handled in the same way, except that now we use $|\psi(\lambda)| \leq A/\lambda^4$, $\lambda \geq 1$. Only the last term needs to be dealt with separately. We write

(28) $$f_y = f_y(x^0, y + \delta^*) + \int_{y+\delta^*}^{y+\lambda\delta^*} f_{yy}(x^0, t) \, dt$$

and substitute this in (26). The contribution of the integral involving $f_y(x^0, y + \delta^*)$ vanishes identically by the orthogonality condition on ψ, and we are reduced to estimating

$$|y|^{-1} \int_1^\infty \left\{ \int_{y+\delta^*}^{y+\lambda\delta^*} |f_{yy}(x^0, t)| \, dt \right\} \lambda^{-3} \, d\lambda.$$

An interchange of the order of integration reduces this to the earlier case (analogous to the right-side of (27)), and disposes of the case $k = 2$.

For general k, if we carry out the differentiation under the integral sign in (24), then we will get various orders derivatives of f. Whenever the total order appearing in f is less than k, (it may be as low as 1), then we write the Taylor expansion of this derivative about the point $(x^0, y + \delta^*)$; carry it up to order k with integral remainder, and then proceed as above.

In fact assume the order of differentiation appearing in f is k_0, with $k_0 < k$. Let g be that partial derivative of f of order k_0. Then we write

$$g(x^0, y + \lambda \delta^*) = \sum_{j=0}^{l-1} \frac{((\lambda - 1)\delta^*)^j}{j!} \frac{\partial^j}{\partial y^j} g \bigg]_{(x^0, y + \delta^*)}$$
$$+ \frac{1}{l!} \int_{\delta^*}^{\lambda \delta^*} (\lambda \delta^* - t)^{l-1} \frac{\partial^l}{\partial t^l} g(x^0, y + t)\, dt$$

$$\text{with}\quad k_0 + l = k$$

Only the integral gives a non-zero contribution, but it is dominated by

$$A(\lambda \delta^*)^{l-1} \int_{\delta^*}^{\lambda \delta^*} \left| \frac{\partial^l}{\partial t^l} g(x^0, y + t) \right| dt,$$

and the argument is then as before.

We should notice that in all these calculations, leading to the proof of (25), the only effect the region D has on the bounds that appear is via its Lipschitz bound M of (22).

3.2.4　The last step of the proof of Theorem 5' will be to remove the restriction that f is C^∞ in D and it and all its partial derivatives are continuous in $\bar{D}$ and bounded.

For this purpose suppose $\eta \in C^\infty(\mathbf{R}^{n+1})$ is non-negative, has total integral 1, and that the support of η lies strictly in the interior of the cone Γ_-. For any $\varepsilon \geq 0$ write $\eta_\varepsilon(u) = \varepsilon^{-n-1}\eta(u/\varepsilon)$, $(u \in \mathbf{R}^{n+1})$, and $f_\varepsilon(u) = \int f(u - v)\eta_\varepsilon(v)\, dv$, where f is given in $L_k^p(D)$. Notice that if $u \in \bar{D}$, the integral involves only $u - v \in D$, and so is well-defined. Moreover since the support of η_ε is strictly inside Γ_-, we see that the integral defines $f_\varepsilon(u)$ in a neighborhood of $\bar{D}$, and f_ε is C^∞ there. It is also clear that

$$\left\| \frac{\partial^\alpha f_\varepsilon}{\partial x^\alpha} \right\|_{L^p(D)} \leq \left\| \frac{\partial^\alpha f}{\partial x^\alpha} \right\|_{L^p(D)}, \qquad |\alpha| \leq k,$$

therefore

(29) $$\|f_\varepsilon\|_{L_k^p(D)} \leq \|f\|_{L_k^p(D)}.$$

Now if $p < \infty$, we know that $f_\varepsilon \to f$ in the norm of $L_k^p(D)$. For $p = \infty$ it suffices to content ourselves with the weaker statement that if $k \geq 1$, then $f_\varepsilon \to f$ in the L^∞_{k-1} norm. (See the closely related facts, §2.1 and §6.2 of Chapter V.)

Now let $\mathfrak{E}_\varepsilon(f) = \mathfrak{E}(f_\varepsilon)$, then we have by (25) and (29) that

(30)
$$\|\mathfrak{E}_\varepsilon(f)\|_{L^p_k(\mathbf{R}^{n+1})} \leq A_{k,n}(M) \|f\|_{L^p_k(D)}.$$

Now clearly $\|f_\varepsilon - f_{\varepsilon'}\|_{L^p_{k-1}(D)} \to 0$, as $\varepsilon, \varepsilon' \to 0$. Therefore $\mathfrak{E}_\varepsilon(f)$ is Cauchy in $L^p_k(\mathbf{R}^{n+1})$, and its limit also satisfies (30), which means we have proved the existence of an operator $\mathfrak{E} = \lim \mathfrak{E}_\varepsilon$, extending our $\mathfrak{E}$ defined originally on C^∞ functions on $\bar{D}$. This is obviously the required extension operator satisfying (25). Theorem 5′ is therefore completely proved.

3.3 The general case. Having disposed of the case of the special Lipschitz domain, we leave the setting of $\mathbf{R}^{n+1}$ and return to $\mathbf{R}^n$. It will be convenient to modify our terminology slightly by referring to rotations of these domains also as special Lipschitz domains. The notion of the Lipschitz bound of such domain is then defined in the obvious manner, and is of course rotation invariant.

Now let D be an open set in $\mathbf{R}^n$, and let ∂D be its boundary. We shall say that ∂D is *minimally smooth*, if there exists an $\varepsilon > 0$, an integer N, an $M > 0$, and a sequence $U_1, U_2, \ldots U_n \ldots$ of open sets so that:

(i) If $x \in \partial D$, then $B(x, \varepsilon) \subset U_i$, for some i; $B(x, \varepsilon)$ is the ball of center x and radius ε.

(ii) No point of $\mathbf{R}^n$ is contained in more than N of the U_i's.

(iii) For each i there exists a special Lipschitz domain D_i whose bound does not exceed M so that

$$U_i \cap D = U_i \cap D_i.$$

Some examples of the above are:

Example 1. Suppose D is a bounded domain in $\mathbf{R}^n$ whose boundary is C^1 embedded in $\mathbf{R}^n$. In this case only finitely many U_i's are needed.

Example 2. D is any open bounded convex set. Again only finitely many U_i's are required.

Example 3. $D \subset \mathbf{R}^1$, and $D = \bigcup_j I_j$, where I_j are disjoint open intervals. The conditions (i)–(iii) are satisfied if there exists a $\delta > 0$, so that length $I_j \geq \delta$, and dist $(I_j, I_k) \geq \delta$, if $j \neq k$. In this example an infinite number of U_i's are required if there are infinitely many I_j. The conditions length $I_j \geq \delta$, dist $(I_j, I_k) \geq \delta$ are also necessary. The reader can easily verify that the condition length $I_j \geq \delta$ is required if we are to have a bounded extension for L_1^1, while the condition dist $(I_j, I_k) \geq \delta$ is needed in order to have a bounded extension for L_1^∞.

3.3.1 We shall prove Theorem 5 by reducing it to Theorem 5' for special Lipschitz domains. The argument however is somewhat tricky.

For any set $U \subset \mathbf{R}^n$ and any $\varepsilon > 0$, denote by $U^\varepsilon = \{x:B(x, \varepsilon) \subset U\}$. Notice that $U^\varepsilon \subset U$ and the condition (i) can be restated by saying $\bigcup_i U_i^\varepsilon \supset \partial D$. Let $\eta(x)$ denote a fixed C^∞ function in $\mathbf{R}^n$ of total integral one, whose support is contained in the unit ball, and denote $\eta_\varepsilon(x) = \varepsilon^{-n}\eta(x/\varepsilon)$. Suppose that χ_i is the characteristic function of $U_i^{3\varepsilon/4}$, and let

$$\lambda_i(x) = (\chi_i * \eta_{\varepsilon/4})(x).$$

The following properties of the λ_i are now evident:
  (a) each λ_i is supported in U_i;
  (b) $\lambda_i(x) = 1$, if $x \in U_i^{\varepsilon/2}$, and so in particular if $x \in U_i^\varepsilon$;
  (c) each $\lambda_i \in C^\infty$ has bounded derivatives of all orders and the bounds of the derivatives of λ_i can be taken to be independent of i. (These depend only on the L^1 norms of the corresponding derivatives of $\eta_{\varepsilon/4}$.)

Consider in addition three other open sets, covering respectively a neighborhood of D, the boundary of D, and that part of the interior of D away from the boundary, namely

$$U_0 = \{x:\text{dist } (x, D) < \varepsilon/4\}$$
$$U_+ = \{x:\text{dist } (x, \partial D) < (3/4)\varepsilon\}$$
$$U_- = \{x \in D:\text{dist } (x, \partial D) > \varepsilon/4\}.$$

Write χ_0, χ_+, and χ_- for the characteristic functions of these sets and regularize the functions as above, to wit $\lambda_0 = \chi_0 * \eta_{\varepsilon/4}$, $\lambda_+ = \chi_+ * \eta_{\varepsilon/4}$, and $\lambda_- = \chi_- * \eta_{\varepsilon/4}$. Then clearly $\lambda_0(x) = 1$, if $x \in \bar{D}$; $\lambda_+(x) = 1$ if dist $(x, \partial D) \leq \varepsilon/2$; and $\lambda_-(x) = 1$ if $x \in D$ and dist $(x, \partial D) \geq \varepsilon/2$. Moreover the supports of λ_0, λ_+, and λ_- lie respectively in the $\varepsilon/2$-neighborhood of D, the ε-neighborhood of ∂D, and in D. Also the functions are bounded in $\mathbf{R}^n$ together with all their partial derivatives. Now set

$$\Lambda_+ = \lambda_0\left(\frac{\lambda_+}{\lambda_+ + \lambda_-}\right) \quad \text{and} \quad \Lambda_- = \lambda_0\left(\frac{\lambda_-}{\lambda_+ + \lambda_-}\right)$$

We see that λ_0 is supported in the set where $\lambda_+ + \lambda_- \geq 1$. Therefore Λ_+ and Λ_- also have all their derivatives bounded in $\mathbf{R}^n$, and $\Lambda_+ + \Lambda_- = 1$ if $x \in \bar{D}$, while $\Lambda_+ + \Lambda_- = 0$ outside the $\varepsilon/2$-neighborhood of D.

Recall that the open sets $U_1, U_2, \ldots U_i, \ldots$ covering the boundary of D had special Lipschitz domains $D_1, D_2, \ldots, D_i, \ldots$ associated with them.

Let $\mathfrak{E}^i$ be the extension operator for $L_k^p(D_i)$ whose properties are given by Theorem 5'.

After all these preliminaries we can finally write down the required extension operator $\mathfrak{E}$ for D. In fact for $f \in L^p(D)$ define

$$(31) \qquad (\mathfrak{E}f)(x) = \Lambda_+(x) \left\{ \frac{\displaystyle\sum_{i=1}^{\infty} \lambda_i(x)\mathfrak{E}^i(\lambda_i f)}{\displaystyle\sum_{i=1}^{\infty} \lambda_i^2(x)} \right\} + \Lambda_-(x)f(x).$$

Observe the following facts:

(d) For x in the support of Λ_+, (more generally if dist $(x, \partial D) \leq \varepsilon/2$), then $x \in U_i^{\varepsilon/2}$ for at least one i, and thus $\sum \lambda_i^2(x) \geq 1$ there. (See (b) above.)

(e) For each x the sum (30) involves at most $N + 1$ non-vanishing terms (because of the condition (ii) on the covering $\{U_i\}$);

(f) The term $\Lambda_-(x)f(x)$ is well-defined since the support of Λ_- is contained in D;

(g) The terms $\mathfrak{E}^i(\lambda_i f)$ are well-defined since the $\lambda_i f$ are given in the special Lipschitz domains D_i;

(h) It is evident that $(\mathfrak{E}f)(x) = f(x)$ for $x \in D$.

In order to prove the basic inequality

$$(32) \qquad \|\mathfrak{E}(f)\|_{L_k^p(\mathbf{R}^n)} \leq A_{kn}(D)\|f\|_{L_k^p(D)}, \quad \text{if} \quad f \in L_k^p(D)$$

we require the following remark.

PROPOSITION. *Suppose* $A(x) = \displaystyle\sum_{i=1}^{\infty} a_i(x)$, *and for each x at most N of the terms* $\{a_i(x)\}$ *are non-vanishing. Then*

$$\|A(x)\|_p \leq N^{1-1/p} \left(\sum_i \|a_i(x)\|_p^p \right)^{1/p} \quad \text{if} \quad p < \infty$$

and

$$\|A(x)\|_\infty \leq N \sup_i \|a_i(x)\|_\infty, \qquad \text{if} \quad p = \infty.$$

The symbol $\|\cdot\|_p$ denotes the standard L^p norm. The case L^∞ of this is trivial as well as the L^1 case. The general case is not much more difficult, and follows from the observation that

$$|A(x)|^p \leq N^{p-1} \sum_{i=0}^{\infty} |a_i(x)|^p$$

which is in turn an obvious consequence of Hölder's inequality.

We prove first (32) when $k = 0$. Using properties (a)–(c) of the λ_i, and then (d) to (h), together with the Proposition and the case $(k = 0)$ of

Theorem 5′ we get, if $p < \infty$,

$$\|\mathfrak{E}f\|_p \leq N^{1-1/p}\left(\sum_i \int_{U_i} |\mathfrak{E}^i(\lambda_i f)|^p \, dx\right)^{1/p} + \left(\int_D |f(x)|^p \, dx\right)^{1/p}$$

$$\leq AN^{1-1/p}\left(\sum_i \int_D |\lambda_i f|^p \, dx\right)^{1/p} + \left(\int_D |f|^p \, dx\right)^{1/p}$$

$$\leq AN\left(\int_D |f|^p \, dx\right)^{1/p} + \left(\int_D |f|^p \, dx\right)^{1/p}, \quad \text{since} \quad \left(\sum_i \lambda_i\right)^{1/p} \leq N^{1/p}.$$

A similar reasoning holds for $p = \infty$. This proves (32) for $k = 0$. A very similar argument works for all k since every fixed partial derivative of the λ_i, $i = 1, 2, \ldots, \Lambda_+$ and Λ_- are all uniformly bounded. The proof of Theorem 5 is therefore concluded.

4. Further results

The following paragraphs §4.1 to 4.5, deal with the restriction of functions in $\mathscr{L}^p_\alpha(\mathbf{R}^n)$ to linear sub-varieties.

4.1 Let f be a locally integrable function on $\mathbf{R}^n$. We shall say that f *can be strictly defined at* x^0, if $\lim\limits_{\varepsilon \to 0} \dfrac{1}{m(B_\varepsilon)} \displaystyle\int_{B_\varepsilon} f(x^0 - t) \, dt$ exists, where B_ε denotes the ball of radius ε centered at the origin. In this case we redefine (if necessary) f to have this value at x^0. If this is done at each point x^0 where this is possible we say that f is *strictly defined*. Thus by the fundamental theorem of Chapter I every locally integrable function can be strictly defined, and agrees with the original function, almost everywhere. For the study of the restrictions of $\mathscr{L}^p_\alpha$ the following lemma is important.

LEMMA. *Suppose* $f \in \mathscr{L}^p_\alpha(\mathbf{R}^n)$, $f = G_\alpha * \varphi$, *with* $\alpha > 0$, $1 \leq p \leq \infty$. *Suppose* x^0 *is a point where the integral* $\int_{\mathbf{R}^n} G_\alpha(x - t)\varphi(t) \, dt = f(x)$ *representing* f *converges absolutely. Then*

$$\frac{1}{m(B_\varepsilon)} \int_{B_\varepsilon} |f(x^0 - t) - f(x^0)| \, dt \to 0, \quad \text{as} \quad \varepsilon \to 0.$$

Thus f can be strictly defined at x^0.

4.2 We consider a linear m-dimensional sub-variety of $\mathbf{R}^n$. It will be no loss of generality to assume that it is the subspace $\mathbf{R}^m$ of $\mathbf{R}^n$ of points whose first m coordinates are arbitrary, and whose last $n - m$ coordinates vanish. Now assume that $\alpha > (n - m)/p$, $1 \leq p \leq \infty$, and $f \in \mathscr{L}^p_\alpha(\mathbf{R}^n)$. Then f can be strictly defined at all points of $\mathbf{R}^m$, except for a subset of $\mathbf{R}^m$ of m-dimensional Lebesgue measure zero. If we denote this restriction by $\mathscr{R}(f)$, then $\mathscr{R}(f) \in L^p(\mathbf{R}^m)$, and the mapping $f \to \mathscr{R}(f)$ of $\mathscr{L}^p_\alpha(\mathbf{R}^n)$ to $L^p(\mathbf{R}^m)$ is continuous, i.e.

$$\|\mathscr{R}f\|_{L^p(\mathbf{R}^m)} \leq A \|f\|_{\mathscr{L}^p_\alpha(\mathbf{R}^n)}.$$

4.3 The $\mathscr{R}(f)$ described above belongs not only to $L^p(\mathbf{R}^m)$, but also to an appropriate Λ space. More precisely, let $\beta = \alpha - (n - m)/p > 0$, and $1 < p < \infty$. If $f \in \mathscr{L}^p_\alpha(\mathbf{R}^n)$, then $\mathscr{R}(f) \in \Lambda^{p,p}_\beta(\mathbf{R}^m)$, and the mapping $f \to \mathscr{R}(f)$ is continuous, that is

$$\|\mathscr{R}(f)\|_{\Lambda_\beta p, p(\mathbf{R}^n)} \le A \|f\|_{\mathscr{L}_\alpha p(\mathbf{R}^n)}.$$

4.4 The converse to §4.3 also holds. Thus the restriction of elements of $\mathscr{L}^p_\alpha(\mathbf{R}^n)$ to $\mathbf{R}^m$ consists of exactly $\Lambda^{p,p}_\beta(\mathbf{R}^m)$, with $\beta = \alpha - \dfrac{(n - m)}{p}$. We give an explicit description of this converse for the case $m = n - 1$. Let φ be a fixed C^∞ function on $\mathbf{R}^{n-1}$ with compact support, such that $\int_{\mathbf{R}^{n-1}} \varphi(x) \, dx = 1$. Let η be a fixed C^∞ function on $\mathbf{R}^1$ with compact support, so that $\eta(0) = 1$. For each locally integrable function f on $\mathbf{R}^{n-1}$ consider its extension to $\mathbf{R}^n$ given by

$$(Ef)(x, y) = \eta(y) \int_{\mathbf{R}^{n-1}} f(x - yt)\varphi(t) \, dt, \qquad (x, y) \in \mathbf{R}^{n-1} \times \mathbf{R}^1 = \mathbf{R}^n.$$

It can then be shown that $f \in \Lambda^{p,p}_\beta(\mathbf{R}^{n-1})$ implies that $Ef \in \mathscr{L}^p_\alpha(\mathbf{R}^n)$, and

$$\|E(f)\|_{\mathscr{L}_\alpha p(\mathbf{R}^n)} \le A \|f\|_{\Lambda_\beta p, p(\mathbf{R}^{n-1})}.$$

Here it is assumed that $\beta = \alpha - 1/p > 0$.

A more precise form of this extension is as follows. Let k be the largest integer not greater than β. Let $\eta_0(y), \ldots, \eta_k(y)$, be C^∞ functions on $\mathbf{R}^1$ with compact support, with the property that $\dfrac{d^l}{dy^l} \eta_j(y) \Big|_{y=0} = \delta_{jl}$, $0 \le j, \, l \le k$. Assume that $f_j \in \Lambda^{p,p}_{\beta-j}(\mathbf{R}^{n-1})$, $0 \le j \le k$. Set

$$F(x, y) = \sum_{j=0}^{k} \eta_j(y) \int_{\mathbf{R}^{n-1}} f_j(x - yt)\varphi(t) \, dt.$$

Then $F \in \mathscr{L}^p_\alpha(\mathbf{R}^n)$,

$$\|F\|_{\mathscr{L}_\alpha p(\mathbf{R}^n)} \le A \left\{ \sum_{j=0}^{k} \|f_j\|_{\Lambda_{\beta-j} p, p(\mathbf{R}^{n-1})} \right\},$$

and $\mathscr{R}\left(\dfrac{\partial^j F(x, y)}{\partial y^j}\right) = f_j(x), j = 0, \ldots, k$, where $\mathscr{R}$ denotes the restriction to the hyperplane $y = 0$. For the above results see Stein [7], and the later treatments in Aronszajn, Mulla, and Szeptycki [1] and Lizorkin [2]. For the background see Gagliardo [1] and Aronszajn and Smith [1].

4.5 Analogous results hold for the space $\Lambda^{p,q}_\alpha(\mathbf{R}^n)$. If $\alpha > \dfrac{(n - m)}{p}$, then the restriction of an element in $\Lambda^{p,q}_\beta(\mathbf{R}^n)$ to $\mathbf{R}^m$ belongs to $\Lambda^{p,q}_\alpha(\mathbf{R}^m)$ where $\beta = \alpha - \dfrac{(n - m)}{p}$. Conversely every function in $\Lambda^{p,q}_\beta(\mathbf{R}^m)$ can be extended to $\mathbf{R}^n$ so that it is an element in $\Lambda^{p,q}_\alpha(\mathbf{R}^n)$. For details see Besov [1], and Taibleson [2].

4.6 Let $\omega(\delta)$, $0 < \delta$, be a regular modulus of continuity as defined in §2.2.3 above. For any closed set F and every non-negative integer k define the space Lip $(k + \omega, F)$ as in §2.3 (equations (16) and (17)), except that the assumption on R_j should read

$$|R_j(x, y)| \leq M |x - y|^{k-|j|} \omega(|x - y|).$$

Then the operator $\mathscr{E}_k$ gives an extension from Lip $(k + \omega, F)$ to Lip $(k + \omega, \mathbf{R}^n)$.

4.7 In the extension theorem of §2.3, assume $\gamma = k$ and in addition

$$R_j(x, y) = o(|x - y|^{k+|j|}),$$

in the sense that for any $\bar{x} \in F$, and any $\varepsilon > 0$, there exists a $\delta > 0$, so that $|\bar{x} - x| < \delta$, $|\bar{x} - y| < \delta$, and $x, y \in F$, implies that $|R_j(x, y)| \leq \varepsilon |x - y|^{k-|j|}$. Then $\mathscr{E}_k(f) \in C^k(\mathbf{R}^n)$. This is essentially the original extension theorem of the type in Whitney [1].

4.8 The first extension theorem for functions in $L_k^p(D)$, where D is a region of the kind treated in §3, was proved by the following argument.

In giving an outline of this argument we shall limit ourselves to a *special Lipschitz domain* $D \subset \mathbf{R}^n$ (as defined in §3.2 for $\mathbf{R}^{n+1}$). For our purpose all we need as a consequence of this definition is the existence of a fixed cone Γ, so that wherever $x \in D$, then $x + \Gamma \subset D$. Now fix $k \geq 1$ in the rest of this discussion. We choose a fixed function φ supported in $-\Gamma$, with the following additional properties: (1) φ has bounded support. (2) Near the origin φ equals a function which is homogeneous of degree $-n + k$. (3) φ is C^∞ in $\mathbf{R}^n - \{0\}$. (4) $\lim_{\varepsilon \to 0} \int_{S^{n-1}} \varepsilon^{n-k} \varphi(\varepsilon x') \, d\sigma(x') = 1/(k - 1)!$. (Observe that in view of the assumed homogeneity of φ near the origin, the integral is actually constant for small ε.)

Let $\psi(\rho y') = \left(\dfrac{\partial}{\partial \rho}\right)^k (\rho^{n-1} \varphi(\rho y'))$, $(y = \rho y', |y| = \rho)$. Then ψ vanishes near the origin, and so is C^∞ everywhere (and of course has bounded support). Let

$$(\mathfrak{E}f)(x) = \int \varphi(\rho y') \rho^{n-1} \left(\frac{\partial}{\partial \rho}\right)^k f(x - \rho y') \, d\rho \, dy'$$

$$- \int \psi(\rho y') f(x - \rho y') \rho^{n-1} \, d\rho \, dy'.$$

The first integral may be rewritten as

$$(-1)^k \int_{\mathbf{R}^n} \varphi(y) \sum_{|\alpha|=k} \frac{k!}{\alpha!} y^\alpha \left(\frac{\partial}{\partial x}\right)^\alpha f(x - y) |y|^{-k} \, dy$$

while the second integral clearly equals

$$\int_{\mathbf{R}^n} \psi(y) f(x - y) \, dy.$$

We remark that f, and $\left(\dfrac{\partial}{\partial x}\right)^{\alpha} f$, $|\alpha| = k$, are given to us only in D. For the purposes of the integral formula defining $\mathfrak{E}f$, we set f and $\left(\dfrac{\partial}{\partial x}\right)^{\alpha} f$ equal to zero outside D.

It can then be verified that $\mathfrak{E}f$ indeed extends f, and

$$\|\mathfrak{E}f\|_{L_k^p(\mathbf{R}^n)} \leq A_{p,k} \|f\|_{L_k^p(D)}, \, 1 < p < \infty.$$

The latter fact is proved by observing that a differentiation of order k applied to the first integral defining $\mathfrak{E}f$ leads essentially to a singular integral operator of the kind treated in §4 of Chapter II. This explains the occurrence of the limitation $1 < p < \infty$. The reader should consult Calderón [4] for further details.

We make two further remarks about this extension. The operator $\mathfrak{E}$ depends on the particular k, and so is not universal, in the sense that the extension given in §3 is. On the other hand it has the interesting property that away from D, $\mathfrak{E}f$ depends only on the part of f near the boundary of D. This means that if $f \in L_k^p(D)$, and f vanishes near the boundary of D, then $\mathfrak{E}f(x) = 0$, if $x \notin D$.

Notes

Sections 1 and 2. The fundamental paper is Whitney [1]. The particular form of Theorem 1, is in Stein [10]; Theorem 2 is in Calderón and Zygmund [7]. See also Glaeser [1].

Section 3. The extension theorem for functions defined in domains with Lipschitz boundaries originates in Calderón [4]. Some of the ideas used were implicit in Sobolov [2]. Calderón's extension does not apply to the limiting cases $p = 1$, or $p = \infty$, since it relies on the L^p boundedness of the singular integrals treated in Chapter II. For the present version, see Stein [10]. Ideas related to the present extension and applied to the case $p = 2$ can be found in Adams, Aronszajn and Smith [1].

CHAPTER VII

Return to the Theory of Harmonic Functions

We return to the theory of harmonic functions to make a deeper study of some of its aspects and in particular with that part dealing with the notion of conjugate harmonic functions first discussed in Chapter III. The main threads of our development will be as follows:

(A) The notion of *non-tangential convergence*. While Theorem 1 in Chapter III guarantees the existence almost everywhere of boundary values for the perpendicular approach for Poisson integrals, the more general non-tangential limits also exist, and this is fundamental in what follows. It allows for a wide extension of the classical Fatou theorem to a purely "local" setup. These matters are dealt with in §1.

(B) *The "area integral" of Lusin*. This object was already pointed out in §2 of Chapter IV, but our attention there was focused on the closely related g and g^* functions.

The basic role of the area integral is due to the fact that it serves to characterize at once both non-tangential convergence and also L^p and H^p norms. The characterization of non-tangential convergence for harmonic functions will be taken up in §2 and will find important application to the study of differentiability properties of functions in Chapter VIII.

(C) H^p *theory for conjugate harmonic functions*. The notion of conjugacy was previously introduced in the generalized Cauchy-Riemann equations in §2.3 of Chapter III, in connection with the Riesz transforms. Here we shall see that by using it, a certain subharmonicity, and the area integral of Lusin we can study the analogue for $p = 1$ of various results for singular integrals operators and multiplier transformations. One of the ideas (characterizing the L^p boundedness of these operators in terms of g, or g^*) was already used in §3 of Chapter IV.

1. Non-tangential convergence and Fatou's theorem

1.1 We shall use the following notation: $\mathbf{R}^{n+1}_+$ is the $n + 1$ dimensional half-space of points (x, y), with $y > 0$, $x \in \mathbf{R}^n$. Its boundary $\{(x, 0)\}$ is identical with $\mathbf{R}^n$. For any $x^0 \in \mathbf{R}^n$, and $\alpha > 0$, $\Gamma_\alpha(x^0)$ will denote the (infinite) cone, $\Gamma_\alpha(x^0) = \{(x, y) \in \mathbf{R}_+^{n+1} : |x - x^0| < \alpha y\}$, whose vertex is at x^0. If $u(x, y)$ is defined at those points in $\mathbf{R}_+^{n+1}$ near a boundary point $(x^0, 0)$, then u has a non-tangential limit (which equals l) at $(x^0, 0)$ if for every $\alpha > 0$ the conditions $(x, y) \in \Gamma_\alpha(x^0)$ and $(x, y) \to (x^0, 0)$ imply that $u(x, y) \to l$.

The basic result about non-tangential convergence for Poisson integrals is contained in the following extension of Theorem 1 of Chapter III (see p. 62).

THEOREM 1. *Suppose $f \in L^p(\mathbf{R}^n)$, $1 \le p \le \infty$, and let $u(x, y)$ be its Poisson integral. Assume that α is a fixed positive number. Then*

(a)
$$\sup_{(x,y) \in \Gamma_\alpha(x^0)} |u(x, y)| \le A_\alpha M(f)(x^0),$$

where Mf is the maximal function of Chapter I, §1. A_α is independent of f.

(b)
$$\lim_{\substack{(x,y) \to (x^0,0) \\ (x,y) \in \Gamma_\alpha(x^0)}} u(x, y) = f(x^0),$$

for almost every x^0, and in particular for every point x^0 in the Lebesgue set of f.

This theorem is actually a relatively easy consequence of the corresponding theorem in Chapter III.

To prove (a), recall that

$$P_y(x) = \frac{c_n y}{(|x|^2 + y^2)^{(n+1)/2}},$$

(Proposition 1, Chapter III). Therefore as is easily seen,

(1) $P_y(x - t) \le A_\alpha P_y(x)$ if $|t| < \alpha y.$

However,

$$u(x^0 - t, y) = \int_{\mathbf{R}^n} P_y(x^0 - t - z, y) f(z)\, dx.$$

Therefore

$$\sup_{|t| < \alpha y} |u(x^0 - t, y)| = \sup_{(x,y) \in \Gamma_\alpha(x^0)} |u(x, y)|$$

$$\le \sup_{|t| < \alpha y} \int_{\mathbf{R}^n} P_y(x^0 - t - z, y)\, |f(z)|\, dz$$

$$\le A_\alpha \sup_y \int_{\mathbf{R}^n} P_y(x^0 - z)\, |f(z)|\, dz$$

$$\le A_\alpha(Mf)(x^0).$$

Thus (a) is proved.

Incidentally, the simple property (1) can in effect be re-expressed in the fact that if $P_1(x) = \varphi(x)$, then $\sup_{|t|<1} \varphi(x - t) \leq A\varphi(x)$ for some constant A independent of x. (With this remark the reader should have no difficulty in formulating and proving a general non-tangential version for the approximations $f * \varphi_\varepsilon$, analogous to Theorem 2 of §2.2, Chapter III.)

To prove (b) assume that x^0 is a point of the Lebesgue set of f. Thus for any $\varepsilon > 0$, there exists a $\delta > 0$ so that

$$\frac{1}{m(B(r))} \int_{B(r)} |f(x^0 - z) - f(x^0)| \, dz < \varepsilon,$$

whenever $r > \delta$; $B(r)$ denotes the ball of radius r about the origin. Now set $g(x) = f(x) - f(x^0)$, if $|x - x^0| \leq \delta$ and $g(x) = 0$ if $|x - x^0| > \delta$. Notice, therefore, that $M(g)(x^0) < \varepsilon$. We have, however,

$$u(x^0 - t, y) - f(x^0) = \int_{\mathbf{R}^n} P_y(z - t)[f(x^0 - z) - f(x^0)] \, dz.$$

So by (1), if $|t| < \alpha y$,

$$|u(x^0 - t, y) - f(x^0)| \leq A_\alpha \int_{\mathbf{R}^n} P_y(z) \, |f(x^0 - z) - f(x^0)| \, dz$$

$$= A_\alpha \left\{ \int_{|z| < \delta} + \int_{|z| \geq \delta} \right\}.$$

Now

$$\int_{|z| \leq \delta} \leq \int_{\mathbf{R}^n} P_y(t) |g(x^0 - z)| \, dz \leq M(g)(x^0) < \varepsilon,$$

by Theorem 1 (part (1)) in Chapter III. However,

$$\int_{|z| \geq \delta} P_y(z) \, |f(x^0 - z) - f(x^0)| \, dz \to 0, \quad \text{as} \quad y \to 0$$

which follows immediately from the fact that

$$\left(\int_{|z| \geq \delta} [P_y(z)]^q \, dz \right)^{1/q} \leq A_\delta y \to 0,$$

for $1 \leq q \leq \infty$, where q is chosen to be successively the exponent dual to p, and then 1. Therefore $\limsup_{|t| < \alpha y, y \to 0} |u(x^0 - t, y) - f(x^0)| \leq A_\alpha \varepsilon$, which proves the required non-tangential convergence at every point of the Lebesgue set of f. Since almost every point is a point of the Lebesgue set (see §1.8 in Chapter I) we have therefore concluded the proof of the theorem.

1.2 Fatou's theorem. We characterize first bounded harmonic functions in $\mathbf{R}_{+}^{n+1}$.

PROPOSITION 1. *Suppose u is given in $\mathbf{R}_{+}^{n+1}$. Then u is the Poisson integral of a function in $L^{\infty}(\mathbf{R}^n)$ if and only if u is harmonic and bounded.*

That the Poisson integral of a bounded function is (harmonic) and bounded follows from the considerations in §2 of Chapter III.

To prove the converse assume that u is harmonic and $|u| \leq M$ in $\mathbf{R}_{+}^{n+1}$.

For each integer k set $f_k(x) = u\left(x, \dfrac{1}{k}\right)$, and let $u_k(x, y)$ be the Poisson

integral of f_k. Finally let $\Delta_k(x, y) = u\left(x, y + \dfrac{1}{k}\right) - u_k(x, y)$. Notice the

following obvious properties of Δ_k. First, Δ_k is harmonic in $\mathbf{R}_{+}^{n+1}$; also

$|\Delta_k| \leq \left| u\left(x, y + \dfrac{1}{k}\right) \right| + |u_k(x, y)| \leq 2M$, so Δ_k is bounded. Finally by

the corollary (on p. 65) in §2.2 of Chapter III, since f_k is continuous and bounded, $u_k(x, y)$ is continuous in $\bar{\mathbf{R}}_{+}^{n+1}$ and $u_k(x, 0) = f_k(x)$; thus Δ_k is continuous in $\bar{\mathbf{R}}_{+}^{n+1}$ and $\Delta_k(x, 0) = 0$. We want to conclude on the basis of these properties that $\Delta_k \equiv 0$. To do this it suffices to show that $\Delta_k(0, 1) = 0$, since our assumptions are invariant under translation in the x variables and dilations jointly in the x and y variables and any point in $\mathbf{R}_{+}^{n+1}$ can be mapped to $(0, 1)$ by a product of those transformations. For fixed $\varepsilon > 0$ consider the function

$$U(x, y) = \Delta_k(x, y) + 2M\varepsilon y + \varepsilon\left[\prod_{j=1}^{n} \cosh\left(\frac{\varepsilon\pi}{4n^{1/2}} x_j\right)\right] \cos\left(\frac{\varepsilon\pi}{4} y\right)$$

which is clearly harmonic in $\mathbf{R}_{+}^{n+1}$, and continuous in $\bar{\mathbf{R}}_{+}^{n+1}$. We restrict

our attention to the cylindrical section $\Sigma = \left\{(x, y) : 0 \leq y \leq \dfrac{1}{\varepsilon}, |x| \leq R\right\}$,

where R is sufficiently large. On that part of the boundary of Σ where

$y = 0$, we have that $U(x, y) \geq 0$ since $\Delta_k = 0$ there; where $y = \dfrac{1}{\varepsilon}$, we

have $U(x, y) \geq 0$ since $|\Delta_k| \leq 2M$. Finally if R is sufficiently large, then again since Δ_k is bounded, we have $U(x, y) \geq 0$ on that part of the boundary. Thus by the maximum principle (see Appendix C) we have that $U(0, 1) \geq 0$, which means that $\Delta_k(0, 1) \geq -(2M + 1)\varepsilon$; a similar conclusion holds with $-\Delta_k$ in place of Δ_k, and as a result $\Delta_k \equiv 0$ as desired. This can be rewritten as

(2) $$u\left(x, y + \frac{1}{k}\right) = \int_{\mathbf{R}^n} P_y(x - t) f_k(t)\, dt.$$

Now recall that

$$\|f_k\|_\infty = \left\|u\left(x, \frac{1}{k}\right)\right\|_\infty \leq M.$$

Therefore by a familiar weak-compactness argument, there is an $f \in L^\infty$, $\|f\|_\infty \leq M$, and a subsequence $\{f_{k'}\}$, so that $f_{k'} \to f$ weakly in the sense that $\int_{\mathbf{R}^n} f_{k'} \varphi \, dt \to \int_{\mathbf{R}^n} f\varphi \, dt$, for any $\varphi \in L^1(\mathbf{R}^n)$. For fixed $(x, y) \in \mathbf{R}^{n+1}_+$, choose $\varphi(t) = P_y(x - t)$. Then (2) becomes in the limit

$$u(x, y) = \int_{\mathbf{R}^n} P_y(x - t)f(t) \, dt$$

which means that u is the Poisson integral of the bounded function f. The proposition is therefore completely proved.

When we combine the proposition and Theorem 1 we immediately get the n-dimensional version of a classical theorem of Fatou.

THEOREM 2.   *Suppose u is harmonic and bounded in $\mathbf{R}^{n+1}_+$. Then u has non-tangential limits at almost every point of the boundary $(\mathbf{R}^n)$ of $\mathbf{R}^{n+1}_+$.*

1.2.1   Proposition 1 quickly leads to a generalization of itself.

COROLLARY.   *Suppose u is harmonic in $\mathbf{R}^{n+1}_+$, and $1 \leq p \leq \infty$. If $\sup_{y > 0} \|u(\cdot, y)\|_{L^p(\mathbf{R}^n)} < \infty$, then u is the Poisson integral of an $f \in L^p(\mathbf{R}^n)$, if $1 < p$. If $p = 1$, u is the Poisson integral of a finite measure.*

This result was stated without proof in §4.2, Chapter III. The easy converse is contained in §2.2 of that chapter. To prove the corollary, assume $p < \infty$. For each $(x, y) \in \mathbf{R}^{n+1}$ let B be the ball whose center is (x, y) and radius is y. By the mean value theorem

$$|u(x, y)|^p \leq \frac{1}{m(B)} \iint_B |u(x', y')|^p \, dx' \, dy'.$$

However, $B \subset \{(x', y'): \ 0 < y' < 2y\}$, and $m(B) = cy^{n+1}$. Therefore

$$|u(x, y)|^p \leq c'y^{-n-1} \int_0^{2y} \int_{\mathbf{R}^n} |u(x', y')|^p \, dx' \, dy',$$

and

$$|u(x, y)| \leq c''y^{-n/p}.$$

For each positive integer k we can therefore apply Proposition 1 (and Theorem 2) and obtain $u\left(x, y + \frac{1}{k}\right) = P_y * f_k$, with $f_k(x) = u\left(x, \frac{1}{k}\right)$.

However, by assumption $\sup_k \|f_k\|_p < \infty$, and thus the familiar weak compactness arguments apply. More specifically, if $p > 1$, there exists an $f \in L^p(\mathbf{R}^n)$ and a subsequence $\{f_{k'}\}$, so that $f_{k'} \to f$ weakly. For $p = 1$ there exists a finite measure $d\mu$ and a subsequence $\{f_{k'}\}$ so that $f_{k'} \to d\mu$ weakly. In either case it follows that $u(x, y) = P_y * f$ or $u(x, y) = P_y * d\mu$ respectively. Observe that $\|f\|_p = \sup_{y>0} \|u(\cdot, y)\|_p$, if $p > 1$, and $\|d\mu\| = \sup_{y>0} \|u(\cdot, y)\|_1$.

1.3 A local version. We define first the appropriate notion of non-tangential boundedness. For any $\alpha > 0$, and $h > 0$, we let $\Gamma_\alpha^h(x^0)$ denote the *truncated cone* $\Gamma_\alpha^h(x_0) = \{(x, y) \in \mathbf{R}_+^{n+1} : |x - x^0| < \alpha y, \ 0 < y < h\}$. If u is defined in $\mathbf{R}_+^{n+1}$, we say that u is *non-tangentially bounded* at x^0 if for some α and h,

$$\sup |u(x, y)| < \infty, \ (x, y) \in \Gamma_\alpha^h(x^0).$$

Notice the non-tangential boundedness at x^0 requires a condition with respect to only one truncated cone whose vertex is x^0, while the existence of the non-tangential limit at x^0 (as described in §1.1 above) requires a condition for *all* cones at x^0.

The main result of this section is the following local analogue of Fatou's theorem (Theorem 2).

THEOREM 3. *Suppose u is harmonic in $\mathbf{R}_+^{n+1}$. Let E be a subset of $\mathbf{R}^n$ and suppose that u is non-tangentially bounded at every $x^0 \in E$. Then u has a non-tangential limit at almost every $x^0 \in E$.*

1.3.1 A preliminary argument allows us to uniformize the situation.

LEMMA. *Let u be continuous in $\mathbf{R}_+^{n+1}$, and suppose it is non-tangentially bounded at every point of the set E, $E \subset \mathbf{R}^n$. Then for $\varepsilon > 0$, there exists a compact set E_1 with:*
(1) $E_1 \subset E, m(E - E_1) < \varepsilon$.
(2) *For any $\alpha > 0$, $h > 0$, there is a bound $M = M(\alpha, h, \varepsilon)$, so that $|u(x, y)| \leq M$, $(x, y) \in \Gamma_\alpha^h(x^0)$, $x^0 \in E_1$.*

Proof. 1st step. By considering only α and h with rational values, we can find an E_0, with $m(E - E_0)$ small (say $m(E - E_0) < \varepsilon/3$), so that $|u(x, y)| \leq M$, $(x, y) \in \Gamma_\alpha^h(x^0)$, $x^0 \in E_0$, with some fixed α and h. We may assume that E_0 is also compact.

2nd step. With this E_0 fixed and k a large integer we shall choose a further subset E_{00}, $E_{00} \subset E_0$, with $m(E_0 - E_{00}) < \varepsilon/3$ so that

$$|u(x, y)| \leq M', \quad (x, y) \in \Gamma_k^k(x^0), \quad x^0 \in E_{00}.$$

To do so we proceed as follows. If $\eta < 1$ and $\varepsilon/3$ are given, there exists a $\delta > 0$, and a subset E_{00} so that $m(E_0 - E_{00}) < \varepsilon/3$ and

$$\frac{m(B(x, r) \cap E_0)}{m(B(x, r))} \geq \eta,$$

for $x \in E_{00}$, and $r \leq \delta$; this is because almost every point of E_0 is a point of density. Our required boundedness on E_{00} will hold if we show that for some δ sufficiently small we have

$$(3) \qquad \Gamma_k^{\bar{\delta}}(x^0) \subset \bigcup_{x' \in E_0} \Gamma_\alpha^h(x'), \quad \text{all} \quad x^0 \in E_{00}.$$

To prove (3) assume for simplicity that $x^0 = 0$. We must therefore consider pairs (x, y) in the cone $|x| < ky$ (with $y < \bar{\delta}$). Fix such a pair (x, y). We want to show that there exists an $x' \in E_0$, so that $|x - x'| < \alpha y$. Suppose, on the contrary, that for this x (and y) there is no such x'. Then the ball $|x - x'| < \alpha y$ is in $^cE_0$. This is a ball of radius αy to be compared with the (large) ball of radius ky (centered at the origin). Since the origin is in E_{00} by assumption, the relative measure in the large ball is at most $1 - \alpha/k$, which would be a contradiction if we had chosen $\eta > 1 - \alpha/k$. Thus for sufficiently small y we can find a required x', and (3) is proved, with $\bar{\delta} = \delta/k$. In view of the continuity of u it is clear that u is bounded for points at a positive distance from $\mathbf{R}^n$. Notice that we can also choose E_{00} to be compact.

Step 3. For each integer k we construct a subset E_{00} of this type. An appropriate intersection of a countable collection of such E_{00} gives the required set E_1.

1.3.2 With E any compact subset of $\mathbf{R}^n$, we define the open region $\mathscr{R}$ by

$$(4) \qquad \mathscr{R} = \bigcup_{x^0 \in E} \Gamma_\alpha^h(x^0).$$

A moment's reflection shows that the above lemma allows one to reduce Theorem 3 to its essential core: *Assume u is harmonic in* $\mathbf{R}_+^{n+1}$ *and* $|u| \leq 1$ *for* $(x, y) \in \mathscr{R}$. *Then for almost every* $x^0 \in E$ *the limit,* $\lim u(x, y)$ *exists as* $(x, y) \to (x^0, 0)$, *and* $(x, y) \in \mathscr{R}$. This we now proceed to prove.

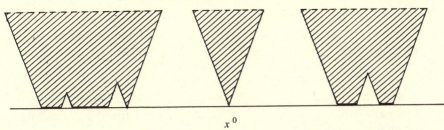

Figure 4.   The region $\mathscr{R}$, which is the union of cones $\Gamma_{\alpha}^{h}(x^{0})$, $x^{0} \in E$.

1.3.3   For each $m > 0$, we define the function φ_{m} on $\mathbf{R}^{n}$ as follows:
$\varphi_{m}(x) = u\left(x, \dfrac{1}{m}\right)$, if $\left(x, \dfrac{1}{m}\right) \in \mathscr{R}$; $\varphi_{m} = 0$ otherwise. Let $\varphi_{m}(x, y)$ be the Poisson integral of φ_{m}, that is $\varphi_{m}(x, y) = (P_{y} * \varphi_{m})(x)$. Define $\psi_{m}(x, y)$ by

(5)
$$u\left(x, y + \frac{1}{m}\right) = \varphi_{m}(x, y) + \psi_{m}(x, y).$$

Now the sequence of functions $\{\varphi_{m}(x)\}$ are uniformly bounded in the $L^{\infty}(\mathbf{R}^{n})$ norm, and thus we can find a $\varphi(x) \in L^{\infty}$ (in fact $|\varphi(x)| \leq 1$), and a subsequence $\{\varphi_{m'}\}$ so that $\varphi_{m'} \to \varphi$ weakly. Let $\varphi(x, y) =$ Poisson integral of φ. Then clearly $\varphi_{m'}(x, y) \to \varphi(x, y)$ at each point $(x, y) \in \mathbf{R}_{+}^{n+1}$; also $u\left(x, y + \dfrac{1}{m}\right) \to u(x, y)$ and so $\psi_{m'}(x, y)$ converges pointwise to $\psi(x, y)$ and we have

(5′)
$$u(x, y) = \varphi(x, y) + \psi(x, y).$$

The function $\varphi(x, y)$ is the Poisson integral of a bounded function and so non-tangential convergence almost everywhere holds for it. The function ψ is constructed to have zero boundary values on E, and this is what we in fact intend to prove about it in terms of a simple inequality. To do this we consider an auxiliary harmonic function $H(x, y)$ with the following properties in the region $\mathscr{R}$. We divide the boundary of $\mathscr{R}$ into two parts: we write $\partial\mathscr{R} = \mathscr{B} = \mathscr{B}_{0} \cup \mathscr{B}_{+}$, where $\mathscr{B}_{0}$ is the part of the region which intersects the boundary hyperplane $\mathbf{R}^{n}$, and $\mathscr{B}_{+}$ is that part of the boundary lying above $\mathbf{R}^{n}$. That is $\mathscr{B}_{0} = \{(x, 0) \in \overline{\mathscr{R}} - \mathscr{R}\}$, and $\mathscr{B}_{+} = \{(x, y) \in \overline{\mathscr{R}} - \mathscr{R}, y > 0\}$.

H will have the following properties:
   (i) H is harmonic in $\mathbf{R}_{+}^{n+1}$
   (ii) $H \geq 0$ in $\mathbf{R}_{+}^{n+1}$, (this is appropriate for $\mathscr{B}_{0}$)
   (iii) $H \geq 2$ on $\mathscr{B}_{+}$
   (iv) as $(x, y) \to (x^{0}, 0)$ non-tangentially, $H(x, y) \to 0$, for almost all $x^{0} \in E$.

The construction of H is simple. Let χ denote the characteristic function of the complement of E. For a constant c to be determined momentarily we write $H(x, y) = c\{(P_y * \chi)(x) + y\}$. The properties (i) and (ii) are obvious; (iv) is an immediate consequence of Theorem 1 of this chapter; the only one that requires further examination is (iii).

Now the boundary $\mathscr{B}_+$ can be further decomposed into its "trivial" part namely that part for which $y = h$, and its non-trivial part, that for which $0 < y < h$. For the trivial part we can always assure (iii) by taking c large enough $\left(c \geq \dfrac{2}{h}\right)$. It remains to consider the non-trivial part of the boundary $\mathscr{B}_+$. It should be observed, and this is also important for later considerations, that the part of the boundary in question is a part of the Lipschitz hypersurface given by $y = \dfrac{1}{\alpha}\, \mathrm{dist}\,(x, E)$. We can therefore conclude that for such points (x, y), the ball B in $\mathbf{R}^n$ whose center is x and which has radius αy lies outside E. Hence

$$(P_y * \chi)(x) = c_n y \int_{\mathbf{R}^n} \frac{\chi(t)}{(|\chi - t|^2 + y^2)^{(n+1)/2}}\, dt \geq c_n y \int_B \cdots dt$$

$$= c_n y \int_{|t| < \alpha y} \frac{1 \cdot dt}{(|t|^2 + y^2)^{(n+1)/2}} = \text{constant.}$$

The fact that the last integral is a constant can be seen by an obvious change of variables. Again taking c large enough we see that all the properties of H have been verified. We shall now prove that for fixed m

(6) $$|\psi_m(x, y)| \leq H(x, y) \quad \text{when} \quad (x, y) \in \mathscr{R}.$$

In view of the harmonic character of ψ_n and H and the maximum principle if this were not so there would exist an $\varepsilon > 0$, and a sequence of points (x_k, y_k) converging to a point on the boundary of $\mathscr{R}$ so that $|\psi_m(x_k, y_k)| \geq H(x_k, y_k) + \varepsilon$. First u and φ_n are both bounded by 1 in absolute value, and therefore $|\psi_m(x, y)| \leq 2$ and so by property (iii) the limit of $\{(x_k, y_k)\}$ cannot be on $\mathscr{B}_+$. However, $\varphi_m(x, y)$ is the Poisson integral of a function which is continuous on an open set containing E, and its values there are $u\left(x, \dfrac{1}{m}\right)$. Thus by Theorem 1, part (b)

$$\lim_k \left\{ u\left(x_k, y_k + \frac{1}{m}\right) - \varphi_m(x_k, y_k) \right\} = \lim_k \psi_m(x_k, y_k) = 0,$$

and again we obtain a contradiction, this time by (ii). We have therefore proved (6), and we can now let $m \to \infty$ through the subsequence $\{m'\}$.

The result is

(6') $$|\psi(x, y)| \leq H(x, y)$$

Property (iv) then gives us the required convergence of ψ (to zero) as $(x, y) \to (x^0, 0)$, $(x, y) \in \mathcal{R}$, for almost every $(x^0, 0) \in E$. This concludes the proof of Theorem 3.

1.3.4 It is interesting to point out that this theorem and many other results of this chapter (in particular Theorem 5 below) do not have valid analogues if non-tangential convergence is replaced by perpendicular convergence to the boundary. See §4.12 below.

2. The area integral

2.1 The theorem we have just proved shows that for a harmonic function in $\mathbf{R}_+^{n+1}$ the properties of non-tangential boundedness and existence of non-tangential limits are almost everywhere equivalent. Looked at from a broad point of view the two properties (boundedness and existence of limits) are not so very different. There is, however, another condition, almost everywhere equivalent with the first two, which has a different character, and finds its expression in a certain quadratic integral introduced by Lusin.

We fix the shape of our typical truncated cone Γ_α^h; (that is we fix α and h). Whenever u is given in $\mathbf{R}_+^{n+1}$ (or sub-region of $\mathbf{R}_+^{n+1}$ which contains the part near the boundary $\mathbf{R}^n$) we define

(7) $$S(u)(x^0) = \left(\iint_{\Gamma_{\alpha(x^0)}^h} |\nabla u|^2 \, y^{1-n} \, dy \, dx \right)^{1/2}.$$

(This is to be compared with the variant occurring in §2.3 of Chapter IV; see p. 89.)

Here

$$|\nabla u|^2 = \left| \frac{\partial u}{\partial y} \right|^2 + \sum_{j=1}^n \left| \frac{\partial u}{\partial x_j} \right|^2,$$

and we have written a double integral in (7) to emphasize that we are dealing with an $n + 1$ dimensional integration in distinction to certain n-dimensional integrations to be carried out below. The reason for the name "area integral" arose when $n = 1$. Then $(S(u)(x^0))^2$ represents the area (points counted with their multiplicity) of the image in $\mathbf{R}^2$ of the triangle $\Gamma_\alpha^h(x^0)$ under the analytic mapping $z \to F(z)$, where the real part of F is u. When $n > 1$, S loses this simple interpretation, but we shall nevertheless continue to use the terminology of the case $n = 1$. The characterization of non-tangential convergence in terms of the area integral is as follows.

THEOREM 4.   *Let u be harmonic in* $\mathbf{R}^{n+1}_{+}$. *Then except for a set of points* x^0 *of zero measure in* $\mathbf{R}^n$, *the following two conditions are equivalent:*
(1) *u has a non-tangential limit at* x^0
(2) $S(u)(x^0) < \infty$

2.1.1   A word of explanation is needed about the statement of the theorem. The proof that $(1) \Rightarrow (2)$ will show in fact that except for a subset of measure zero, at the points x^0 where the non-tangential limit exists we have $S(u)(x^0) < \infty$, no matter what choice of the parameters α and h we make that determine the shape of the truncated cone.

Conversely suppose that we have $S(u)(x^0) < \infty$ for x^0 in a given set, where the truncated cones may vary from point to point. Then the proof of the theorem will show that u has non-tangential limits at almost every point of the set.

In particular it follows that assuming the finiteness of S for variable cones is almost everywhere equivalent with assuming the finiteness of S for all truncated cones. (This last point, however, is quite elementary and could be proved directly if one wished.)

2.2.1   The proof of the theorem will depend on an application of Green's theorem to the region $\mathscr{R}$ defined in §1.3.2, (p. 202). The boundary of this region has just barely the required regularity to utilize Green's theorem and so it is convenient to approximate $\mathscr{R}$ by a family of regions $\{\mathscr{R}_\varepsilon\}$ with very smooth boundaries. The family $\partial\mathscr{R}_\varepsilon$ have a certain uniform smoothness which reflects the minimal smoothness of $\partial\mathscr{R}$.

LEMMA.   *There exists a family of regions* $\{\mathscr{R}_\varepsilon\}$, $\varepsilon > 0$, *with the following properties:*

(α) $\overline{\mathscr{R}_\varepsilon} \subset \mathscr{R}$, $\mathscr{R}_{\varepsilon_1} \subset \mathscr{R}_{\varepsilon_2}$ *if* $\varepsilon_2 < \varepsilon_1$.
(β) $\mathscr{R}_\varepsilon \to \mathscr{R}$ *as* $\varepsilon \to 0$ (i.e. $\cup \mathscr{R}_\varepsilon = \mathscr{R}$).
(γ) *the boundary* $\partial\mathscr{R}_\varepsilon = \mathscr{B}_\varepsilon$ *of the region is the union of two parts* $\mathscr{B}^1_\varepsilon \cup \mathscr{B}^2_\varepsilon$, *so that* $\mathscr{B}^2_\varepsilon$ *is a portion of the hyperplane* $y = h - \varepsilon$ *and*
(δ) $\mathscr{B}^1_\varepsilon$ *is a portion of the hypersurface* $y = \alpha^{-1}\delta_\varepsilon(x)$ *where* $\delta_\varepsilon \in C^\infty$, *and*

$$\left|\frac{\partial\delta_\varepsilon}{\partial x_j}\right| \leq 1, \qquad j = i, \ldots, n.$$

Set $\delta(x) = \text{dist}\,(x, E)$. Then clearly $|\delta(x) - \delta(x')| \leq |x - x'|$, if $x, x' \in \mathbf{R}^n$. Suppose $\varphi \in C^\infty(\mathbf{R}^n)$, φ has compact support, is positive, and

$$\int_{\mathbf{R}^n} \varphi(x)\,dx = 1.$$

Write $\tilde{\delta}_\eta(x) = \varphi_\eta * \delta$, with $\varphi_\eta(x) = \eta^{-n}\varphi(x/\eta)$. Thus it is clear that $\tilde{\delta}_\eta \in C^\infty$ and $\left|\dfrac{\partial \tilde{\delta}_\eta}{\partial x_j}\right| \leq 1$; also $\tilde{\delta}_\eta \to \delta$ uniformly as $\eta \to$. It is then easy to see that with an appropriate choice of η and η' (η' is a constant > 0) the functions $\delta_\varepsilon(x) = \tilde{\delta}_\eta(x) + \eta'$ and the corresponding regions $\mathcal{R}_\varepsilon = \{(x, y): \delta_\varepsilon(x) < \alpha y, \ 0 < y < h - \varepsilon\}$, satisfy all the conclusions of our lemma.

2.2.2 For fixed α and h we have our typical truncated cone $\Gamma_\alpha^h = \Gamma_\alpha^h(0)$. It will be necessary in what is done below to consider another truncated cone strictly containing it. We fix it by fixing β and k, with $\beta > \alpha$, and $k > h$. Then clearly $\Gamma_\beta^k \supset \Gamma_\alpha^h$, and the only point that the boundaries of these cones have in common is their common vertex.

LEMMA. *Suppose u is harmonic in Γ_β^k.*
  (i) *If $|u| \leq 1$ in Γ_β^k, then $|y\nabla u| \leq c$ in Γ_α^h.*

  (ii) *If $\iint_{\Gamma_\beta^k} |\nabla u|^2 y^{1-n} \, dx \, dy \leq 1$, then $|y\nabla u| \leq c$ in Γ_α^h. c is a constant which depends only on α, β, h, k and the dimension n.*

Observe that there exists a constant c_1, $c_1 > 0$, with the property that if (x, y) is any point of the smaller cone, Γ_α^h, the ball B whose center is (x, y) and has radius $c_1 y$ lies in Γ_β^k. Now by the mean-value property of harmonic functions we can assert the following inequalities (see Appendix C):

$$|(\nabla u)(x, y)| \leq c_2 r^{-1} \sup_{(x,y)\in B} |u(x, y)|$$

and

$$|(\nabla u)(x, y)|^2 \leq c_2 r^{-n-1} \iint_B |\nabla u|^2 \, dx' \, dy'$$

where $r = $ radius of B ($= c_1 y$).

The first of these two inequalities gives conclusion (i) immediately. To prove the second conclusion observe that for $(x', y') \in B$, the value y' remains comparable to the radius of B and more particularly $r \geq c_3 y'$, and so $y \geq c_4 y'$. Thus

$$y^2 |\nabla u|^2 \leq c_4 \iint_B |\nabla u|^2 y^{1-n} \, dx' \, dy \leq c_4 \iint_{\Gamma_\beta^k} |\nabla u|^2 y^{1-n} dx' dy.$$

The lemma is therefore proved.

2.3 **Proof of the direct part, (1) $\Rightarrow$ (2).** Suppose α and h are given. It suffices to show that at almost every point x^0 of a given compact set E, we have $\int \int_{\Gamma^h(x^0)} |\nabla u|^2 y^{1-n} \, dx \, dy < \infty$, if we assume that

$$(8) \qquad \sup_{x^0 \in E} \sup_{(x,y) \in \Gamma_\beta^k(x^0)} |u(x, y)| \le 1$$

for some fixed $\beta > \alpha$, $k > h$. Indeed assuming that u has non-tangential limits at a given set, we can always find compact subsets whose measure is arbitrarily close to the given set, and where the uniformity leading to (8) would hold, (after multiplication by a suitable non-zero constant).

We therefore write $\widetilde{\mathscr{R}} = \bigcup_{x^0 \in E} \Gamma_\beta^k(x^0)$, (in the same way as

$$\mathscr{R} = \bigcup_{x^0 \in E} \Gamma_\alpha^h(x^0)).$$

Our assumption becomes, therefore, that $|u| \le 1$ in $\widetilde{\mathscr{R}}$.

In order to show that $S(u)(x^0) < \infty$ almost everywhere in E it suffices to show that $\int_E S^2(u)(x^0) \, dx^0 < \infty$. This integral equals

$$\iint \left(\int_E \psi(x^9, x, y) \, dx^0 \right) y^{1-n} |\nabla u(x, y)|^2 \, dx \, dy$$

where ψ is the characteristic function of the set $\{|(x - x^0| < \alpha y, 0 < y < h\}$. However

$$\int_E \psi(x^0, x, y) \, dx^0 \le \int_{|x^0 - x| < \alpha y} dx^0 = cy^n,$$

thus we want to see why

$$(9) \qquad \iint_{\mathscr{R}} y \, |\nabla u(x, y)|^2 \, dx \, dy < \infty.$$

We replace now $\mathscr{R}$ by the approximating regions $\mathscr{R}_\varepsilon$ and then (9) is equivalent with

$$(9') \qquad \iint_{\mathscr{R}_\varepsilon} y \, |\nabla u|^2 \, dx \, dy \le A < \infty$$

with A independent of ε. To evaluate (9') we use Green's theorem in the form

$$(10) \qquad \iint_{\mathscr{R}_\varepsilon} (A \, \Delta B - B \, \Delta A) \, dx \, dy = \int_{\mathscr{B}_\varepsilon} \left(A \frac{\partial B}{\partial n_\varepsilon} - B \frac{\partial A}{\partial n_\varepsilon} \right) d\tau_\varepsilon$$

for the region with smooth boundary $\partial \mathscr{R}_\varepsilon = \mathscr{B}_\varepsilon$, where $\dfrac{\partial}{\partial n_\varepsilon}$ indicates the directional derivative along the outward normal and $d\tau_\varepsilon$ is the element of "area" of $\partial \mathscr{R}_\varepsilon$. Now let $B = \dfrac{u^2}{2}$, and $A = y$, then $\Delta B = |\nabla u|^2$ and $\Delta A = 0$. Since $\partial \mathscr{R}_\varepsilon \subset \widetilde{\mathscr{R}}$ the estimate (i) of the lemma holds there. That is, $\left|\dfrac{\partial B}{\partial n_\varepsilon}\right| \leq |u|\,|\nabla u|$, and so $\left|\dfrac{A\partial B}{\partial n_\varepsilon}\right| \leq c$ on $\mathscr{B}_\varepsilon$. Similarly, since $\left|\dfrac{\partial y}{\partial n_\varepsilon}\right| \leq 1$ we

have $\left|\dfrac{B\partial y}{\partial n_\varepsilon}\right| \leq \dfrac{1}{2}$. Altogether then the integrand of the integral on $\mathscr{B}_\varepsilon$ in
(10) is uniformly bounded, and so

$$\left| \int_{\mathscr{B}_\varepsilon} \left(\frac{A\partial B}{\partial n_\varepsilon} - \frac{B\partial A}{\partial n_\varepsilon} \right) d\tau_\varepsilon \right| \leq c \int_{\mathscr{B}_\varepsilon} d\tau_\varepsilon.$$

However,

$$\int_{\mathscr{B}_\varepsilon} d\tau_\varepsilon = \int_{\mathscr{B}_\varepsilon^1} d\tau_\varepsilon + \int_{\mathscr{B}_\varepsilon^2} d\tau_\varepsilon$$

where $\mathscr{B}_\varepsilon^2$ is a portion of the hyperplane $y = h - \varepsilon$, and $\mathscr{B}_\varepsilon^1$ is a portion of
the surface $y = \alpha^{-1}\delta_\varepsilon(x)$. On $\mathscr{B}_\varepsilon^2$ we have $d\tau_\varepsilon = dx$, while on $\mathscr{B}_\varepsilon^1$

$$d\tau_\varepsilon = \left(\sqrt{1 + \alpha^{-2}\sum_{j=1}^{n} \left(\frac{\partial \delta_\varepsilon}{\partial x_j} \right)^2} \right) dx \leq \sqrt{1 + \alpha^{-2}n}\, dx$$

by property (δ) of the lemma in §2.2.1. Since in any case $\mathscr{R}$ and thus $\mathscr{B}_\varepsilon$
is contained in a fixed compact set, it then follows that $\int_{\mathscr{B}_\varepsilon} d\tau_\varepsilon \leq$ constant $< \infty$, and so (9) is proved.

2.4 Proof of the converse part, $(2) \Rightarrow (1)$. We may reduce our
assumptions to the following. We can assume that for certain β and k
(which may be small), and a given bounded set E_0 we have

$$\iint_{\Gamma_\beta^k(x^0)} |\nabla u|^2 y^{1-n} dx\, dy \leq 1, \quad \text{for}\quad x^0 \in E_0.$$

Now let E be chosen as follows: $E \subset E_0$, E is compact and $m(E_0 - E)$ is
small; also there exists an $\eta > 0$, so that if $x^0 \in E_0$ then

$$m\{\{|x - x^0| < r\} \cap E_0\} \geq \tfrac{1}{2}m\{|x - x^0| < r\},$$

for $0 < r < \eta$. At the end of the proof we let $m(E_0 - E) \to 0$.
   Such choices of E are possible since almost every point of E_0 is a point
of density of E_0.
   Now suppose α and h are fixed with $\alpha < \beta$, $h < k$. We shall then study
u on $\mathscr{R} = \bigcup_{x^0 \in E} \Gamma_\alpha^h(x^0)$. We will now try to reverse the logical chain used
in the direct part, and our first objective is the proof of (9).
   Since we assumed that E_0 was bounded, we get

$$\text{(11)} \qquad \int_{E_0} \left\{ \iint_{\Gamma_\beta^k} |\nabla u|^2 y^{1-n}\, dx\, dy \right\} dx^0 < \infty.$$

It is now necessary to estimate $\int_{E_0} \tilde{\psi}(x_0, x, y)\, dx^0$ from below for $(x, y) \in \mathscr{R}$. Here $\tilde{\psi}$ is the characteristic function of $\{|x - x_0| < \beta y, 0 < y < k\}$. Now $(x, y) \in \mathscr{R}$ means there exists a $z \in E$, so that $|x - z| < \alpha y, 0 < y < h$. Hence we see that

$$\int_{E_0} \tilde{\psi}(x_0, x, y)\, dx^0 \geq \int_{E_0 \cap \{|x^0 - z| < (\beta - \alpha)y\}} dx^0.$$

By the property of relative density not less than $\frac{1}{2}$, which was assumed for E, we get that the second integral exceeds cy^n, for some constant c. Substituting this in (11) then gives (9). But (9) is equivalent with (9′) and in turn this is the same as

$$(12) \qquad \left| \int_{\mathscr{B}_\varepsilon} \left(y \frac{\partial u^2}{\partial n_\varepsilon} - u^2 \frac{dy}{\partial n_\varepsilon} \right) d\tau_\varepsilon \right| \leq c < \infty.$$

However, $\mathscr{B}_\varepsilon = \mathscr{B}_\varepsilon^1 \cup \mathscr{B}_\varepsilon^2$, and $\mathscr{B}_\varepsilon^2$ is at a strictly positive distance from the set $\{y = 0\}$, and is also compact. Thus the contribution in (12) coming from $\mathscr{B}_\varepsilon^2$ is bounded and hence we have

$$(12') \qquad \left| \int_{\mathscr{B}_\varepsilon^1} \left(y \frac{\partial u^2}{\partial n_\varepsilon} - u^2 \frac{\partial y}{\partial n_\varepsilon} \right) d\tau_\varepsilon \right| \leq c < \infty.$$

Now on $\mathscr{B}_\varepsilon^1$, $\frac{\partial y}{\partial n_\varepsilon} \leq -\alpha(\alpha^2 + n^2)^{-1/2}$. In fact $\frac{\partial}{\partial n_\varepsilon}$ is the outward normal derivative to the surface whose equation is $F_\varepsilon(x, y) \equiv \alpha y - \delta_\varepsilon(x) = 0$. The direction of $\frac{\partial}{\partial n_\varepsilon}$ is then given by the unit vector with the same direction as

$$\left(-\frac{\partial F_\varepsilon}{\partial y}, \frac{\partial F_\varepsilon}{\partial x_1}, \dots, \frac{\partial F_\varepsilon}{\partial x_n} \right) = \left(-\alpha, \frac{\partial \delta_\varepsilon}{\partial x_j}, \dots, \frac{\partial \delta_\varepsilon}{\partial x_n} \right).$$

Since $\left| \frac{\partial \delta_\varepsilon}{\partial x_j} \right| \leq 1$ we see that $\frac{\partial y}{\partial n_\varepsilon} \leq -\alpha(\alpha^2 + n^2)^{-1/2}$.

Next let $\mathscr{I}_\varepsilon^2 = \int_{\mathscr{B}_\varepsilon^1} u^2\, d\tau_\varepsilon$. According to what we have just said

$$\mathscr{I}_\varepsilon^2 \leq c_1 \int_{\mathscr{B}_\varepsilon^1} |u|\, y \left| \frac{\partial u}{\partial n_\varepsilon} \right| d\tau_\varepsilon + c_2, \quad \text{by (12′)}.$$

Now $\mathscr{B}_\varepsilon^1 \subset \tilde{\mathscr{R}} = \bigcup_{x^0 \in E} \Gamma_\beta^k(x^0)$, thus by the conclusion (ii) in the lemma of §2.2.2 we see that $y \left| \frac{\partial u}{\partial n_\varepsilon} \right| \leq y\, |\nabla u| \leq c$ there. Hence

$$\int_{\mathscr{B}_\varepsilon^1} |u|\, y \left| \frac{\partial u}{\partial n_\varepsilon} \right| d\tau_\varepsilon \leq c \int_{\mathscr{B}_\varepsilon^1} |u|\, d\tau_\varepsilon \leq c \mathscr{I}_\varepsilon \left(\int_{\mathscr{B}_\varepsilon^1} d\tau_\varepsilon \right)^{1/2} \leq c_3 \mathscr{I}_\varepsilon.$$

Altogether this gives $\mathscr{I}_\varepsilon^2 \leq c_3 \mathscr{I}_\varepsilon + c_2$, and hence $\mathscr{I}_\varepsilon$ is bounded in ε. We have therefore

(13) $$\int_{\mathscr{B}_\varepsilon^2} u^2 \, d\tau_\varepsilon \leq \text{constant.}$$

2.4.1 We continue with the proof of the converse direction (2) $\Rightarrow$ (1). We seek to majorize the function u by another, v, whose non-tangential behavior is known to us. We proceed as follows. The surface $\mathscr{B}_\varepsilon^1$ is a portion of the surface $y = \alpha^{-1}\delta_\varepsilon(x)$. Let $f_\varepsilon(x)$ be the function defined on $y = 0$ which is the projection on $y = 0$ of the restriction of u to $\mathscr{B}_\varepsilon^1$, and zero otherwise. That is

$$f_\varepsilon(x) = u(x, \alpha^{-1}\delta_\varepsilon(x))$$

for those $(x, 0)$ lying below $\mathscr{B}_\varepsilon^1$, and $f_\varepsilon(x) = 0$ otherwise. Since clearly $d\tau_\varepsilon \geq dx$, we have that

$$\int_{R^n} |f_\varepsilon(x)|^2 \, dx \leq \int_{\mathscr{B}_\varepsilon^1} |u(x)|^2 \, d\tau_\varepsilon \leq c.$$

Let now $v_\varepsilon(x, y)$ be the Poisson integral of $|f_\varepsilon|$. We shall show that for two appropriate constants c_1 and c_2

(14) $$|u(x, y)| \leq c_1 v_\varepsilon(x, y) + c_2, \quad \text{with} \quad (x, y) \in \mathscr{R}_\varepsilon$$

By the maximum principle for harmonic functions it suffices to show that the above inequality holds for points on the boundary $\mathscr{B}_\varepsilon$ of $\mathscr{R}_\varepsilon$. We have $\mathscr{B}_\varepsilon = \mathscr{B}_\varepsilon^1 \cup \mathscr{B}_\varepsilon^2$, where $\mathscr{B}_\varepsilon^2$ is a portion of the hyperplane $y = h - \varepsilon$ lying in a fixed sphere. Since $v_\varepsilon \geq 0$ we can satisfy (14) on $\mathscr{B}_\varepsilon^2$ by choosing c_2 large enough (and independent of ε).

It remains to consider points $(x, y) \in \mathscr{B}_\varepsilon^1$. Since $\mathscr{B}_\varepsilon^1 \subset \mathscr{R} = \bigcup_{x^0 \in E} \Gamma_\alpha^h(x^0)$ we can find a constant $c > 0$ so the ball B of center (x, y) and radius cy lies entirely in $\bigcup_{x^0 \in E} \Gamma_{\alpha^*}^{h^*}(x^0)$, where $\alpha < \alpha^* < \beta$ and $h < h^* < k$. Recall that for the cones $\Gamma_\beta^k(x^0)$ we have the property

$$\iint_{\Gamma_\beta^k(x^0)} |\nabla u|^2 \, y^{1-n} \, dx \, dy \leq 1.$$

Therefore by the lemma in §2.2.2 we have that $y |\nabla u| \leq c$, for all points in the ball B. Now $|u(p_1) - u(p_2)| \leq |p_1 - p_2| \sup_l |\nabla u|$ where l is the line segment joining (p_1, p_2); thus if $p_1 = (x, y)$, and p_2 is any other point of the ball B, $|p_1 - p_2| \leq$ radius of $B = cy$ and hence

(15) $$|u(p_1) - u(p_2)| \leq c_2.$$

Next suppose S_ε is that portion of the surface $\mathscr{B}_\varepsilon^1$ which is contained in the ball B. Write $|S_\varepsilon| = \int_{\mathscr{B}_\varepsilon^1 \cap B} d\tau_\varepsilon$. Then because of (15) we clearly have

$$|u(p_1)| \le \frac{1}{|S_\varepsilon|} \int_{S_\varepsilon} |u_\varepsilon(p_2)| \, d\tau_\varepsilon(p_2) + c_2$$

Because $d\tau_\varepsilon \ge dx$ and B has radius cy it is clear that $|S_\varepsilon| \ge ay^n$, where a is an appropriate constant, $a > 0$. Recalling the definition of $f_\varepsilon(x)$ and the fact that $d\tau_\varepsilon \le c \, dx$, we then get

$$|u(p_1)| = |u(x, y)| \le by^{-n} \int_{|z-x| < cy} |f_\varepsilon(z)| \, dz + c_2.$$

The Poisson kernel P_y has the property that

$$P_y(z) \ge c_1^{-1} by^{-n}, \quad \text{for} \quad |z| < cy$$

(for an appropriate constant c_1).

If we substitute in the above, since v_ε is the Poisson integral of $|f_\varepsilon|$ we get

$$|u(x, y)| \le c_1 v_\varepsilon(x, y) + c_2 \quad \text{for} \quad (x, y) \in \mathscr{R}_\varepsilon.$$

In view of the uniform boundedness in the L^2 norm of the $|f_\varepsilon|$ we can select a subsequence $|f_{\varepsilon'}|$ which converges weakly to an f in $L^2(\mathbf{R}^n)$. Let v denote the Poisson integral of f; thus $v_{\varepsilon'}(x, y)$ converges pointwise to $v(x, y)$, for $(x, y) \in \mathbf{R}_+^{n+1}$. Finally since $\mathscr{R}_\varepsilon \to \mathscr{R}$ we get from (14) that

$$(14') \qquad |u(x, y)| \le c_1 v(x, y) + c_2, \qquad (x, y) \in \mathscr{R}.$$

In that v is the Poisson integral of an L^2 function, it is by Theorem 1 non-tangentially bounded at almost every point of $\mathbf{R}^n$ and hence almost everywhere at E. Because of $(14')$ the same is true for u, and therefore in view of Theorem 3, u has non-tangential limits at almost every point of E.

2.5 Application to non-tangential convergence. Let $u_0, u_1, \ldots, u_n$ be a system of conjugate harmonic functions in the sense of Chapter III, §2.3. For simplicity of notation we write $y = x_0$. Then these $n + 1$ functions satisfy the equations

$$\sum_{j=0}^{n} \frac{\partial u_j}{\partial x_j} = 0,$$

and

$$\frac{\partial u_j}{\partial x_k} = \frac{\partial u_k}{\partial x_j}, \qquad 0 \le k, j \le n.$$

Our purpose now will be to prove the following theorem which is itself implied by the theorem for the area integral just proved.

THEOREM 5. *The following two statements are equivalent for any set* $E \subset \mathbf{R}^n$:

(a) u_0 *has a non-tangential limit at almost every point of E.*

(b) $u_1, u_2, \ldots, u_n$ *all have non-tangential limits at almost every point of E.*

The reader will have no difficulty in deducing the following consequence of Theorem 5.

COROLLARY. *Suppose H is harmonic in* $\mathbf{R}_+^{n+1}$, *and* $\dfrac{\partial^k}{\partial y^k} H$ *has non-tangential limits at almost every point of a set E, $E \subset \mathbf{R}^n$. Then the same is true of* $P\left(\dfrac{\partial}{\partial x}\right) H$, *where* $P\left(\dfrac{\partial}{\partial x}\right)$ *is a homogeneous polynomial in*

$$\frac{\partial}{\partial x_0}, \frac{\partial}{\partial x_1}, \ldots, \frac{\partial}{\partial x_n},$$

of degree k.

Because of the interpretation of conjugate harmonic functions in terms of Riesz transforms (as described in Chapter III, §2.2), Theorem 5 can be viewed in the following light. Namely, it is a local version, dealing with behavior almost everywhere in arbitrary sets for Riesz transforms; the global version, the one giving $L^p(\mathbf{R}^n)$ inequalities, was of course treated in Chapters II and III.

The case $n = 1$ is exceptional in the proof of Theorem 5, since then clearly $|\nabla u_0|^2 = |\nabla u_1|^2$, and thus $S(u_0)(x) = S(u_1)(x)$. In that case, therefore, the theorem is an immediate consequence of Theorem 4. However, in the general case the fact that $S(u_0)(x)$ dominates the $S(u_j)(x)$, $j = 1, \ldots, n$, requires an extra argument, which we shall now give.

2.5.1 A lemma.

LEMMA. *Let* Γ_α^h *and* Γ_β^k *be a pair of truncated cones (for simplicity we assume that their vertices are at the origin). Suppose* $\alpha < \beta$, $h < k$, *and so* Γ_α^h *is strictly contained in* Γ_β^k. *Assume that*

$$\iint_{\Gamma_\beta^k} \left|\frac{\partial u}{\partial y}\right|^2 y^{1-n} \, dx \, dy < \infty.$$

Then

$$\iint_{\Gamma_\alpha^h} \left|\frac{\partial u}{\partial x_j}\right|^2 y^{1-n} \, dx \, dy < \infty, \qquad j = 1, \ldots, n.$$

We proceed as follows. We let ρ denote any unit vector of direction determined by the smaller cone Γ_α^h; that is ρ is any unit vector so that $\rho s \in \Gamma_\alpha^h$ for some $s > 0$. Let s_0 denote the least upper bound of s so that $\rho s \in \Gamma_\alpha^h$. Then clearly we always have $h \le s_0 \le h^*$ where h^* is a constant which depends on the cone Γ_α^h only. For any function U defined in Γ_α^h we denote its restriction to the ray determined by ρ as U_ρ; more precisely $U_\rho(s) = U(\rho s)$. We shall show

$$\int_0^h s \left| \frac{\partial u_\rho(s)}{\partial x_j} \right|^2 ds \le A < \infty$$

where the constant A is independent of the direction ρ in question. An integration over all relevant ρ will then give our conclusion.

We write

(16)
$$\frac{\partial u}{\partial x_j} = -\int_y^h \frac{\partial^2 u}{\partial x_j \, \partial y} \, (x, \tau) \, d\tau + \frac{\partial u}{\partial x_j} \, (x, h)$$

The second term on the right side is harmless since it is uniformly bounded; (it represents values of $\dfrac{\partial u}{\partial x_j}$ taken at points strictly away from the boundary of the larger cone Γ_α^k). We therefore need to deal with only the integral on the right side. For this we use the following estimate which is a consequence of the mean-value theorem (see Appendix C).

(17)
$$\left| \frac{\partial^2 u}{\partial x_j \, \partial y} \, (x, \tau) \right|^2 \le \frac{cr^{-2}}{m(B)} \int\int_B \left| \frac{\partial u}{\partial y} \right|^2 dx' \, dy'.$$

Here B is a ball centered at the point (x, τ) whose radius is r. We now choose a constant c_1 with the property that if $(x, \tau) \in \Gamma_\alpha^h$ then the ball B of radius $r = c_1 \tau$ lies entirely in the larger cone Γ_β^k. With this choice (17) becomes

(17')
$$\left| \frac{\partial^2 u}{\partial x_j \, \partial y} \, (x, \tau) \right| \le c_2 \tau^{-(n+3)/2} \left(\int\int_B \left| \frac{\partial u}{\partial y} \right|^2 dx' \, dy' \right)^{1/2}$$

Now call S_τ the "layer" in the larger cone Γ_β^k given by

$$S_\tau = \{(x, y) : |x| < \beta y, \ \tau - c_1 \tau < y < \tau + c_1 \tau\}.$$

Then clearly since $B \subset \Gamma_\beta^k$, we have $B \subset S_\tau$ and hence if we write

$$\mathscr{I}_\tau = \int\int_{S_\tau} \left| \frac{\partial u}{\partial y} \right|^2 dx' \, dy'$$

the inequality (17') becomes

(17'')
$$\left| \frac{\partial^2 u}{\partial x_j \, \partial y} \, (x, \tau) \right| \le c_2 \tau^{-(n+3)/2} \mathscr{I}_\tau^{1/2}.$$

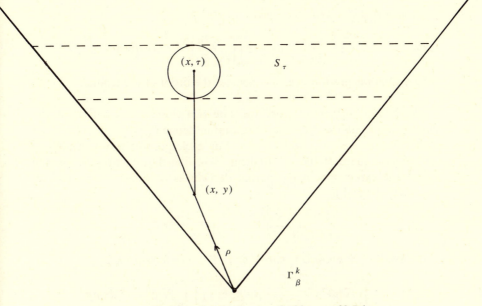

Figure 5. The situation in the proof of lemma §2.5.1.

We call θ the angle that ρ makes with the y-axis. Since this direction is contained in the cone $|x| < \alpha y$, $y > 0$, it follows that $\cos \theta \geq a_0 = (1 + \alpha^2)^{-1/2} > 0$. With this observation we can insert (17″) in (16). The result is

$$\left| \frac{\partial u_\rho(s)}{\partial x_j} \right| \leq c_3 \int_{sa_0}^h \tau^{-(n+3)/2} \mathscr{I}_\tau^{1/2} \, d\tau + c_4.$$

Now use the Hardy inequality (see Appendix A) and we get

$$\int_0^h s \left| \frac{\partial u_\rho(s)}{\partial x_j} \right|^2 ds \leq c_5 \int_0^h \tau^{-n} \mathscr{I}_\tau \, d\tau + c'.$$

However,

$$\int_0^h \tau^{-n} \mathscr{I}_\tau \, d\tau = \int_0^h \tau^{-n} \left\{ \int\!\!\int_{S_\tau} \left| \frac{\partial u}{\partial y} \right|^2 dx' \, dy' \right\} d\tau$$

$$\leq \int\!\!\int_{\Gamma_\beta^k} \left| \frac{\partial u}{\partial y} \right|^2 \left\{ \int \tau^{-n} \chi(\tau, x', y') \, d\tau \right\} dx' \, dy'.$$

where χ is the characteristic function of the layer S_τ in Γ_β^k. But

$$\int \tau^{-n} \chi(\tau, x', y) \, d\tau \leq \int_{\tau - c_1\tau < y < \tau + c_1\tau} \tau^{-n} \, d\tau = \int_{y/(1+c_1)}^{y/(1-c_1)} \tau^{-n} \, d\tau = c' y^{-n+1}.$$

Inserting this in the above gives

$$\int_0^h s \left| \frac{\partial u_\rho(s)}{\partial x_j} \right|^2 ds \leq c \iint_{\Gamma_\beta^k} \left| \frac{\partial u}{\partial y} \right|^2 y^{1-n} \, dx \, dy + c$$

and a final integration over ρ concludes the proof of the lemma.

2.5.2 For further reference we state two closely related inequalities. In this connection we let Γ_α denote the infinite cone $\Gamma_\alpha = \{(x, y) : |x| < \alpha y\}$. We consider also Γ_β, with $\beta > \alpha$. For each positive k we denote by $|\nabla^k u|^2$ the square of the kth gradient of u, namely that positive definite quadratic expression in the partial derivatives of order k which is recursively defined by

$$|\nabla^k u|^2 = \sum_{j=0}^n \left| \nabla^{k-1} \frac{\partial u}{\partial x_j} \right|^2, \qquad (x_0 = y).$$

LEMMA 1. *Suppose u is harmonic in Γ_β, then for each $k \geq 1$*

$$\iint_{\Gamma_\alpha} |\nabla^k u|^2 y^{2k-n-1} \, dx \, dy \leq c_{\alpha,\beta,k} \iint_{\Gamma_\beta} |\nabla u|^2 y^{1-n} \, dx \, dy.$$

LEMMA 2. *Suppose u is harmonic in Γ_β, and $|\nabla u| \to 0$ as $y \to \infty$, for $(x, y) \in \Gamma_\beta$. Then for each $k \geq 1$*

$$\iint_{\Gamma_\beta} |\nabla u|^2 y^{1-n} \, dx \, dy \leq c_{\beta,k} \iint_{\Gamma_\beta} |\nabla^k u|^2 y^{2k-n-1} \, dx \, dy.$$

To prove Lemma 1 we use the inequality

$$(18) \qquad |\nabla^k u(x, \tau)|^2 \leq c_k \frac{r^{-2(k-1)}}{m(B)} \iint_B |\nabla u|^2 \, dx' \, dy'$$

where B is a ball of radius $r = c_1 \tau$ centered at (x, τ); the rest is then as in §2.5.1.

To prove Lemma 2 we observe that $|\nabla u| \to 0$ implies by (18) that $|\nabla^k u| \to 0$ as $y \to \infty$, for any proper sub-cone of Γ_β. Now if $k > 1$, then

$$(19) \qquad \frac{\partial u}{\partial x_j}(x, y) = \frac{-1}{(k-2)!} \int_y^\infty \frac{\partial^k}{\partial y^k \, \partial x_j} u(x, \tau)(y - \tau)^{k-2} \, d\tau$$

By the inequalities for integral operators with homogeneous kernels (see Appendix A) we get from (19) that

$$\int_{|x|/\beta}^\infty \left| \frac{\partial u}{\partial x_j}(x, y) \right|^2 y^{1-n} \, dy \leq c_k \int_{|x|/\beta}^\infty \left| \frac{\partial^k u}{\partial y^{k-1} \, \partial x_j}(x, y) \right|^2 y^{2k-n-1} \, dy$$

and a final integration in x gives the desired result, for $\dfrac{\partial u}{\partial x_j}$. A similar argument works for $\dfrac{\partial u}{\partial y}$.

The lemma in §2.5.1 as well as those in §2.5.2 tell, among other things, that as far as the basic properties of the area integral are concerned we could have used higher derivatives as well as first derivatives. However, it is important to note that we could not have used the zero derivative. In fact, observe that the analogue for the area integral for $k = 0$ is infinite even when u is a constant!

3. *Application of the theory of H^p spaces*

We intend to study further the notion of conjugacy given by the generalized Cauchy-Riemann equations

$$(18) \qquad \sum_{j=0}^{n} \frac{\partial u_j}{\partial x_j} = 0, \quad \text{and} \quad \frac{\partial u_j}{\partial x_k} = \frac{\partial u_k}{\partial x_j}, \qquad x_0 = y.$$

One of the tools will be a certain result of subharmonicity which is subsumed in the following lemma.

3.1 A subharmonic property of the gradient. We write $F = (u_0, u_1, \ldots, u_n)$, where the u_j satisfy the Cauchy-Riemann equations (18) in some open set. (Observe that this implies that in some neighborhood of each point of this set the vector F is the gradient of a harmonic function H in that neighborhood.) For certain technical reasons it will be convenient to assume that the u_j take their values in a fixed finite-dimensional inner product space (i.e., Hilbert space) over the real numbers. We write $|\cdot|$ for the norm in that space and $u_i \cdot u_i'$ for the inner product of the u_i with u_i'. We write also

$$|F| = \left(\sum_{j=0}^{n} |u_j|^2 \right)^{1/2}, \quad \text{and} \quad \Delta = \sum_{j=0}^{n} \frac{\partial^2}{\partial x_j}.$$

LEMMA. *Suppose $|F| > 0$ at a point. If $q \geq \dfrac{n-1}{n}$, then $\Delta(|F|^q) \geq 0$ there. More precisely, if $q > \dfrac{n-1}{n}$, then*

$$(19) \qquad C_q |F|^{q-2} |\nabla F|^2 \geq \Delta(|F|^q) \geq c_q |F|^{q-2} |\nabla F|^2$$

*for two positive constants C_q and c_q.**

---

* The proof will show that we may take $C_q = q(q - 1)$, $c_q = q$, if $q \geq 2$ and $C_q = q$, $c_q = q \left[1 + (q - 2) \left(\dfrac{n}{n + 1} \right) \right]$ if $q \leq 2$. The condition $q \geq \dfrac{n-1}{n}$ is exactly the condition $c_q \geq 0$. Observe also that $C_2 = c_2 = 2$.

To understand the thrust of this lemma we remark that the assertion that $\Delta |F|^q \geq 0$, if $q \geq 1$, is a consequence of the fact that each u_j is harmonic, and does not otherwise depend on the differential relations (18) for the u_j. The particular usefulness of this lemma is that it holds for some $q < 1$. The actual range of q given, $q \geq \dfrac{n-1}{n}$ is best possible as simple examples show.

The inequality (19) should also be compared to the following two simpler identities. First

$$(20) \qquad \Delta u^q = q(q-1)u^{q-2}|\nabla u|^2,$$

whenever u is harmonic and $u > 0$. We have already used this identity in Chapter IV, §2.

The second identity is

$$(21) \qquad \Delta |F|^q = \frac{q^2}{2}|F|^{q-2}|\nabla F|^2 = q^2|F|^{q-2}|F'|^2,$$

in the case when $n = 1$ and $F = u_0 - iu_1$ is a holomorphic function of $x_0 + ix_1$.

3.1.3 To prove the lemma we use the notation $F \cdot G = u_0 \cdot v_0 + u_1 \cdot v_1 + \cdots + u_n \cdot v_n$, where $F = (u_0, u_1, \ldots, u_n)$, $G = (v_0, v_1, \ldots, v_n)$; also $F_{x_j} = \dfrac{\partial F}{\partial x_j}$; then $|\nabla F|^2 = \sum\limits_{j=0}^{n}|F_{x_j}|^2$. Now $\dfrac{\partial}{\partial x_j}(F \cdot G) = F_{x_j} \cdot G + F \cdot G_{x_j}$; therefore

$$\frac{\partial}{\partial x_j}|F|^q = \frac{\partial}{\partial x_j}(F \cdot F)^{q/2} = q|F|^{q-2}(F_{x_j} \cdot F).$$

Hence,

$$\frac{\partial^2}{\partial x_j^2}|F|^q = q(q-2)|F|^{q-4}(F_{x_j} \cdot F)^2 + q|F|^{q-2}\{|F_{x_j}|^2 + F_{x_j x_j} \cdot F\}.$$

Adding with respect to j and taking into account the fact that $F_{x_0 x_0} + F_{x_1 x_1} \cdots + F_{x_n x_n} = 0$ we obtain

$$\Delta |F|^q = q|F|^{q-4}\left\{(q-2)\sum_j (F_{x_j} \cdot F)^2 + |F|^2 \sum_j |F_{x_j}|^2\right\}.$$

It is therefore only a matter of comparing

$$(22) \qquad (q-2)\sum_j (F_{x_j} \cdot F)^2 + |F|^2 \sum_j |F_{x_j}|^2$$

with $|F|^2|\nabla F|^2 = |F|^2 \sum\limits_j |F_{x_j}|^2$.

By Schwarz's inequality $\sum_j (F_{x_j} \cdot F)^2 \leq |F|^2 \sum_j |F_{x_j}|^2$ and so (22) is certainly dominated by $|F|^2 |\nabla F|^2$, if $q \leq 2$, and $(q-1)|F|^2 |F|^2$ if $q \geq 2$. This shows that $\Delta |F|^q \leq C_q |F|^{q-2} |\nabla F|^2$. Similarly the bound of $\Delta |F|^q$ from below is simple when $q \geq 2$, and does not use the full thrust of the Cauchy-Riemann equations (18). We now assume that $q < 2$. Then the heart of the matter is contained in the following improvement over Schwarz's inequality

(23)
$$\sum_j (F_{x_j} \cdot F)^2 \leq \left(\frac{n}{n+1}\right) |F|^2 \sum_j |F_{x_j}|^2$$

which we shall prove momentarily. Assuming (23) we see that (22) is bounded from below by

$$|F|^2 \sum_j |F_{x_j}|^2 \left\{ 1 + (q-2)\left(\frac{n}{n+1}\right)\right\}.$$

Inserting this in the above gives $\Delta |F|^q \geq c_q |F|^{q-2} |\nabla F|^2$, with $c_q = q\left\{1 + (q-2)\left(\frac{n}{n+1}\right)\right\}$, if $q \leq 2$. The lemma will then be proved once we have established the inequality (23), which makes full use of the generalized Cauchy-Riemann equations.

3.1.2 We observe first that it suffices to prove (23) in the special case when the functions $u_0, u_1, \ldots, u_n$ are real-valued, that is when the Hilbert spaces in question are one-dimensional. The passage to the general case considered is then effected by introducing any orthonormal basis.

Let now $\mathcal{M} = (m_{jk})$ be an $(n+1) \times (n+1)$ matrix with (real) entries m_{jk}. We will consider two norms on such matrices. First the ordinary norm $\|\mathcal{M}\| = \sup |\mathcal{M}(F)|$, taken over all vectors F of length 1. Also the Hilbert-Schmidt norm, $\|\|\mathcal{M}\|\|$, given by $\|\|\mathcal{M}\|\|^2 = \sum_{j,k} |m_{jk}|^2$. By Schwarz's inequality it is clear that $\|\mathcal{M}\| \leq \|\|\mathcal{M}\|\|$. Suppose now that we assume that the matrix $\mathcal{M}$ is symmetric and has trace zero. Then we can strengthen the above inequality to read

(23′)
$$\|\mathcal{M}\|^2 \leq \left(\frac{n}{n+1}\right) \|\|\mathcal{M}\|\|^2.$$

To prove (23′) we observe that both norms are orthogonal invariants and after a proper choice of basis in the $n+1$ dimensional Euclidean space we can assume that the symmetric matrix $\mathcal{M}$ is diagonal, with diagonal entries $\lambda_0, \lambda_1, \ldots, \lambda_n$. Since the trace of $\mathcal{M}$ vanishes we get $\sum_{j=0}^{n} \lambda_j = 0$.

Therefore $\lambda_{j_0} = -\sum_{j \neq j_0} \lambda_j$; so by Schwarz's inequality $\lambda_{j_0}^2 \leq n \sum_{j \neq j_0} \lambda_j^2$, and consequently

$$\sup_j \lambda_j^2 = \lambda_{j_0}^2 \leq \left(\frac{n}{n+1}\right) \sum_j \lambda_j^2.$$

This is (23′) in case of a diagonal matrix, and thus by what has been said already (23′) is proved for any symmetric matrix of trace zero. With $F = (u_0, \ldots, u_n)$, let $m_{jk} = \dfrac{\partial u_j}{\partial x_j}$. Then the generalized Cauchy-Riemann equations (18) state exactly that the matrix $\mathcal{M} = (m_{jk})$ is symmetric and has trace zero. It is then easy to see that (23′) implies (23) and the lemma is completely proved.

3.2 H^p spaces; in particular H^1. In analogy with the classical theory we define the n-dimensional form of the Hardy space H^p as follows. Suppose $F = (u_0, u_1, \ldots, u_n)$ satisfies the generalized Cauchy-Riemann equations (18) in $\mathbf{R}_+^{n+1}$. For $p > 0$ we say that $F \in H^p$ if

$$(24) \qquad \sup_{y > 0} \left(\int_{\mathbf{R}^n} |F(x, y)|^p \, dx \right)^{1/p} < \infty$$

We shall write $\|F\|_p$ for the quantity appearing above. It is a norm when $p \geq 1$.

Because of the lemma in §3.1 it will be shown that some of the classical theory ($n = 1$) of H^p spaces extends to n-dimensions when $p > \dfrac{n-1}{n}$, and what is most interesting for our purposes, is that this contains the case of $p = 1$. To understand better the implications of the latter point we consider first the H^p spaces when $1 < p < \infty$.

Suppose therefore that $F \in H^p$. Then by the corollary in §1.2.1 there exist $f_0, f_1, f_2, \ldots, f_n$, each in $L^p(\mathbf{R}^n)$, so that $u_j(x, y)$ is the Poisson integral of f_j, $j = 0, \ldots, n$. Also by §4.4 of Chapter III $f_j = R_j(f_0)$, where $R_1, R_2, \ldots, R_n$ are the Riesz transforms. Conversely, suppose $f_0 \in L^p(\mathbf{R}^n)$, and let $f_j = R_j(f_0)$, and $u_j(x, y)$ be the Poisson integrals of f_j, $j = 0, \ldots, n$. Then $F = (u_0, u_1, \ldots, u_n) \in H^p$; moreover $\|f_0\|_p \leq \|F\|_p \leq A_p \|f_0\|_p$.

To summarize: for $1 < p < \infty$ the space H^p is naturally equivalent with the space $L^p(\mathbf{R}^n)$.

For $p = 1$ this identity no longer holds and we may view the space H^1 as a substitute for $L^1(\mathbf{R}^n)$. Our purpose will be to show that in this context various results which break down for L^1 have correct versions for H^1. In this way the H^1 theorems may be thought to be in some sense analogous and complementary to the results involving weak-type (1, 1) which occur in Chapters I and II. Our first theorem of this kind deals in effect with the maximal function.

THEOREM 6. *Let $F \in H^1$. Then* $\lim_{y \to 0} F(x, y) = F(x)$ *exists almost everywhere and in* $L^1(\mathbf{R}^n)$ *norm. Also*

$$\int_{\mathbf{R}^n} \sup_{y > 0} |F(x, y)| \, dx \leq A \sup_{y > 0} \int_{\mathbf{R}^n} |F(x, y)| \, dx = \|F\|_1.$$

Before we come to the proof of this theorem we shall formulate several corollaries which will indicate more clearly what is actually involved.

Let $d\mu_0$ be a finite measure on $\mathbf{R}^n$. We shall say that the Riesz transform $R_j(d\mu_0)$ is also a measure, say $d\mu_j$, if the identity

$$(25) \qquad \hat{\mu}_j(x) = i \frac{x_j}{|x|} \hat{\mu}_0(x)$$

holds, where $\hat{\mu}_0$ and $\hat{\mu}_j$ denote respectively the Fourier transforms of the measure $d\mu_0$ and $d\mu_j$. This is of course consistent with the usual definition in view of identity (8) in Chapter III, (see p. 58).

As a special case of this definition we can assume that $d\mu_0 = f_0 \, dx$ where $f_0 \in L^1(\mathbf{R}^n)$. We say then analogously that $R_j(f_0) \in L^1(\mathbf{R}^n)$ if there exists an $f_j \in L^1(\mathbf{R}^n)$ so that $\hat{f}_j(x) = i \frac{x_j}{|x|} \hat{f}_0(x)$.

COROLLARY 1. *Suppose $d\mu_0$ is a finite measure and all its Riesz transforms $R_j(d\mu)$ are also finite measures, $R_j(d\mu_0) = d\mu_j, j = 1, \ldots, n$. Then there exist $L^1(\mathbf{R}^n)$ functions $f_0, f_1, \ldots, f_n$, so that $d\mu_j = f_j \, dx, j = 0, \ldots, n$.*

The space H^1 is naturally isomorphic with the space of $L^1(\mathbf{R}^n)$ functions f_0 which have the property that $R_j(f_0) \in L^1(\mathbf{R}^n), j = 1, \ldots, n$. The H^1 norm is then equivalent with $\|f_0\|_1 + \sum_{j=1}^{n} \|R_j(f_0)\|_1$.

This space of f_0 consists in effect of the "real parts" of the boundary values of F in H^1. It is tempting to refer to this Banach space as H^1 also, whenever this does not cause confusion with the parent space of F initially defined.

COROLLARY 2. *Suppose $f_0 \in L^1(\mathbf{R}^n)$ and $R_j(f_0) \in L^1(\mathbf{R}^n), R_j(f_0) = f_j, j = 1, \ldots, n$. Then*

$$\sum_{j=0}^{n} \int_{\mathbf{R}^n} \sup_{y > 0} |u_j(x, y)| \, dx \leq A \sum_{j=0}^{n} \|f_j\|_1.$$

3.2.1 We prove Theorem 6 and its corollaries. To begin with let us suppose, as we did earlier, that if $F = (u_0, \ldots, u_n)$, each of the u_j's take their values in a fixed finite-dimensional Hilbert space; we call this inner-product space V_1. We shall also need another finite dimensional Hilbert space V_2 and we consider $V = V_1 \oplus V_2$, their direct sum; V_1 and V_2 are orthogonal complements in V.

We next find a fixed function in $\mathbf{R}_+^{n+1}$, $\Phi(x, y) = (v_0(x, y), v_1(x, y), \ldots, v_n(x, y))$, so that the v_j's take their values in V_2 and

(i) The v_j satisfy the Cauchy-Riemann equations (18).

(ii) $|\Phi(x, y)| = c \, |(x, y + 1)|^{-n-1} = c(|x|^2 + (y + 1)^2)^{-(n)+1/2}$.

To do this let V_2 denote the standard coordinate space of $(n + 1)$ dimensions; let $H(x, y)$ be the harmonic function $|(x, y + 1)|^{-n+1} = (|x|^2 + (y + 1)^2)^{-(n-1)/2}$.* For every j, $0 \leq j \leq n$ set

$$v_j(x, y) = \left(\frac{\partial^2 H}{\partial x_j \, \partial x_k}\right)^n_{k=0}.$$

Then

$$|\Phi(x, y)|^2 = \sum_{j=0}^{n} \sum_{k=0}^{n} \left|\frac{\partial^2 H}{\partial x_j \, \partial x_k}\right|^2.$$

It is therefore easy to verify (i) and (ii) with $c^2 = (n^2 - 1)(n^2 - n)$.

Next, for every $\varepsilon > 0$ we define F_ε by

(26) $$F_\varepsilon(x, y) = F(x, y + \varepsilon) + \varepsilon\Phi(x, y).$$

If we write $F_\varepsilon(x, y) = (u_0^\varepsilon(x, y), \ldots, u_n^\varepsilon(x, y))$, then

$$u_j^\varepsilon(x, y) = u_j(x, y + \varepsilon) + \varepsilon v_j(x, y),$$

and so the u_j^ε take their values in $V_1 \oplus V_2 = V$. Observe also that the components u_j^ε of F_ε satisfy the Cauchy-Riemann equations, and that F_ε is continuous in $\overline{\mathbf{R}}_+^{n+1}$.

Now because of our assumptions on F, and in view of the corollary in §1.2.1 of this chapter, we can assert that each $u_j(x, y)$ is the Poisson integral of a finite measure. Thus it is easy to see that for every fixed ε each $u_j(x, y + \varepsilon)$ tends to zero as $|(x, y)| \to \infty$, for $(x, y) \in \overline{\mathbf{R}}_+^{n+1}$. The same is true for the components of Φ (by property (ii)) and so $|F_\varepsilon(x, y)| \to 0$ as $|(x, y)| \to \infty$ in $\overline{\mathbf{R}}_+^{n+1}$. In addition to this we have $|F_\varepsilon(x, y)|^2 = |F(x, y + \varepsilon)|^2 + \varepsilon^2 |\Phi(x, y)|^2 > 0$, because of the orthogonality of V_1 and V_2 in V. Thus $|F_\varepsilon|^q$ is smooth, and we can apply the lemma in §3.1 with $q = \dfrac{n-1}{n}$, obtaining $\Delta(|F_\varepsilon|^q) \geq 0$ everywhere. Set

$$g_\varepsilon(x) = |F_\varepsilon(x, 0)|^q = (|F(x, \varepsilon)|^2 + \varepsilon^2 |\Phi(x, 0)|^2)^{q/2}.$$

---

* We assume here and in the rest of this chapter that $n > 1$. The argument for $n = 1$ needs certain slight modifications.

Then

(27) $$\int_{\mathbf{R}^n} (g_\varepsilon(x))^p \, dx = \int_{\mathbf{R}^n} |F_\varepsilon(x, 0)| \, dx \le \|F\|_1 + \varepsilon \|\Phi\|_1$$

where $p = 1/q$, and it is important that as a result $p > 1$. Let $g_\varepsilon(x, y)$ be the Poisson integral of g_ε. We claim that

(28) $$|F_\varepsilon(x, y)|^q \le g_\varepsilon(x, y), \, (x, y) \in \overline{\mathbf{R}}_+^{n+1}.$$

To verify (28) observe that both F_ε and g_ε are continuous in $\overline{\mathbf{R}}_+^{n+1}$ and F_ε vanishes at infinity. We have $\Delta(|F_\varepsilon|^q - g_\varepsilon) \ge 0$, and therefore in view of the maximum principle of Appendix C it suffices to verify (27) on the boundary, $y = 0$, for which the equality sign holds. Therefore (28) is proved.

We now select a subsequence of the family $\{g_\varepsilon(x)\}$, $\varepsilon \to 0$, which converges weakly to a function g in $L^p(\mathbf{R}^n)$. Because of (27) we have $\|g\|_p^p \le \|F\|_1$. If $g(x, y)$ denotes the Poisson integral of g then (28) leads to

(28') $$|F(x, y)|^q \le g(x, y)$$

However, by the maximal theorem for Poisson integrals (see Chapter III, p. 62) we have $\sup_{y>0} |F(x, y)|^q \le \sup_{y>0} g(x, y) \le (Mg)(x)$, and therefore

$$\int_{\mathbf{R}^n} \sup_{y>0} |F(x, y)| \, dx \le \int_{\mathbf{R}^n} (Mg(x))^p \, dx \le A_p^p \int_{\mathbf{R}^n} (g(x))^p \, dx \le A_p^p \|F\|_1.$$

This proves the main conclusion of Theorem 6. That $\lim_{y \to 0} F(x, y)$ exists almost everywhere follows because the Poisson integral of a finite measure has this property (see §4.1 of Chapter III). The almost everywhere convergence can also be proved by appealing to (28') which shows that the $F(x, y)$ is almost everywhere non-tangentially bounded, and then by using Theorem 3 of this chapter.

Finally the convergence in the L^1 norm is a consequence of the almost everywhere convergence and the maximal inequality just proved which shows that $|F(x, y)|$ is majorized by a fixed integrable function.

To prove Corollary 1 we merely notice that if $u_j(x, y)$ are the Poisson integrals of $d\mu_j$ and $d\mu_j = R_j(d\mu_0)$, then $F = (u_0, \ldots, u_n)$ satisfies the Cauchy-Riemann equations. Also $\sup_{y>0} \|u_j(x, y)\|_1 = \|d\mu_j\| < \infty$, and so $F \in H^1$. Let $f_j(x) = \lim_{y \to 0} u_j(x, y)$, where the existence of this limit (in the L^1 norm) is guaranteed by the theorem. If $\hat{\mu}_j$ and $\hat{f}_j$ are the Fourier transforms of $d\mu_j$ and f_j respectively, we have $(\mu(x, y))^\wedge = \hat{\mu}_j(x)e^{-2\pi|x|y}$. Thus $\hat{\mu}_j(x)e^{-2\pi|x|y} \to \hat{f}_j(x)$, as $y \to 0$; therefore $\hat{\mu}_j(x) = \hat{f}_j(x)$ and $d\mu_j = f_j \, dx$.

Corollary 2 follows from the above once we notice that if $d\mu_j = f_j(x)\,dx$, then $\|d\mu_j\| = \|f_j\|_1$.

3.3   The area integral and H^1. If $f \in L^p(\mathbf{R}^n)$, we have studied in Chapter IV, and also in part in the present chapter, three related auxiliary expressions, namely

$$g(f)(x), \qquad S(f)(x), \qquad \text{and} \qquad g_\lambda^*(f)(x).$$

These were defined as follows. If $u(x, y)$ is the Poisson integral of f, then

$$g(f)(x) = \left(\int_0^\infty y\,|\nabla u(x, y)|^2\,dy \right)^{1/2}$$

and

$$g_\lambda^*(f)(x) = \left(\int_0^\infty \int_{\mathbf{R}^n} |\nabla u(x - t, y)|^2 \left(\frac{y}{|t| + y} \right)^{n\lambda} y^{1-n}\,dt\,dy \right)^{1/2}$$

To define the area integral S we need to fix the cone entering in its definition. In what follows it will be convenient to fix this cone as the usual right circular cone. Thus we take S to be given by

$$S(f)(x) = \left(\iint_{|t| \leq y} |\nabla u(x - t, y)|^2 y^{1-n}\,dt\,dy \right)^{1/2}.$$

We observed in Chapter IV that $g(f)(x) \leq cS(f)(x) \leq c_\lambda g_\lambda^*(f)(x)$, and proved in Theorems 1 and 2 that

$$B_p \|f\|_p \leq \|g(f)\|_p \leq c_\lambda \|g_\lambda^*(f)\|_p \leq A_{p,\lambda} \|f\|_p$$

if $1 < p < \infty$, when $\lambda \geq 2$.

Our purpose will be to formulate the extension of these facts for $p = 1$ in the context of the H^1 spaces, and then to apply this in §3.4 below.

Let $F \in H^p$, $1 \leq p$, and define $S(F)(x)$ and $g_\lambda^*(F)(x)$ in analogy to the above, namely

$$S(F)(x) = \left(\iint_{|t| \leq y} |\nabla F(x - t, y)|^2 y^{1-n}\,dt\,dy \right)^{1/2}$$

$$g_\lambda^*(F)(x) = \left(\iint |\nabla F(x - t, y)|^2 \left(\frac{y}{|t| + y} \right)^{n\lambda} y^{1-n}\,dt\,dy \right)^{1/2}.$$

Again, of course, $S(F)(x) \leq c_\lambda g_\lambda^*(F)(x)$. Our results are as follows:

THEOREM 7.   *If $F \in H^p$, $1 \leq p < \infty$, then $\|g_\lambda^*(F)\|_p \leq A_{p,\lambda}\|F\|_p$, as long as $p > \dfrac{2}{\lambda}$. In particular $\|S(F)\|_p \leq A_p \|F\|_p$.*

THEOREM 8.   *Suppose $F \in H^p$, $1 \leq p < \infty$. Then $B_p \|F\|_p \leq \|S(F)\|_p$.*

It can be shown that results of this kind are also valid for $p > \dfrac{n-1}{n}$, (see §4.9 below), but in its present formulation the theorem and proof are already fairly typical of the general case. The interest in these theorems is for us the case $p = 1$, since the case in that part of the theorem when $p > 1$ is contained in the results of Chapter IV. In the arguments that follow we therefore limit ourselves to $p = 1$.

3.3.1 Proof of Theorem 7. For the proof of this and the succeeding theorems it will be very convenient to restrict our attention to an appropriate sub-class of H^p which we now define.

Let H_0^p consist of all F such that

  (i) $F \in H^p$
  (ii) F is continuous in $\overline{\mathbf{R}}_+^{n+1}$; it is also rapidly decreasing, that is pF is bounded in $\overline{\mathbf{R}}_+^{n+1}$ for every polynomial p in the $x_1, \ldots, x_n$ and y
  (iii) Property (ii) holds also for every order partial derivative of F in the $x_1, \ldots, x_n$ and y.

What will be important for our purposes is the following fact.

LEMMA. H_0^p *is dense in* H_p, $1 \le p < \infty$.

When $1 < p < \infty$ the proof of the lemma can be given by a rather straightforward limiting argument since the Riesz transforms R_j are continuous in L^p. The lack of continuity for $p = 1$ complicates the situation for that basic special case, and it is just this case which concerns us here. However, to come directly to the presentation of both Theorems 7 and 8 we shall postpone the proof of the lemma to §3.3.3 below.

Assuming the truth of this lemma, we are therefore faced with the task of showing that

$$(29) \qquad \|g_\lambda^*(F)\|_1 \le A_\lambda \|F\|_1, \ \lambda > 2, \ F \in H_0^1.$$

It will be useful at this stage to make a further reduction of our problem by replacing F by $F + \varepsilon\Phi$, in analogy to the proof of §3.2.1. It is to be recalled that the components of the Φ are orthogonal to those of F, that they also satisfy the Cauchy-Riemann equations, and that $|\Phi(x, y)| = c\,|(x, y + 1)|^{-n-1}$. (For later purposes it is to be observed that $|\nabla\Phi| = c'\,|(x, y + 1)|^{-n-2}$.)

The main point of introducing the perturbation $\varepsilon\Phi$ is of course to eliminate the zeroes of F; in fact we have $|F + \varepsilon\Phi|^2 = |F|^2 + \varepsilon^2\,|\Phi|^2 > 0$. Our objective will therefore be (29) with F replaced by $F + \varepsilon\Phi$.

Our argument will be a modification of the presentation for $p > 1$ given in §2 of Chapter IV, and in particular in §2.5.2 of that chapter. Let us draw the parallel in detail.

Lemma 1 (p. 86) will be replaced by the inequality

$$(30) \qquad C_1 |F|^{-1} |\nabla F|^2 \geq \Delta(|F|) \geq c_1 |F|^{-1} |\nabla F|^2$$

which is merely a special case of the lemma in §3.1 of the present chapter.

Lemma 2 (p. 87) will be restated with a slight modification as follows. Suppose G is continuous in $\overline{\mathbf{R}}^+_{n+1}$, G is of class C^2 in $\mathbf{R}^{n+1}_+$, $y \Delta G \in L^1(\mathbf{R}^{n+1}_+, dy\,dx)$, $|G(x, y)| \leq |(x, y)|^{-n-\varepsilon}$, and

$$|\nabla G(x, y)| \leq A\,|(x, y)|^{-n-1-\varepsilon},$$

with $\varepsilon > 0$. Then

$$(31) \qquad \iint_{\mathbf{R}^{n+1}_+} y\,\Delta G\,dx\,dy = \int_{\mathbf{R}^n} G(x, 0)\,dx.$$

Finally for the maximal Lemma 3, and its variant inequality (24), (see p. 92), we shall have

For any fixed μ, μ sufficiently close to 1, $\mu < 1$,

$$(32) \quad |F(x - t, y)| \leq (1 + |t|/y)^{n/\mu} F^*_\mu(x), \text{ with } \int_{\mathbf{R}^n} F^*_\mu(x)\,dx \leq A_\mu \, \|F\|_1.$$

In fact by the inequality (24) of Chapter IV we have

$$|g(x - t, y)| \leq A\left(1 + \frac{|t|}{y}\right)^n M(g)(x),$$

where $g(x, y)$ is the Poisson integral of an arbitrary function $g(x)$. We now invoke the majorization (28′), namely $|F(x, y)|^q \leq g(x, y)$ with $g(x) = |F(x, 0)|^q$ where we choose $q = \mu < 1$. Then

$$|F(x - t, y)| = A^{1/\mu}(1 + |t|/y)^{n/\mu} M^{1/\mu}(g),$$

and (32) is proved with

$$F^*(x) = A^{1/\mu} M^{1/\mu}(g)(x),$$

when we recall that if $p = 1/q$, then $\|g\|_p^p \leq \|F\|_1$.

Now whatever fixed λ, $\lambda > 2$ we start with, we can find a $\lambda' > 1$, and a $\mu < 1$, μ sufficiently close to 1, so that $\lambda' = \lambda - 1/\mu$. We set

$$I^*(x) = \int_{\mathbf{R}^{n+1}_+} y^{1-n}\left(\frac{y}{y + |t|}\right)^{\lambda' n} \Delta G(x - t, y)\,dt\,dy,$$

where $G(x, y) = |F(x, y) + \varepsilon \Phi(x, y)|$. Observe (as in §2.5.2 of Chapter

IV) that

$$\int_{\mathbf{R}^n} I^*(x)\, dx = c_{\lambda'} \iint_{\mathbf{R}^{n+1}_+} y\, \Delta G(t, y)\, dt\, dy = c_{\lambda'} \int_{\mathbf{R}^n} G(t, 0)\, dt$$

$$= c_{\lambda'} \int_{\mathbf{R}^n} |F(x, 0) + \varepsilon\Phi(x, 0)|\, dx,$$

as soon as we verify that G satisfies the requirements given for (31). Now since $F \in H^1_0$, it is rapidly decreasing together with all its partial derivatives; also $|\Phi(x, y)| = c\, |(x, y + 1)|^{-n-1}$ and $|\nabla\Phi| = c'\, |(x, y + 1)|^{-n-2}$. Therefore by (30) for $F + \varepsilon\Phi$ in place of F we have

$$\Delta\, |F + \varepsilon\Phi| \le A\, |(x, y + 1)|^{-n-3}.$$

This shows $y\, \Delta G \in L^1(\mathbf{R}^{n+1}_+, dy\, dx)$. Similarly

$$|G(x, y)| \le A\, |(x, y + 1)|^{-n-1}$$

and $|\nabla G(x, y)| \le A\, |(x, y + 1)|^{-n-2}$.

Next by (30) again

$$g_\lambda(F + \varepsilon\Phi)^2(x) \le c_1^{-1} \iint_{\mathbf{R}^{n+1}_+} y^{1-n}\left(\frac{y}{y + |t|}\right)^{\lambda n} G\, \Delta G\, dt\, dy.$$

The latter is majorized by $c_1^{-1} I^*(x) F^*_{\mu,\varepsilon}(x)$, because of (32); here

$$\sup_{t,y} \left(\frac{y}{y + |t|}\right)^{n/\mu} G(x - t, y) = F^*_{\mu,\varepsilon}(x)$$

and

$$\|F^*_{\mu,\varepsilon}\|_1 \le A_\mu \|F + \varepsilon\Phi\|_1.$$

Thus by Schwarz's inequality

$$\int_{\mathbf{R}^n} g^*_\lambda(F + \varepsilon\Phi)(x)\, dx \le c_1^{-1/2} c_{\lambda'}^{1/2} \|F + \varepsilon\Phi\|_1.$$

Letting $\varepsilon \to 0$ proves our desired inequality (29) and thus Theorem 7.

3.3.2 *Proof of Theorem 8.* In addition to the operator S we want to consider a whole family of variants, defined for any q, with $q > \dfrac{n-1}{n}$. We set

$$\mathfrak{S}_q(F)(x) = \left(\iint_{\Gamma(x)} y^{1-n} \Delta(|F|^q)(y, t)\, dy\, dt\right)^{1/q}$$

where $\Gamma(x)$ is our basic cone,

$$\Gamma(x) = \{(t, y) : |x - t| < y\}.$$

$\mathfrak{S}_q(F)$ is well-defined, since

$$\Delta(|F|)^q \geq 0 \quad \text{for} \quad q \geq \frac{n-1}{n}$$

by the lemma in §3.1. We make the following observations about $\mathfrak{S}_q$.

(α) $$\mathfrak{S}_2(F) = \sqrt{2}\, S(F)$$

because $\Delta |F|^2 = 2\,|\nabla F|^2$, by the lemma in §3.1. In addition $\mathfrak{S}_q$ satisfies a certain convexity property in q, namely

(β) $$\mathfrak{S}_q(F) \leq c(\mathfrak{S}_{q_0}(F))^{1-\theta}(\mathfrak{S}_{q_1}(F))^\theta$$

whenever $q_j > (n-1)/n$, and $1/q = (1-\theta)/q_0 + \theta/q_1$ where $0 \leq \theta \leq 1$. c is a constant which depends on q_0, q_1 and θ, but is independent of F. Inequality (β) is a consequence of Hölder's inequality and of the lemma in §3.1. This lemma states, in effect, that $\Delta\,|F|^q$ is comparable with

$$|F|^{q-2}\,|\nabla F|^2$$

and so, in effect, $\mathfrak{S}_q(F)$ behaves like a qth norm of $|F|$.

After these preliminaries we come to the main point. In proving the inequality

(33) $$B\,\|F\|_1 \leq \|S(F)\|_1, \quad F \in H^1$$

we shall make the same reduction we did in the proof of Theorem 7. That is, we assume that in place of F we have $F + \varepsilon\Phi$, where $F \in H_0^1$, and Φ is as above. In order to simplify the notation we shall assume that F already has the required form. Write now $G(x, y) = |F(x, y)|$ and we apply the identity $\iint_{R_+^{n+1}} y\,\Delta G\,dx\,dy = \int_{R^n} G(x, 0)\,dx$ again, which we already saw was perfectly legitimate. In fact

$$\int_{R^n} \mathfrak{S}_1(F)\,dx = \int_{R^n} \left(\iint_{\Gamma(x)} \Delta\,|F(t, y)|\,y^{1-n}\,dy\,dt \right) dx$$

$$= c \iint_{R_+^{n+1}} y\,\Delta\,|F(x, y)|\,dx\,dy = c \int_{R^n} |F(x, 0)|\,dx.$$

Thus, $\|F\|_1 = c^{-1}\,\|\mathfrak{S}_1(F)\|_1$ and hence by (β) we have using Hölder's inequality

(34) $$\|F\|_1 \leq c'\,\|(\mathfrak{S}_2(F))^{\theta'}(\mathfrak{S}_\eta(F))^{1-\theta'}\|_1 \leq c'\,\|\mathfrak{S}_2(F)\|_1^{\theta'}\,\|\mathfrak{S}_\eta(F)\|_1^{1-\theta'}$$

where η is an arbitrary, but fixed exponent in the range $\dfrac{n-1}{n} < \eta < 1$, and θ' is chosen accordingly.

If we can prove the inequality

(35) $$\|\mathfrak{S}_\eta(F)\|_1 \leq c\,\|F\|_1$$

then substituting this in (34) gives $\|F\|_1 \le c' \|\mathfrak{S}_2(F)\|_1$, which is (33). We turn our attention therefore to (35).

The question whether $\mathfrak{S}_\eta(F) \in L^1(\mathbf{R}^n)$ is the same as that whether $\mathfrak{S}_\eta(F)^\eta \in L^{1/\eta}(\mathbf{R}^n)$, and since $\eta < 1$, the exponent $1/\eta$ is greater than one. Let r be the index conjugate to $1/\eta$, namely $1/r + \eta = 1$. Then what we need to do is estimate

$$\sup_\varphi \int_{\mathbf{R}^n} \mathfrak{S}_\eta(F)^\eta(x)\varphi(x)\, dx$$

where φ ranges over an (appropriate) dense subset of elements in $L^r(\mathbf{R}^n)$, with $\|\varphi\|_r \le 1$.

For our φ we choose the non-negative C^∞ functions on $\mathbf{R}^n$, each with compact support, and so that $\|\varphi\|_r \le 1$. Now

$$\mathfrak{S}_\eta(F)^\eta(x) = \iint_{\Gamma(x)} \Delta(|F|^\eta(t, y))y^{1-n}\, dt\, dy$$

$$= \iint_{\mathbf{R}^{n+1}_+} \psi(x, t, y) \Delta(|F|^\eta(t, y))y^{1-n}\, dt\, dy$$

with $\psi(x, t, y)$ the characteristic function of the cone $\Gamma(x) = \{|x - t| < y\}$. If $P_y(x)$ is the Poisson kernel then $\psi(x, t, y)y^{-n} \le cP_y(x)$, therefore

(36) $$\int_{\mathbf{R}^n} \mathfrak{S}_\eta(F)^\eta(x)\varphi(x)\, dx \le c \iint_{\mathbf{R}^{n+1}_+} \varphi(x, y) \Delta |F(x, y)|^\eta\, y\, dx\, dy$$

with $\varphi(x, y)$ the Poisson integral of φ, that is $\varphi(x, y) = (P_y * \varphi)(x)$. We next observe the following differential identity: if $\Delta(A) = 0$

$$A\,\Delta(B) = \Delta(AB) - 2\sum_{j=0}^n \frac{\partial A}{\partial x_j}\frac{\partial B}{\partial x_j}$$

Set $A = \varphi(x, y)$, $B = |F(x, y)|^\eta$, $(y = x_0)$, then the right side of (36) is majorized by a constant multiple of

$$\iint_{\mathbf{R}^{n+1}_+} \Delta(\varphi |F|^\eta)y\, dy\, dx + 2\iint_{\mathbf{R}^{n+1}_+} |\nabla |F|^\eta| \cdot |\nabla\varphi|\, y\, dy\, dx.$$

The integral on the left of the above can be evaluated by the identity (31) whose application is again legitimate in view of the assumed form of F. Its value is then $\int_{\mathbf{R}^n} |F(x, 0)|^\eta\, \varphi(x, 0)\, dx$, and this is majorized by

$$\|F\|_1^\eta \|\varphi\|_r \le \|F\|_1^\eta.$$

The integral on the right equals a constant multiple of

(37) $$\int_{\mathbf{R}^n} \left\{ \iint_{\Gamma(x)} |\nabla |F|^\eta(t, y)| \cdot |\nabla\varphi(t, y)|\, y^{1-n}\, dt\, dy \right\} dx.$$

Now $|\nabla |F|^{\eta}| \le$ constant $|F|^{\eta-1} |\nabla F|$. Also by the maximal inequality ((32), p. 226) we have $\sup\limits_{(t,y)\in\Gamma(x)} |F(t, y)| \le cF^*(x)$, with $\int_{\mathbf{R}^n} F^*(x)\,dx \le A \|F\|_1$. Finally $\Delta |F|^{\eta} \ge$ constant $|F|^{\eta-2} |\nabla F|^2$. Altogether then, using Schwarz's inequality, we see that (37) is majorized by a constant multiple of

$$\int_{\mathbf{R}^n} (F^*(x))^{\eta/2}(\mathfrak{S}_\eta(F))^{\eta/2}S(\varphi)\,dx$$

To this integral we apply Hölder's inequality in the form

$$\int_{\mathbf{R}^n} A_1 A_2 A_3\,dx \le \|A_1\|_{p_1} \|A_2\|_{p_2} \|A_3\|_{p_3},$$

where $1/p_1 + 1/p_2 + 1/p_3 = 1$, with $p_1 = p_2 = 2/\eta$, and $p_3 = r$ (recall $1/r + \eta = 1$). The majorant is then $\|F^*\|_1^{\eta/2} \|\mathfrak{S}_\eta(F)\|_1^{\eta/2} \|S(\varphi)\|_r$ which in turn is dominated by a constant multiple of $\|F\|_1^{\eta/2} \|\mathfrak{S}_\eta(F)\|_1^{\eta/2}$; this is true on two counts, first since $\|S(\varphi)\|_r \le A \|\varphi\|_r \le A$ by the results of Chapter III (here $r > 1$), and also $\|F^*\|_1 \le A \|F\|_1$ as already pointed out.

In summary we have obtained the following:

$$\|\mathfrak{S}_\eta(F)\|_1^{\eta} = \sup_\varphi \int_{\mathbf{R}^n} \mathfrak{S}_\eta(F)^{\eta}\varphi\,dx$$

$$\le \|F\|_1^{\eta} + A \|F\|_1^{\eta/2} \|\mathfrak{S}_\eta(F)\|_1^{\eta/2}$$

This implies inequality (35) and therefore inequality (33), giving the desired conclusion for our F of the special form $F + \varepsilon\Phi$, with $F \in H_0^1$. The limiting passage to general F in H^1 is then routine.

3.3.3 *Proof of the density lemma.*

We dispose here of the lemma which was stated without proof in §3.3.1 above. We are going to prove that H_0^1 is dense in H^1.

We suppose $f \in L^1(\mathbf{R}^n)$, and f has the property that the $R_j(f) = f_j$ all belong to $L^1(\mathbf{R}^n)$. (The assumption means of course that $i\dfrac{x_j}{|x|}\hat{f}(x)$ each are Fourier transforms of L^1 functions; observe that as a consequence $\hat{f}(0) = \hat{f}_j(0) = 0, j = 1, \ldots, n$.)

Our proof of the lemma will be in two steps. First we shall see that for each f of the above kind we can find a sequence $\{f^{(k)}\}_k \in L^1$, so that $\hat{f}^{(k)}$ has support which is compact and at a positive distance from the origin, and so that $f^{(k)} \to f$ and $R_j(f^{(k)}) \to R_j(f)$ in the L^1 norm, as $k \to \infty$.

Let us choose a fixed C^∞ function in $\mathbf{R}^n$ with compact support, Φ, with the additional property that $\Phi(x) = 1$, for $|x| \le 1$. For each $\delta > 0$, define the transformation T_δ on $L^1(\mathbf{R}^n)$ by $(T_\delta f)^\wedge(x) = \Phi(x/\delta)\hat{f}(x)$.

Clearly

$$T_\delta(f)(x) = \delta^n \int_{\mathbf{R}^n} f(x - y)\varphi(\delta y) \, dy$$

with $\varphi = \Phi$. It is also to be observed that $\|T_\delta(f)\|_1 \leq A \|f\|_1$ with A independent of δ or f. Consider $T_N(I - T_\varepsilon)f$. As $N \to \infty$ and $\varepsilon \to 0$ it converges to f in $L^1(\mathbf{R}^n)$ norm, if f belongs to the closed subspace L^1_0 of $L^1(\mathbf{R}^n)$ consisting of functions whose Fourier transform vanishes at the origin. To prove this assertion it suffices, in view of the uniform boundedness of the operators $T_N(I - T_\varepsilon)$, to verify it for a dense subset of this subspace. An appropriate such subset consists of $L^1(\mathbf{R}^n)$ functions whose Fourier transform has support which is compact and is at a positive distance from the origin. For those f, obviously $T_N(I - T_\varepsilon)f = f$ for N sufficiently large and ε sufficiently small. That such f are dense in the closed subspace L^1_0 can be proved directly by elementary computation, or one can appeal to Wiener's theorem characterizing the maximal ideals of $L^1(\mathbf{R}^n)$. In any case then we take $T_k(I - T_{1/k})f = f^{(k)}$, so $f^{(k)} \to f$ in L^1 norm. Also $R_j(f^{(k)}) = T_k(I - T_{1/k})R_j(f)$, so $R_j(f^{(k)}) \to R_j(f)$ and we have achieved the first step.

We may assume now that $f \in L^1(\mathbf{R}^n)$ and that the support of $\hat{f}$ is a compact set K disjoint from the origin. Consider a standard regularization of $\hat{f}$ given by $\hat{f} * k^n\psi(kx)$ where ψ is C^∞ with compact support, $\int_{\mathbf{R}^n} \psi \, dx = 1$. Observe that the $\hat{f} * k^n\psi(kx)$ belong to C^∞, and if k is sufficiently large they have a common support contained in a compact set K' which is at a positive distance from the origin. If $\Psi^\wedge(x) = \psi(x)$ then $\Psi(0) = 1$ and $\hat{f} * k^n\psi(kx)$ is the Fourier transform of $f(x)\Psi(x/k) = f_k$ which clearly converges to f in $L^1(\mathbf{R}^n)$ norm. We claim that $R_j(f_k) \to R_j(f)$ also. In fact for each compact set K' of the type described there is a C^∞ function $m_j(x)$ so that $m_j(x) = i\dfrac{x_j}{|x|}$ for $x \in K'$. Let M_j be the L^1 function determined by $\hat{M}_j(x) = m_j(x)$. Then

$$M_j * f_k = R_j(f_k).$$

The convergence of $R_j(f_k)$ to $R_j(f)$ in the $L^1(\mathbf{R}^n)$ norm is then obvious. It is also apparent that the element of H^1 whose boundary values are $(f_k, R_1(f_k), \ldots, R_n(f_k))$ is in fact in H^1_0. This proves the lemma.

It will be useful to record the essence of the lemma as follows. Consider the Banach space $\{f \in L^1(\mathbf{R}^n); R_j(f) \in L^1(\mathbf{R}^n), j = 1, \ldots, n\}$, with norm

$$\|f\| = \|f\|_1 + \sum_{j=1}^n \|R_j(f)\|_1.$$ Then we have proved above:

COROLLARY. *The collection of f whose Fourier transforms are C^∞ and which have compact support strictly disjoint from the origin is dense in the whole Banach space.*

Let us call by H_{00}^1 the corresponding subspace of F in H^1. We have of course $H_{00}^1 \subset H_0^1 \subset H^1$, and we have shown that H_{00}^1 is dense in H^1.

3.4 Multiplier transformations in H^1. After all the exertions of the previous pages in dealing with the functions $S(F)$ and $g_\lambda^*(F)$ we come now to some results which indicate that these efforts were justified. We intend to show that many of the singular integral operators (more broadly: multiplier transformations) studied in Chapters II, III, and IV extend to bounded operators on H^1.

We require a definition. Let $m(x)$ be a function defined on $\mathbf{R}^n$. Suppose that whenever $F \in H^1$ we can find another element of H^1, namely $\tilde{F}$, with the property that if $\lim_{y \to 0} F(x, y) = F(x, 0) = (f_0(x), f_1(x), \ldots, f_n(x))$, and $\tilde{F}(x, 0) = (\tilde{f}_0(x), \ldots, \tilde{f}_n(x))$, we have

$$(38) \qquad (\tilde{f}_j)^\wedge(x) = m(x)\hat{f}_j(x), \qquad j = 0, 1, \ldots, n.$$

The function m will then define a mapping T_m of H^1 to itself, given by $\tilde{F} = T_m(F)$. If T_m is bounded on H^1 [*] we shall say that m is a multiplier for H^1. The matter can be put in another way. We can say that m is a multiplier for H^1 if there exists a constant A with the following property: whenever $f_0, f_1, \ldots, f_n$ are $L^1(\mathbf{R}^n)$ and $f_j = R_j(f_0)$, then there exists $L^1(\mathbf{R}^n)$ functions $\tilde{f}_0, \tilde{f}_1, \ldots, \tilde{f}_n$ defined by (38) so that

$$\sum_{j=0}^n \|\tilde{f}_j\|_1 \leq A \sum_{j=0}^n \|f_j\|_1.$$

Our theorem is as follows:

THEOREM 9. *Suppose that $m(x)$ is of class $C^{(n+1)}$ in the complement of the origin of $\mathbf{R}^n$. Assume that*

$$(39) \qquad \sup_{0 < R < \infty} R^{2|\alpha|-n} \int_{R \leq |x| \leq 2R} \left|\left(\frac{\partial}{\partial x}\right)^\alpha m(x)\right|^2 dx \leq B$$

for every differential monomial $\left(\dfrac{\partial}{\partial x}\right)^\alpha$, where $\alpha = (\alpha_1, \alpha_2, \ldots, \alpha_n)$ and $|\alpha| = \alpha_1 + \alpha_2 \cdots + \alpha_n \leq n + 1$. Then m is a multiplier for H^1; that is,

$$(40) \qquad \|T_m(F)\|_1 \leq A \|F\|_1.$$

---

* The assumption that T_m is bounded on H^1 is strictly speaking unnecessary since it follows automatically from the closed graph theorem and the assumption that T_m is defined on all of H^1.

Among the operators that are covered by this theorem are the following:
(i) The operators of the kind

$$f \to \lim_{\varepsilon \to 0} \int_{|y| \geq \varepsilon} \frac{\Omega(y)}{|y|^n} f(x - y)\, dy$$

if Ω is a sufficiently smooth homogeneous function of degree 0 with vanishing mean-value on the unit sphere. Operators of this type arose in §4 of Chapter II, and contain the class of operators described in Theorem 6 (p. 75) of Chapter III. The latter class includes of course the Riesz transforms and the algebra they generate.

(ii) A substantial sub-class of the multipliers arising in the L^p multiplier theorem (Theorem 3 and its corollary) in Chapter IV (see p. 96).

Incidentally the proof will show that $A \leq CB$, where C is some absolute constant; thus the theorem leads directly to a slight extension of itself where the condition of $C^{(n+1)}$ is relaxed to some condition involving differentiality in the L^2 context. We shall however, not pursue this refinement.

3.4.1 The theorem will be a direct consequence of the following lemma:

LEMMA. *Suppose F belongs to the dense subspace* H_{00}^1. *Then* $T_m(F) \in H^1$ *and*

$$(41) \qquad S(T_m(F))(x) \leq A' g_\lambda^*(F)(x), \quad \text{with} \quad \lambda = \frac{2n + 2}{n}$$

The lemma is proved in much the same way as the corresponding lemma in §3.2 of Chapter IV.

First since $F \in H_{00}^1$, each $\hat{f}_j$ is C^∞ and has compact support away from the origin. Hence in view of the fact that m is $C^{(n+1)}$ away from the origin it follows that $m(x)\hat{f}_j(x)$ is of class $C^{(n+1)}$ and has compact support, and therefore is the Fourier transform of an L^1 function. Thus by definition (38) $T_m(F) \in H^1$.

Let us define the harmonic function $M(x, y)\, y > 0$, by

$$M(x, y) = \int_{\mathbf{R}^n} e^{-2\pi i x \cdot t} e^{-2\pi |t| y} m(t)\, dt.$$

Then following the argument in §3.3 of Chapter IV we have

$$\tilde{F}(x, y) = (T_m F)(x, y) = \int_{\mathbf{R}^n} M(t, y_1) F(x - t, y_2)\, dt, \qquad y = y_1 + y_2,$$

and therefore

$$(42) \qquad |\nabla^{(k+1)} \tilde{F}(x, y)| \leq \int_{\mathbf{R}^n} |\nabla^k M(t, y/2)|\, |\nabla F(x - t, y/2)|\, dt$$

where ∇^k denotes the kth gradient.

The conditions (39) may be reinterpreted in terms of M. Following the same reasoning used in Chapter IV we get

(43) $$|\nabla^k M(t, y)| \leq B' y^{-n-k}$$

(43') $$\int_{\mathbf{R}^n} |t|^{2k} |\nabla^k M(t, y)|^2 \, dt \leq B' y^{-n}, \quad \text{for} \quad k = n + 1.$$

In proving the inequality (41) it suffices by translation invariance to consider the origin only. Inserting (43) and (43') in (42) and using Schwarz's inequality gives (with $k = n + 1$)

$$|\nabla^{(k+1)} \tilde{F}(x, y)|^2 \leq A y^{-n-2k} \int_{|t| \leq 2y} |\nabla F(x - t, y/2)|^2 \, dt$$

$$+ A y^{-n} \int_{|t| \geq 2y} |\nabla F(x - t, y/2)|^2 |t|^{-2k} \, dt$$

$$= I_1(x, y) + I_2(x, y)$$

Thus

$$\iint_{|x| \leq y} |\nabla^{k+1} \tilde{F}(x, y)|^2 \, y^{2k-n+1} \, dx \, dy \leq \sum_{j=1}^{2} \iint_{|x| \leq y} I_j(x, y) y^{2k-n+1} \, dx \, dy$$

To evaluate

$$\iint_{|x| \leq y} I_1(x, y) y^{2k-n+1} \, dx \, dy$$

observe that $|t| \leq 2y$ and $|x| \leq y$ implies that $|x - t| \leq 3y$. Thus a simple computation shows that this integral is dominated by a constant multiple of

$$\iint_{|x'| \leq 6y} |\nabla F(x', y)|^2 \, y^{1-n} \, dx' \, dy.$$

Similarly the integral

$$\iint_{|x| \leq y} I_2(x, y) y^{2k-n+1} \, dx \, dy$$

is dominated by a constant multiple of

$$\iint_{|x'| \geq 2y} |\nabla F(x', y)|^2 \, y^{1-n} \left(\frac{y}{|x'|}\right)^{2k} \, dx' \, dy$$

Both of these results are of course majorized by a constant multiple of $(g_\lambda^*(F)(0))^2$, where $\lambda = \dfrac{2k}{n}$. Since $k = n + 1$, we get $\lambda = \dfrac{2n + 2}{n}$. Altogether we have verified that

$$\iint_{|x| \leq y} |\nabla^{(k+1)} \tilde{F}(x, y)|^2 \, y^{2k-n+1} \, dx \, dy \leq A'[g_\lambda^*(F)(0)]^2.$$

We now invoke Lemma 2 in §2.5.2 of the present chapter (see p. 216). As a result we get

$$S(\tilde{F})(0) = S(T_m F)(0) \leq A g_\lambda^*(F)(0)$$

and after a translation by an arbitrary x this is (41), and the lemma is proved.

The proof of Theorem 9 is then concluded as follows. Theorems 7 and 8 now show immediately that $\|T_m(F)\|_1 \leq A \|F\|_1$, whenever $F \in H_{00}^1$, with A independent of F.

The bounded operator T_m defined on H_{00}^1 then has an abstract extension as a bounded operator to all of H^1. It is then a trivial matter to see by a limiting argument that this extension satisfies the defining property (38). Theorem 9 is therefore completely proved.

4. Further results

4.1 The results of §1 have analogues when the upper half-plane $\mathbf{R}_+^{n+1}$ is replaced by the unit ball B^{n+1} in $\mathbf{R}^{n+1}$ with its boundary the unit sphere S^n. Let $\mathscr{P}(x, y)$ be the spherical Poisson kernel $\mathscr{P}(x, y) = c_n \dfrac{1 - |x|^2}{|x - y|^{n+1}}$, $(|x| < 1, |y| = 1)$, and $d\sigma(y)$ be the induced Lebesgue measure on S^n. For every $f \in L^p(S^n, d\sigma)$, its Poisson integral is $u(x) = \int_{S^n} \mathscr{P}(x, y)f(y)\, d\sigma(y)$. One then has the following. Suppose u is harmonic in B^{n+1}.

(a) u is the Poisson integral of an L^p, $1 < p \leq \infty$ function if and only if
$$\sup_{0 < r < 1} \left(\int_{S_n} |u(ry)|^p\, d\sigma(y)\right)^{1/p} < \infty$$

(b) u is the Poisson integral of a finite measure on S if and only if
$$\sup_{0 < r < 1} \int_{S^n} |u(ry)|\, d\sigma(y) < \infty$$

(c) u is the Poisson integral of a finite positive measure on S^n if and only if $u \geq 0$ in B^{n+1}.

Under any of the three above conditions the (appropriately defined) non-tangential limits of u exist almost everywhere on S^n.

4.2 Suppose $u(x, y)$ is harmonic in $\mathbf{R}_+^{n+1}$. Then $u(x, y) \geq 0$ if and only if it is of the form

$$u(x, y) = \int_{\mathbf{R}^n} P_y(x - t)\, d\mu(t) + ay, \qquad a \geq 0$$

where $d\mu$ is non-negative Borel measure for which

$$\int_{\mathbf{R}^n} \frac{d\mu(t)}{(1 + |t|^2)^{(n+1)/2}} < \infty. \quad \text{(Hint: use §4.1(c)).}$$

4.3 The fact that non-tangential boundedness implies at almost every point the existence of non-tangential limits (Theorem 3) has been generalized in several directions.

(a) It suffices to assume that at the points in question, the given harmonic function is non-tangentially bounded from below. (Carleson [1])

(b) These results can be extended to domains which have a Lipschitz boundary. (R. Hunt and Wheeden [1])

4.4 Let $d\mu$ be any non-negative measure on $\mathbf{R}_+^{n+1}$ with the property that $\mu(Q) \leq c(\text{diam } Q)^n$ for any cube Q in $\mathbf{R}_+^{n+1}$ which touches the boundary, $\mathbf{R}^n$. Let $u(x, y)$ be the Poisson integral of an $L^p(\mathbf{R}^n)$ function. Then

$$\left(\iint_{\mathbf{R}_+^{n+1}} |u(x, y)|^p \, d\mu \right)^{1/p} \leq cA_p \left(\int_{\mathbf{R}^n} |f(x)|^p \, dx \right)^{1/p}, \qquad 1 < p \leq \infty.$$

See Carleson [2]; also Hörmander [4].

That this result is in reality a consequence of the usual maximal theorem (Theorem 1, Chapter I) can be seen as follows. Let $\Phi(x, y)$ and $\varphi(x)$ be non-negative functions in $\mathbf{R}_+^{n+1}$ and $\mathbf{R}^n$ related by the non-tangential inequality $\sup_{|x-x'|<y} \Phi(x', y) \leq \varphi(x)$. Then $\mu\{(x, y): \Phi > \alpha\} \leq cm\{x: \varphi > \alpha\}$ for each α, and as a result

$$\iint_{\mathbf{R}_+^{n+1}} \Phi^p \, d\mu \leq c \int_{\mathbf{R}^n} \varphi^p(x) \, dx$$

Once this is observed, we need only take $\Phi(x, y) = |u(x, y)|$, $\varphi(x) = AM(f)(x)$. The non-tangential inequality $\sup_{|x-x'|<y} \Phi(x', y) \leq \varphi(x)$ is then contained in Theorem 1, and the result follows from the L^p inequality for $M(f)$.

4.5 There is another type of maximal inequality for Poisson integrals. It has some of the features of the g_λ^* function in that there is a critical L^p class depending on λ. Define $\mathscr{M}_\lambda$ by

$$\mathscr{M}_\lambda(f)(x) = \sup_{y>0} \left(\int_{\mathbf{R}^n} |u(x - t, y)|^2 \, y^{-n} \left(\frac{y}{|t| + y} \right)^{n\lambda} dt \right)^{1/2}$$

Notice that $\mathscr{M}_\lambda(f)(x) \geq c_\lambda M(f)(x)$, if $f \geq 0$. Let $1 < \lambda \leq 2$.

(a) If $p = 2/\lambda$ the mapping $f \to \mathscr{M}_\lambda(f)$ is of weak-type (p, p)

(b) If $p > 2/\lambda$

$$\|\mathscr{M}_\lambda(f)\|_p \leq A_{p,\lambda} \|f\|_p$$

(c) If $p < 2/\lambda$, there exists an $f \in L^p(\mathbf{R}^n)$ so that $\mathscr{M}_\lambda(f)(x) = \infty$ everywhere. See Stein [4].

4.6 Let H be the harmonic function $\dfrac{|(x, y)|^{-n+1}}{-n + 1}$, $n > 1$, (with $|(x, y)| = (y^2 + x_1^2 \cdots + x_n^2)^{1/2}$).

Set $F = \nabla H = \left(\dfrac{\partial H}{\partial y}, \dfrac{\partial H}{\partial x_1}, \ldots \dfrac{\partial H}{\partial x_n} \right)$. Then $|F| = |(x, y)|^{-n}$, and $\Delta |F|^q = nq[nq - n + 1] |(x, y)|^{-nq-2}$. Thus if $q > 0$, $\Delta |F|^q \geq 0$ only if $q \geq \left(\dfrac{n-1}{n} \right)$.

4.7 The L^p inequalities for fractional integration (see §1.2 in Chapter VI) are valid for $p = 1$ in the context of H^1. If $f \in L^1(\mathbf{R}^n)$ and $R_j(f) \in L^1(\mathbf{R}^n)$, $j = 1, \cdots, n$. Then

$$I_\alpha(f) = \frac{1}{\gamma(\alpha)} \int_{\mathbf{R}^n} \frac{f(y)}{|x - y|^{n-\alpha}} \, dy \in L^q(\mathbf{R}^n)$$

and $\|I_\alpha(f)\|_q \leq A_\alpha(\|f\|_1 + \sum_{j=1}^{n} \|R_j(f)\|_1)$ with $1/q = 1 - \alpha/n$, and $0 < \alpha < n$. See Stein and Weiss [2].

4.8 Suppose $f \geq 0$, $f \in L^1(\mathbf{R}^n)$. Define $R_j(f)$ by

$$\lim_{\varepsilon \to 0} c_n \int_{|y| \geq \varepsilon} \frac{y_j}{|y|^{n+1}} f(x - y) \, dy,$$

which we know exists almost everywhere (see Theorem 4, Chapter I).

Suppose each $R_j(f)$ is integrable on each compact set. Then $|f| \log (2 + |f|)$ is also integrable on each compact set. (*Hint:* use Corollary 2 of Theorem 6 in §3.2 of this chapter, together with §5.2(c) in Chapter I. For details see Stein [12].)

4.9 Theorems 6 to 9 have extensions to H^p, for $p > \dfrac{n-1}{n}$ by much the same methods as those given here for $p = 1$.

(a) If $F \in H^p$, then $\lim_{y \to 0} F(x + iy) = F(x)$ exists almost everywhere,

$$\int_{\mathbf{R}^n} |F(x + iy) - F(x)|^p \, dx \to 0,$$

as $y \to 0$, and $\int_{\mathbf{R}^n} \sup_{y>0} |F(x + iy)|^p \, dx \leq A_p^p \|F\|_p^p$. Stein and Weiss [2].

(b) $B_p \|F\|_p \leq \|S(F)\|_p \leq A_p \|F\|_p$. Calderón [6], Segovia [1] and also Gasper [1]

(c) $\qquad\qquad \|g_\lambda^*(F)\|_p \leq A_{p,\lambda} \|F\|_p$, if $p > 2/\lambda$

(d) m is a multiplier of H^p if $|m(x)| \leq B$ and

$$\sup_{0 < R < \infty} R^{2|\alpha|-n} \int_{R \leq |x| \leq 2R} \left| \left(\frac{\partial}{\partial x} \right)^\alpha m(x) \right|^2 dx \leq B$$

where $|\alpha| \leq k$, and $k > n/p$, $(p \leq 2)$. For (c) and (d) see Stein [9].

4.10 Suppose $a(x)$ is defined on $\mathbf{R}^1$ and $|a(x) - a(y)| \le M |x - y|$, $x, y \in \mathbf{R}^1$.

Let $T_\varepsilon(f)(x) = \displaystyle\int_{|x-y| \ge \varepsilon} \left\{ \dfrac{a(x) - a(y)}{(x - y)^2} \right\} f(y)\, dy.$ Then $\| T_\varepsilon(f) \|_p \le A_p \| f \|_p$,
$1 < p < \infty$, with A_p independent of ε. This result can be proved as an application of Theorem 8. For details and further results of this kind see Calderón [6].

4.11 Let $u(x, y)$ be harmonic in $\mathbf{R}^{n+1}_+$. Then at *an individual* point x^0 the three conditions

  (1) u is non-tangentially bounded at x^0;

  (2) u has a non-tangential limit at x^0; and

  (3) $\iint_{\Gamma(x^0)} |\nabla u|^2 \, y^{1-n} \, dx \, dy < \infty$

may all be independent, except that of course (2) implies (1).

To see this take the case $n = 1$. Consider the three functions $u(x, y)$ given by $e^{-i/z}$, $e^{i\gamma \log z}$, and $(\log z)^{1-\delta}$ where γ is real, $1/2 < \delta < 1$; each are single valued and holomorphic in $z = x + iy$ for $(x, y) \in \mathbf{R}^2_+$. Let $x^0 = 0$. For the first we have (2) (and (1)) but not (3); for the second we have (1), but not (2) and (3); and for the third function we have (3), but not (2) or (1).

4.12 That many of the results of this chapter do not hold when the non-tangential approach is replaced by perpendicular convergence to the boundary can be seen by the use of the following theorem.

THEOREM. *Let E be a set of the first category (and possibly of full measure) in* $\mathbf{R}^1$. *Suppose $\Phi(x, y)$ is any continuous function in $\mathbf{R}^2_+$. Then there exists an analytic function $F(z)$, $z = x + iy$ in $\mathbf{R}^2_+$ so that* $\lim_{y \to 0} \{ F(x + iy) - \Phi(x, y) \} = 0$ *for each $x \in E$.* (See Bagemihl and Seidel [1]).

  (a) By the aid of this theorem, choosing an appropriate Φ which is bounded but oscillates as $y \to 0$, we see that the analogue of Theorem 3 in §1.3 for perpendicular approach is not valid. Similarly by choosing a Φ which in addition is purely real-valued, we see that the analogue of Theorem 5 in §2.5 also fails.

  (b) For fixed $\Psi(y)$, let $\Phi(x, y) = \Psi'(y)$, and suppose that $f(z)$ is analytic in $\mathbf{R}^2_+$, with the property that $f'(x + iy) - \Psi'(y) \to 0$, as $y \to 0$, $x \in E$. Then clearly $f(x + iy) - \Psi(y) \to$ limit, as $y \to 0$, $x \in E$. If we choose $\Psi(y) = ye^{i/y}$, we see that $f(x + iy)$ can tend to a limit almost everywhere, as $y \to 0$, but $\int_0^\varepsilon y \, |f'(x + iy)|^2 \, dy = \infty$ almost everywhere. Conversely, if we choose $\Psi(y) = (\log 1/y)^\delta$, $(0 < \delta < 1/2)$ for $y \le 1$, then we see that $\int_0^\varepsilon y \, |f'(x + iy)|^2 \, dy$ can be finite almost everywhere, without $\lim_{y \to 0} f(x + iy)$ existing almost everywhere.
This shows that the perpendicular analogue of the area theorem (in §2) is also not valid.

Notes

Section 1. The local version of Fatou's theorem (when $n = 1$) was proved by Privalov using complex methods. See Zygmund [8, Chapter XIV]. The general version, and proof, given here are taken from Calderón [1].

Section 2. The area integral introduced by Lusin when ($n = 1$), and in this case Theorem 4 is due to Marcinkiewicz and Zygmund and Spencer (see Zygmund [8, Chapter XIV]). For the case of general n see Stein [5], but one direction of the implication had been proved earlier by Calderón [2]. Theorem 5, dealing with non-tangential convergence of conjugate harmonic functions (which will be a key tool in Chapter VIII) goes back to Plessner, when $n = 1$. The argument using complex methods can be found in Zygmund [8, Chapter XIV]. The general case is in Stein [5], as well as the lemma in §2.5.1 used to prove it.

Section 3. The classical theory of H^p spaces may be found in, for example, Zygmund [8, Chapter VII], and K. Hoffman [1]. The n-dimensional real theory was begun in Stein and G. Weiss [2], and is based on a variant of the lemma in §3.1. Theorem 7 was announced in Stein [9]. The argument of Theorem 8 is a simplification of the original proof given by Segovia [1]. For the chain of ideas leading up to Theorem 8 and the reasoning presented here see also Zygmund [8, Chapter XIV], and the significant refinement in Calderón [6].

Theorem 9, which extends to H^1 many of the singular integral theorems, is in Stein [9]. For the extension of the theory of H^p spaces to systems generalizing (18) see Calderón and Zygmund [8], Stein and G. Weiss [2], and *Fourier Analysis*, Chapter VI.

CHAPTER VIII

Differentiation of Functions

In the present chapter we want to bring together various techniques developed in this monograph to study differentiability properties of functions of several variables. In keeping with our approach we shall not aim at the greatest generality, but we will instead pick out certain salient features of a theory which has not yet reached maturity. We shall be concerned with the following problems.

(A) What are conditions that guarantee that a function has derivatives almost everywhere? This is a special case of a further problem which is central in our considerations here.

(B) What are the conditions on a function relative to a given measurable subset E of $\mathbf{R}^n$, that guarantee that this function is differentiable at almost every point of E?

We adopt the following viewpoint with respect to the second problem. We look first for the appropriate *global* analogue of our problem. Here the result is usually stated in terms of identities or inequalities that involve norms of function spaces and are valid everywhere (or almost everywhere) for all of $\mathbf{R}^n$. The derived *local* analogue, which is often deeper, is then given by similar conditions and conclusions, but relative to an arbitrary measurable set E.

We have already dealt with one example of this global and local pairing in Chapter VII (in §1.2), where we considered two versions of Fatou's theorem on boundary values of harmonic functions.

We have made it our goal in this chapter to present three local theorems giving conditions for differentiability. Their global analogues are the following: the differentiability almost everywhere of Lipschitz functions; the boundedness of singular integral transformations on $L^q(\mathbf{R}^n)$, $1 < q < \infty$; and the characterization of the L^q class of a function in terms of the L^q class of the g-function or Lusin integral.

In addition to the global results already alluded to, the main technical ingredients turn out to be as follows.

(i) Generalizations of the ordinary definition of differentiability.

240

(ii) A basic splitting lemma into "good" and "bad" parts which in some sense may be viewed as the local analogue of the corresponding global splitting used in the study of singular integrals in Chapter II.

(iii) An argument of *desymmetrization* which allows one to go from symmetric assumptions to non-symmetric conclusions.

The ideas (i) and (ii) are presented in §1 and §2 and (iii) is detailed in §4.

We add a pedantic comment. It is in the nature of the subject matter considered here that in principle we cannot exclude the possibility that non-measurable sets enter at certain stages of the argument. While this is an awkward point it does not impede the development that follows. We shall return to this in greater detail in §3.1.1 later, but let us at this point already make the following convention. When the words "function" or "set" are written below we will mean Lebesgue measurable function or Lebesgue measurable set, unless there is an explicit qualification to the contrary.

1. *Several notions of pointwise differentiability*

1.1 Suppose that f is defined in an open neighborhood of a set E in $\mathbf{R}^n$. Let $x^0 \in E$. We shall say that f has an *ordinary derivative* (or *is differentiable*) at x^0, if there exists a linear function $\Lambda = \Lambda_{x^0}$ so that

$$(1) \qquad f(x^0 + y) = f(x^0) + \Lambda(y) + o(|y|)$$

as $|y| \to 0$, or what is the same that

$$\sup_{|y| \le r} \frac{|f(x^0 + y) - f(x^0) - \Lambda(y)|}{|y|}$$

tends to zero with r.

On the other hand if f is (say) locally integrable we can define its first partial derivatives in the *weak sense* (the distribution sense) as in Chapter V, §2.1, by saying that $\dfrac{\partial f}{\partial x_k} = f_k$, $k = 1, \ldots, n$, with the f_k locally integrable if

$$(2) \qquad \int_{\mathbf{R}^n} f \frac{\partial \varphi}{\partial x_k} \, dx = -\int_{\mathbf{R}^n} f_k \varphi \, dx$$

for every smooth function φ whose support is compact and is strictly inside the domain of definition of f.

The first important question that arises is whether as a consequence of the second definition f has a derivative in the pointwise sense (1) for almost every x^0.

In the case of one dimension the answer is of course yes since such f are locally absolutely continuous on $\mathbf{R}^1$ (see §6.1 of Chapter V). The situation however differs when the dimension is greater than one. The example in §6.3 of Chapter V shows that the $\dfrac{\partial f}{\partial x_k}$ may exist in the weak sense and be in $L^p(\mathbf{R}^n)$, with $p \leq n$; at the same time f may fail to have a derivative in the sense of (1) at *every* point x^0, since f may be unbounded near every x^0. We are therefore motivated to relax the requirement in the definition of the derivative so as to conform with the type of behavior we can expect of f. We shall say that f has a derivative at x^0 *in the L^q sense*, with $1 \leq q < \infty$, if

$$(3) \quad \left(h^{-n} \int_{|y| \leq h} |f(x^0 + y) - f(x^0) - \Lambda(y)|^q \, dy \right)^{1/q} = o(h), \quad \text{as} \quad h \to 0.$$

This is clearly a generalization of the initial definition (1).

The following result clarifies the significance of the notion just introduced. We assume that $n > 1$.

THEOREM 1.   *Suppose f is a locally integrable function given in an open set Ω, with the property that the $\dfrac{\partial f}{\partial x_j}, j = 1, \ldots, n$ exist weakly there, and f and the $\dfrac{\partial f}{\partial x_j}$ belong locally to $L^p(\mathbf{R}^n)$, with $j = 1, \ldots n$.*

(a) *If $n < p$, f has an ordinary derivative (in the sense (1)) for almost every x^0, when f has been suitable modified on a set of measure zero.*

(b) *If $1 < p < n$, then f has a derivative in the L^q sense with $1/q = 1/p - 1/n$, for almost every x^0.*

1.2   Proof of Theorem 1.   After multiplying by a smooth function with compact support we may assume that f itself has compact support and that f and the $\dfrac{\partial f}{\partial x_j}$ are in $L^p(\mathbf{R}^n)$, with the latter taken in the sense of distributions; this means that $f \in L_1^p(\mathbf{R}^n)$. We can also suppose that $p < \infty$, since the result in the case $p = \infty$ is then a consequence of the case $n < p < \infty$.

If f were in $\mathscr{D}$ we would have the identity

$$(4) \qquad\qquad f = I_1\left(\sum_{j=1}^{n} R_j\left(\frac{\partial f}{\partial x_j} \right) \right).$$

(See §2.3 of Chapter V.)

Now our f is in $L_1^p(\mathbf{R}^n)$, so it can be approached in the norm of that space by a sequence $\{f_m\}$ of such elements (see §2.1 of Chapter V);

since each $\dfrac{\partial f}{\partial x_j}$ also belongs to $L^{p'}(\mathbf{R}^n)$ for every $p' \leq p$, the proof of the approximation argument also shows that

$$\frac{\partial f_m}{\partial x_j} \to \frac{\partial f}{\partial x_j} \quad \text{in} \quad L^{p'}(\mathbf{R}^n).$$

Recall also that the R_j are continuous from $L^{p'}(\mathbf{R}^n)$ to itself, if $1 < p' < \infty$ and I_1 is continuous from $L^{p'}(\mathbf{R}^n)$ to $L^q(\mathbf{R}^n)$ if $1/q' = 1/p' - 1/n$, $1 < p' < n$, (see §1.2 of Chapter V). Thus altogether we have that the identity (4) holds for our f. For our purposes this can be restated as follows.

(5) $\quad f(x) = \displaystyle\int_{\mathbf{R}^n} \frac{g(y)}{|x - y|^{n-1}} \, dy, \quad \text{with} \quad g \in L^{p'}(\mathbf{R}^n), \qquad 1 < p' \leq p.$

Here we have set

$$g(x) = \frac{1}{\gamma(1)} \sum_{j=1}^{n} \left(R_j \frac{\partial f}{\partial x_j} \right)(x).$$

Of course the integral (5) converges absolutely for almost every x, (this can be seen by taking $p' < n$). Redefine f to be the value of this integral wherever it converges absolutely, and for other points of x (if there be such) we may define f arbitrarily.

Next, since $g \in L^p(\mathbf{R}^n)$ the following two properties hold for almost every x^0:

(6) $\qquad \dfrac{1}{r^n} \displaystyle\int_{|y| \leq r} |g(x^0 - y) - g(x^0)|^p \, dy \to 0, \quad \text{as} \quad r \to 0.$

(7) $\qquad \displaystyle\lim_{\varepsilon \to 0} \int_{|y| \geq \varepsilon} \frac{y_j}{|y|^{n+1}} g(x^0 - y) \, dy \quad \text{exists.}$

The property (6) is merely a restatement of §5.7 of Chapter I; (7) is a special case of the theorem of the existence almost everywhere of singular integrals given in §4.5 of Chapter II.

It is at an x^0 where (6) and (7) hold simultaneously that we shall prove the existence of the appropriate derivative of f.

In order to simplify the notation we shall assume that $x^0 = 0$ and $g(x^0) = 0$. This can be achieved first by an appropriate translation and then by replacing $g(x)$ by $g(x) - g(x^0)\varphi(x)$, where φ is a smooth function with compact support and with $\varphi(x^0) = 1$. It is clear that the result with g replaced by $g - g(x^0)\varphi$ implies the desired result for g.

After these preliminaries are out of the way we see that our assumptions are reduced to

(6′) $\qquad \dfrac{1}{r^n} \displaystyle\int_{|y| \leq r} |g(y)|^p \, dy \to 0, \quad \text{as} \quad r \to 0$

and

$$(7') \qquad \alpha_j = \lim_{\varepsilon \to 0} (n-1) \int_{|y| \geq \varepsilon} g(y) \frac{y_j}{|y|^{n+1}} \, dy \quad \text{exists.}$$

We set $\Lambda(x) = \sum_{j=1}^{n} \alpha_j x_j$, with $x = (x_1, \ldots, x_n)$. We remark that in view of the oft-mentioned estimation at infinity, namely that $g \in L^{p'}(\mathbf{R}^n)$, $1 < p' \leq p$, and the majorization (6') near the origin, then the integral $\int_{\mathbf{R}^n} \frac{g(y)}{|y|^{n-1}} \, dy$ which represents $f(0)$ converges absolutely.

It will be our purpose to find the appropriate measure of smallness of

$$f(x) - f(0) - \sum_{j=1}^{n} \alpha_j x_j, \quad \text{as} \quad |x| \to 0.$$

By (5) we have

$$f(x) - f(0) = \int_{|y| \leq 2|x|} g(y) |x-y|^{-n+1} \, dy - \int_{|y| \leq 2|x|} g(y) |y|^{-n+1} \, dy$$

$$+ \int_{|y| \geq 2|x|} g(y) \{|x-y|^{-n+1} - |y|^{-n+1}\} \, dy$$

$$= A_x - B_x + C_x.$$

1.2.1 Study of the integral C_x. We deal with

$$C_x = \int_{|y| > 2|x|} g(y) \{|x-y|^{-n+1} - |y|^{-n+1}\} \, dy;$$

it will be the dominant term of the difference $f(x) - f(0)$. We shall see that as

$$|x| \to 0, \qquad C_x = \sum_{j=1}^{n} \alpha_j x_j + o(|x|).$$

Now the Taylor development of $|x-y|^{-n+1}$ in powers of $x_1, \ldots, x_n$ shows that

$$|x-y|^{-n+1} - |y|^{-n+1} = (-n+1) \sum_{j=1}^{n} x_j \frac{y_j}{|y|^{n+1}} + R(x, y),$$

where

$$|R(x, y)| \leq A \frac{|x|^2}{|y|^{n+1}}, \quad \text{for} \quad |y| \geq 2|x|.$$

Therefore

$$C_x = (-n+1) \sum_{j=1}^{n} x_j \int_{|y| \geq 2|x|} g(y) \frac{y_j}{|y|^{n+1}} \, dy + R_x,$$

where

$$|R_x| \leq A |x|^2 \int_{|y| \geq 2|x|} \frac{|g(y)| \, dy}{|y|^{n+1}} \, .$$

For any $\delta > 0$ we have

$$|x| \int_{|y| \geq 2|x|} \frac{|g(y)|\, dy}{|y|^{n+1}} = |x| \int_{\delta \geq |y| \geq 2|x|} \frac{|g(y)|\, dy}{|y|^{n+1}} + |x| \int_{|y| \geq \delta} \frac{|g(y)|\, dy}{|y|^{n+1}}.$$

The first integral is majorized by a constant multiple of $\delta^{-n} \int_{|y| \leq \delta} |g(y)|\, dy$, which tends to zero with δ by (6'). Once δ is fixed the second integral clearly tends to zero with $|x|$. Altogether we see therefore that

$$(8) \qquad C_x = \sum_{j=1}^{n} \alpha_j x_j + o(|x|), \quad \text{as} \quad |x| \to 0.$$

1.2.2 Majorization of A_x when $n < p < \infty$. Let us consider $A_x = \int_{|y| \leq 2|x|} g(y) |x - y|^{-n+1}\, dy$. We apply Hölder's inequality with the exponents p and r, $1/p + 1/r = 1$; the hypothesis $n < p$ is equivalent with $(n - 1)r < n$, and that $|y|^{-n+1}$ is locally in $L^r(\mathbf{R}^n)$. Thus since $|x - y| \leq 3|x|$ in this integral we have

$$A_x \leq \left(\int_{|y| \leq 2|x|} |g(y)|^p\, dy \right)^{1/p} \left(\int_{|y| \leq 3|x|} y^{(-n+1)r}\, dy \right)^{1/r}$$

$$= o(|x|^{n/p}) \times \{ C\, |x|^{(n+(-n+1)r)/r} \}$$

$$= o(|x|).$$

To summarize

$$(9) \qquad A_x = o(|x|) \quad \text{as} \quad |x| \to 0.$$

1.2.3 Majorization of A_x when $1 < p < n$. For each $n > 0$, let $\chi_h(y)$ be the characteristic function of the ball $|y| \leq h$. Then assuming $|x| \leq h$, we have

$$|A_x| = \left| \int_{|y| \leq 2|x|} \frac{g(y)}{|x - y|^{n-1}}\, dy \right| \leq \int_{\mathbf{R}^n} \frac{|g(y)|\, \chi_{2h}(y)}{|x - y|^{n-1}}\, dy.$$

Hence by the L^p theorem on fractional integration in §1.2 of Chapter V we have

$$\|\chi_h A_x\|_q \leq A\, \|g \chi_{2h}\|_p, \quad \text{where} \quad 1/q = 1/p - 1/n.$$

That is,

$$\int_{|x| \leq h} |A_x|^q\, dx \leq A^q \left(\int_{|y| \leq 2h} |g(y)|^p\, dy \right)^{q/p} = o(h^{nq/p}) = o(h^{n+q}),$$

by property (6').

We obtain therefore

$$(10) \qquad \left(\frac{1}{h^n} \int_{|x| \leq h} |A_x|^q\, dx \right)^{1/q} = o(h), \quad \text{as} \quad h \to 0.$$

1.2.4   Majorization of B_x.   Since

$$B_x = \int_{|y| \le 2x} \frac{g(y)}{|y|^{n-1}} \, dy,$$

we have that

$$|B_x| \le \sum_{j=-1}^{\infty} 2^{-j-1} \int_{|x| \le |y| \le 2^{-j}x} \le \sum_{j=-1}^{\infty} (2^{-j}|x|)^{-n+1} o((2^{-j}|x|)^n)$$

by property (6'). Thus

$$B_x = o(|x|) \sum_{j=-1}^{\infty} 2^{-j},$$

and therefore

(11) $B_x = o(|x|)$ as $|x| \to 0.$

The combination of (8), (9), (10), and (11) then shows that

$$f(x) - f(0) - \sum_{j=1}^{n} \alpha_j x_j = o(|x|)$$

if $n < p$, and

$$\left(h^{-n} \int_{|x| \le h} \left| f(x) - f(0) - \sum_{j=1}^{n} \alpha_j x_j \right|^q dx \right)^{1/q} = o(h) \quad \text{if} \quad 1 < p < n.$$

This proves our theorem.

1.2.5   The theorem is also valid in the case $p = 1$ although the proof must be modified. This and other generalizations are stated in §6 below.

2. *The splitting of functions*

2.1   Derivatives in the harmonic sense.   We come now to one of the main techniques used, a certain splitting of functions into their "good" and "bad" parts. Since we shall also be appealing to the theory of harmonic functions in an essential way it will be important to be able to achieve this splitting in the context of a notion of differentiability deriving from that theory. The idea we have in mind is in fact still more general than the definitions given in §1, and is described as follows.

Let f be a locally integrable function defined in an open set Ω. For a fixed $x^0 \in \Omega$ we modify f outside a bounded open set containing x^0, by setting it equal to zero. For the resulting f (now in $L^1(\mathbf{R}^n)$) we take its Poisson integral, $u(x, y) = P_y * f$. We shall say that f has a *harmonic derivative* at x^0, if u and $\dfrac{\partial u}{\partial x_1}, \dfrac{\partial u}{\partial x_2}, \ldots, \dfrac{\partial u}{\partial x_m}$ have non-tangential limits at

x_0. In this case, we define

$$f(x^0) = \lim_{(x,y)\to(x^0,0)} u(x, y), \quad \text{and} \quad \frac{\partial f}{\partial x_j}(x^0) = \lim_{(x,y)\to(x^0,0)} \frac{\partial u}{\partial x_j}(x, y)$$

where the variable point (x, y) converges to $(x^0, 0)$ non-tangentially.

We need to make the following clarifying remarks about this definition.

(i) The notion of harmonic derivative is well defined in the sense that it is independent of the modification we have subjected f to strictly away from x^0. In particular it is easy to verify that if f is in $L^1(\mathbf{R}^n)$ and vanishes in a neighborhood of x^0, then u and all its partial derivatives with respect to the x_j have non-tangential limits zero at x^0.

(ii) If f has derivative in the ordinary sense at x^0, or more generally in the L^q sense (as defined by (3)) then f has a derivative in the harmonic sense and the values of these derivatives coincide at x^0. To see this it suffices to assume that f has a derivative in the L^1 sense. For appropriate constants $A_0, \alpha_1, \ldots, \alpha_n$ we have that if

$$f(x^0 + y) = A_0 + \sum_{j=1}^{n} \alpha_j y_j + \varepsilon(y),$$

then

$$\int_{|y|\leq r} |\varepsilon(y)|\, dy = o(r^{n+1}), \quad \text{as} \quad r \to 0.$$

Observe that it is immediate from this that if we take $f(x^0)$ to be A, then x^0 is a Lebesgue point for f, and hence by Theorem 1 of Chapter VII (see page 197) it then follows that u has A as non-tangential limit at x^0. Next

$$\frac{\partial u}{\partial x_j}(x, y) = \int_{\mathbf{R}^n} P_y^j(t) f(x - t)\, dt,$$

with

$$P_y^j(t) = \frac{\partial}{\partial t_j} P_y(t), \quad j = 1, \ldots, n.$$

Thus

$$\frac{\partial u}{\partial x_j}(x^0 + h, y) = \int P_y^j(t + h) f(x^0 - t)\, dt$$

$$= \int P_y^j(t + h) A\, dt + \sum_{k=1}^{n} \int P_y^j(t + h)\alpha_k t_k\, dt$$

$$+ \int P_y^j(t + h)\varepsilon(t)\, dt.$$

The first term vanishes and the second group of terms equals α_j. This can be seen by a simple argument of passing the derivative $\dfrac{\partial}{\partial t_j}$ from $P_y(t)$ to A

and $\sum \alpha_k t_k$ respectively. That leaves only the last term; for this we take $|h| < \alpha y$ for some α (this is the non-tangential condition) and rewrite the integral as

$$(12) \quad \int_{|t| \le y} P_y^j(t + h)\varepsilon(t)\, dt + \int_{1 \ge |t| \ge y} P_y^j(t + h)\varepsilon(t)\, dt$$

$$+ \int_{|t| \ge 1} P_y^j(t + h)\varepsilon(t)\, dt.$$

Since

$$P_y(t) = \frac{cy}{(|t|^2 + y^2)^{(n+1)/2}},$$

it is clear that $\dfrac{\partial P_y(t + h)}{\partial t_j}$ is bounded by the following two expressions, if $|h| < \alpha y$; first cy^{-n-1}, and secondly $cy/|t|^{n+2}$. We insert the first estimate for P_j in the first integral and we get therefore that it is bounded by a constant multiple of $y^{-n-1} \int_{|t| \le y} |\varepsilon(t)|\, dt$, which tends to zero with y. Similarly the second integral is bounded by a constant multiple of

$$y \int\limits_{1 \ge |t| \ge y} \frac{|\varepsilon(t)|}{|t|^{n+2}}\, dt \le \sum y \int\limits_{2^{k+1}y \ge |t| \ge 2^k y} \frac{|\varepsilon(t)|\, dt}{|t|^{n+2}},$$

where the sum is taken over all non-negative integral k so that $2^{k+1}y \le 2$. By assumption on ε we get that each term of this sum is

$$y(2^k y)^{-n-2}\, o((2^{k+1}y)^{n+1}) = 2^{-k}o(1);$$

and therefore this integral tends to zero with y. The third integral trivially converges to zero with y, since $\varepsilon(t)$ is the sum of term which is bounded by $c\,|t|$ and a function in $L^1(\mathbf{R}^n)$. This proves our assertion.

(iii) Our last comment about the notion of harmonic derivative is this. Our definition does not require anything about the behavior of $\dfrac{\partial}{\partial y} u(x, y)$. In fact the existence of a non-tangential limit of $\dfrac{\partial u}{\partial y}$ at a given x^0 is *not* a consequence of ordinary differentiability at x^0. If however we had the differentiability (ordinary, in the L^q sense, or harmonic) for a set of x^0 of positive measure, we could conclude that for almost all such x^0 the non-tangential limits of $\dfrac{\partial u}{\partial y}$ exist. We shall return to this deeper fact momentarily.

2.2   The splitting.   Our theorem is as follows:

THEOREM 2.   *Suppose f is a given locally integrable function, and that for every point x^0 in a set of finite measure E, f has a derivative in the*

harmonic sense. Then given any $\varepsilon > 0$ we can find a compact set F, so that
$F \subset E$, $m(E - F) < \varepsilon$, *and we can also write f as $f = g + b$ where*

(1) $g \in L_1^{\infty}(\mathbf{R}^n)$

(2) b *is zero on F.*

Observe that g is the "good" function. In view of Theorem 1 it has an ordinary derivative almost everywhere. (By definition it also has essentially bounded first derivatives in $\mathbf{R}^n$ taken in the weak sense.) While b is the "bad" function, it has the useful redeeming feature that it vanishes on the set F.

We shall now give the proof of Theorem 2. After reducing to a bounded subset of E if necessary (for simplicity we call this subset also E), we may assume that f vanishes outside a neighborhood of this set; thus $f \in L^1(\mathbf{R}^n)$. Let $u(x, y)$ be the Poisson integral of f, and then according to our assumptions $\dfrac{\partial u}{\partial x_1}, \ldots, \dfrac{\partial u}{\partial x_n}$, have non-tangential limits at every point of E. It then follows by Theorem 5 of Chapter VII, page 213, that $\dfrac{\partial u}{\partial y}$ has non-tangential limits for almost every x^0 in E. Also u has non-tangential limits at almost every $x^0 \in E$ as we have seen (in particular at points of the Lebesgue set of f). Let us fix the parameters α and h and consider the truncated cones $\Gamma_\alpha^h(x^0) = \{(x, y) : |x - x^0| < \alpha y, 0 < y < h\}$. Then in view of what has been said and because of the uniformization lemma in §1.3.1 of Chapter VII (page 201) we can conclude that we can find a compact set F with the property that $F \subset E$, $m(E - F) < \varepsilon$ and u, $\dfrac{\partial u}{\partial y}$ and $\dfrac{\partial u}{\partial x_1}, \ldots, \dfrac{\partial u}{\partial x_n}$ are all uniformly bounded in $\mathscr{R} = \displaystyle\bigcup_{x^0 \in F} \Gamma_\alpha^h(x^0)$. We can also assume that u has the non-tangential limit $f(x^0)$ at every point $x^0 \in F$. Now in place of the truncated cones $\Gamma_\alpha^h(x^0)$ consider the infinite cones $\Gamma_\alpha(x^0) = \{(x, y) : |x - x^0| < \alpha y\}$. Since u is the Poisson integral of an L^1 function, it follows also that u, $\dfrac{\partial u}{\partial y}$, $\dfrac{\partial u}{\partial x_1}, \ldots, \dfrac{\partial u}{\partial x_n}$ are all uniformly bounded in the half-space $y \geq h$, and thus in addition to $\mathscr{R}$ are also bounded in $\tilde{\mathscr{R}} = \displaystyle\bigcup_{x^0 \in F} \Gamma_\alpha(x^0)$.

Let us now look more carefully at $\tilde{\mathscr{R}}$. $\tilde{\mathscr{R}}$ is an open subset of $\mathbf{R}^{n+1}$ whose boundary is given by the hypersurface $y = \alpha^{-1}\delta(x)$ with $\delta(x) = \text{dist}\,(x, F)$, being the distance from x to F. Since $\delta(x)$ clearly satisfies the condition $|\delta(x) - \delta(x^1)| \leq |x - x^1|$ we see that $\tilde{\mathscr{R}}$ is a special Lipschitz domain in the terminology of §3.2 of Chapter VI, and therefore the extension theorem in §3.2 applies to u restricted to $\tilde{\mathscr{R}}$. More explicitly let

$U(x, y)$ denote the restriction of u to $\tilde{\mathscr{R}}$. Then of course U is C^∞ in $\tilde{\mathscr{R}}$ and U and its first partial derivatives are bounded in $\tilde{\mathscr{R}}$; thus $U \in L_1^\infty(\tilde{\mathscr{R}})$. Let $\mathfrak{G}(U)$ be the extension of U to all of $\mathbf{R}^{n+1}$. We have therefore that $\mathfrak{G}(U) \in L_1^\infty(\mathbf{R}^{n+1})$. Finally set g to be the restriction of $\mathfrak{G}(U)$ to $\mathbf{R}^n$. Since the restriction of a function in $L_1^\infty(\mathbf{R}^{n+1})$ is in $L_1^\infty(\mathbf{R}^n)$ (this is an immediate consequence of §6.2(a) in Chapter V), we see that $g \in L_1^\infty(\mathbf{R}^n)$.

Because u has a non-tangential limit equal to $f(x^0)$ at every point $(x^0, 0)$ in $\mathbf{R}^{n+1}$, where $x^0 \in F$, so does U; and since $\mathfrak{G}(U)$ is continuous it follows that $\mathfrak{G}(U)(x^0, 0) = f(x^0)$, for every $x^0 \in F$. Hence the same is true of g, its restriction to $\mathbf{R}^n = \{(x, 0\}$; that is $g(x^0) = f(x^0)$ for each $x^0 \in F$. Writing then $b(x) = f(x) - g(x)$, we have achieved the desired conclusion.

3. *A characterization of differentiability*

3.1 Boundedness of difference quotient. Our first application of these techniques will be to prove a theorem which characterizes almost everywhere the notion of differentiability.

The reader should not overlook the striking analogy of the theorem stated below and the local version of Fatou's theorem (Theorem 3, p. 201 in Chapter VII). The application of this theorem on harmonic functions is a main idea in the proof that follows.

THEOREM 3. *Let f be given in an open neighborhood of a given set E, then f is differentiable (in the ordinary sense) at almost every point of E if and only if*

$$(13) \qquad f(x^0 + y) - f(x^0) = O(|y|) \quad as \quad |y| \to 0,$$

for almost every $x^0 \in E$. It is of course not assumed that the constant appearing in the "O" above is uniform in x^0.

The fact that differentiability at a given x^0 implies (13) is trivial, so we turn to the converse.

If we restrict ourselves to a subset E_0 of E where f is bounded and (13) applies, then it is apparent that outside an open neighborhood of this subset f is still bounded, and we can modify f so as to vanish outside this neighborhood. It will suffice to take E_0 to be bounded and the appropriate neighborhood of E_0 to be bounded also. Then it will be enough to show that the modified f is differentiable almost everywhere in E_0. For ease of notation replace E_0 by E; and call the f which is modified also f. Then this $f \in L^1(\mathbf{R}^n)$ and we let u denote its Poisson integral. Just as in the argument of (ii) of §2.1 we see that at every point x^0 where (13) holds the $\dfrac{\partial u}{\partial x_j}$ are non-tangentially bounded at x^0. This is merely a repetition of the argument on page 248, with the appropriate o's replaced by O's. Therefore by the

Theorems 1 and 3 of Chapter VII the non-tangential limits of u and $\frac{\partial u}{\partial x_1}, \frac{\partial u}{\partial x_2}, \ldots, \frac{\partial u}{\partial x_n}$ exist at almost every point of E; thus f has a harmonic derivative at almost every point of E. We can now invoke the splitting theorem of §2.2 above. Therefore for every $\varepsilon > 0$ there exists a compact set F, $F \subset E$, $m(E - F) < \varepsilon$, and a splitting $f = g + b$, with $g = f$ on F. Since $g \in L_1^\infty(\mathbf{R}^n)$, then Theorem 1 in §1.1 shows that g has an ordinary derivative for every point in $\mathbf{R}^n$ and hence it suffices to show that b is differentiable at almost every $x \in F$. However b vanishes at all points of F and thus we get in place of (13)

(13') $\quad b(x^0 + y) = O(|y|)$, as $|y| \to 0$ for almost every $x^0 \in F$.

We wish now to uniformize the relation (13'). For every integer k therefore let F_k be the set

$$F_k = \{x^0 : |b(x^0 + y)| \le k |y|, |y| \le 1/k\}.$$

Clearly $\bigcup_{k=1}^{\infty} F_k$ contains all points where (13') holds and we are reduced to considering what happens for $x^0 \in F_k$.

3.1.1 We must digress now because we have reached an unpleasant point in the argument; namely, the set F_k since it is given by a continuum of inequalities is not necessarily Lebesgue measurable. One could have gotten around this difficulty by assuming f to be everywhere continuous to begin with, but this assumption is an artificial one in the general context of our problems.

We proceed instead by the observation that the statement "almost every point of E is a point of density of E" holds under suitable modification for non-measurable sets. In fact for any not-necessarily-measurable set E, denote by $m_e(E)$ its *exterior* (or *outer*) measure; by definition $m_e(E) = \inf m(F)$, $E \subset F$, where F ranges over measurable sets. Clearly for any set E there exists a measurable set $\tilde{E}$, so that $m_e(E) = m(\tilde{E})$ and more generally $m_e(E \cap F) = m(\tilde{E} \cap F)$ for every measurable set F.

Next we say that x^0 is a point of (*exterior*) density of E if

$$\frac{m_e(E \cap B(x^0, r))}{m(B(x^0, r))} \to 1 \quad \text{as} \quad r \to 0$$

where $B(x^0, r)$ is the ball centered at x^0 of radius r. Obviously x^0 is a point of exterior density of E if and only if it is a point of density of $\tilde{E}$. Thus except for a subset of E of Lebesgue measure zero, the points of E are points of exterior density of E.

3.1.2   To conclude the proof of the theorem we need therefore only prove that at every point x^0 which is a point of exterior density of F_k we have $b(x^0 + y) = o(|y|)$ as $|y| \to 0$. For then b is differentiable at this point of F_k; and by taking the union over k, b is differentiable except for a set of measure zero of F.

The argument now is merely a repetition of the ideas already anticipated in Chapter I in the proof of Proposition 2 (in §2.2, page 13). For a point x^0 of exterior density of F_k and for a given $\varepsilon > 0$, we consider the "small" ball of center $x + y$, of radius $\varepsilon |y|$; also we consider the "large" ball of center x and radius $|y| + \varepsilon |y|$. Now by that argument of point of density there exists a $z \in F_k$ belonging to the small ball, if $|y|$ is small enough. This means that within a distance of $\varepsilon |y|$ from the point $x^0 + y$ we can find a point of F_k. Thus $|b(x + y)| \le k\varepsilon |y|$, by the definition of F_k, and since ε is arbitrary this shows that $b(x + y) = o(|y|)$ as $y \to 0$.

3.2   A refinement of the splitting.   Before we come to the next application we need to refine the conclusions of the splitting theorem in §2.2. Let $1 \le q < \infty$, and suppose that f has at every point of the set E a derivative in the L^q sense (as defined in §1.1), and thus f has at every point of E a derivative in the harmonic sense. Let $\varepsilon > 0$. According to the theorem in §2.2 we can find a set F, $F \subset E$, $m(E - F) < \varepsilon$ and $f = g + F$, so that $g \in L_1^\infty(\mathbf{R}^n)$ and $b(x^0) = 0$ for $x^0 \in F$. We can state more:

COROLLARY (OF THEOREM 2).   *For almost every* $x^0 \in F$

$$(14) \qquad \int_{|y| \le 1} \frac{|b(x^0 + y)|}{|y|^{n+1}} \, dy < \infty.$$

Proof.   In view of the fact that $g \in L_1^\infty(\mathbf{R}^n)$, we get that g has at almost every point an ordinary (and thus also L^q) derivative. Hence b, which vanishes on F, has almost everywhere a derivative in the L^q sense on F; that is in particular

$$(15) \qquad \frac{1}{r^n} \int_{|y| \le r} |b(x^0 + y)|^q \, dy = O(r^q), \quad \text{as} \quad r \to 0$$

for almost every $x^0 \in F$.

We can also assume that after an easy modification (of the type made in the proof of Theorem 3) that $f \in L^q(\mathbf{R}^n)$. Thus after a further trivial modification of g we can suppose that $b \in L^q(\mathbf{R}^n)$.

Now let F_k be the *closed* set given by

$$F_k = \left\{ x^0 : \frac{1}{r^n} \int_{|y| \le r} |b(x^0 + y)|^q \, dy \le kr^q, \, 0 < r \le 1/k \right\}.$$

(Observe that in this case the sets F_k are automatically measurable!)

We shall show that if $\delta(x)$ denotes the distance of x from the closed set F_k we have the inequality

$$(16) \qquad \int_{|y| \leq 1} \frac{|b(x^0 + y)| \, dy}{|y|^{n+1}} \leq c \int_{|y| \leq 1} \frac{\delta(x^0 + y) \, dy}{|y|^{n+1}} \, , \qquad x^0 \in F$$

and therefore our corollary will follow from the theorem about the finiteness of the integral of Marcinkiewicz given in §2.3 of Chapter I, page 14.

To carry out the argument we write the complement of F_k as a "disjoint" union of cubes $\{Q_j\}$ whose diameters are comparable to their distances from F_k, in accordance with Theorem 1 of Chapter VI.

We then have

$$\int_{|y| \leq 1} \frac{|b(x^0 + y)| \, dy}{|y|^{n+1}} = \int_{{}^cF_k \cap \{|x^0 - y| \leq 1\}} \frac{|b(y)| \, dy}{|x^0 - y|^{n+1}}$$

$$= \sum_j \int_{Q_j \cap \{|x^0 - y| \leq 1\}} \frac{|b(y)| \, dy}{|x^0 - y|^{n+1}} \, .$$

Now if $x^0 \in F_k$, then the quantity $\dfrac{1}{|x_0 - y|^{n+1}}$ is roughly constant in y, as y ranges in Q_j; that is

$$\sup_{y \in Q_j} \frac{1}{|x^0 - y|^{n+1}} \leq c_1 \inf_{y \in Q_j} \frac{1}{|x^0 - y|^{n+1}} \, .$$

Also each Q_j is contained in a ball B_j centered at some point of F_k, with radius of B_j comparable to the diameter of Q_j. Thus by the definition of F_k it follows that

$$\int_{Q_j} |b(y)|^q \, dy \leq c_2 m(Q_j)^{1+q/n}$$

and then by Hölder's inequality

$$\int_{Q_j} |b(y)| \, dy \leq c_3 m(Q_j)^{1+1/n}.$$

Finally the diameter of Q_j is comparable to the distance of any point in Q_j from F_k. Thus $\int_{Q_j} |b(y)| \, dy \leq c_4 m(Q_j) \, \delta(y)$, $y \in Q_j$. Altogether then

$$\int_{Q_j \cap \{|x^0 - y| < 1\}} \frac{|b(y)| \, dy}{|x^0 - y|^{n+1}} \leq c \int_{Q_j \cap \{|x^0 - y| < 1\}} \frac{\delta(y)}{|x^0 - y|^{n+1}} \, .$$

Adding in j we obtain (16) and hence our corollary.

REMARK.   The argument above also proves that:

(17)
$$\int_{|y|\leq 1} \frac{|b(x^0 + y)|^q \, dy}{|y|^{n+q}} < \infty, \quad \text{for almost every } x^0 \in F.$$

The integral (17) is majorized by

$$\int_{|y|\leq 1} \frac{(\delta(x^0 + y))^q}{|y|^{n+q}} \, dy,$$

as the reasoning of the corollary shows.

At every point x^0 where (17) holds we evidently also have the conclusion

(18)
$$\frac{1}{r^n}\int_{|y|\leq r} |b(x^0 + y)|^q \, dy = o(r^q), \quad r \to 0.$$

This has the interpretation that not only does b vanish on F, but also at almost every point of F the first derivatives of b in the L^q sense are zero.

3.3   Preservation under the action of singular integrals.   We next consider a local analogue of the facts concerning the boundedness of singular integrals on operators $L^q(\mathbf{R}^n)$. We shall show that the pointwise notion of a derivative in the L^q sense is stable almost everywhere under the action of appropriate singular integrals transformations.

We deal with the following class of operators; their kernels $K(x)$ satisfy:

(i) K is of class C^1 away from the origin
(ii) $|K(x)| \leq A/|x|^n$, $|\nabla K| \leq A/|x|^{n+1}$, $x \neq 0$
(iii)   if $T_\varepsilon(f) = \int_{|y|\geq\varepsilon} K(y)f(x - y) \, dy$, then for some fixed q,

$$\|T_\varepsilon(f)\|_q \leq A_q \|f\|_q,$$

for $f \in L^q(\mathbf{R}^n)$, where A_q is independent of ε. We also assume that $T_\varepsilon f$ converges in the L^q norm to a limit Tf, as $\varepsilon \to 0$.

Among the examples of such transformation are those with $K(x) = \dfrac{\Omega(x)}{|x|^n}$ where Ω is homogeneous of degree 0, is of class C^1 on the unit sphere, and its mean-value on that sphere vanishes. This class of course includes the Riesz transforms, and the higher Riesz transforms of Chapter III.

THEOREM 4.   *Let* $1 < q < \infty$. *Suppose* $f \in L^q(\mathbf{R}^n)$, *and* f *has a derivative in the* L^q *sense at every point of a set* E. *If* T *is a singular integral transformation of the above kind, then* Tf *has a derivative in the* L^q *sense at almost every point of* E.

It is natural to ask whether the notion of the ordinary derivative is also stable under the action of singular integral operators. However this is not so even for $n = 1$; see §6.8 below.

3.3.1 Proof. Let us remark first that if f vanishes in a fixed neighborhood of a given point x^0, then it is easy to see that Tf, (after a suitable correction on a set of measure zero) has an ordinary derivative at every point of that neighborhood, and in particular at x^0. This remark shows that the question of whether Tf is differentiable at x_0 depends only on the behavior of f near x_0. We can therefore assume that the set E is bounded. Using the splitting theorem we write $f = g + b$, where $g \in L_1^\infty(\mathbf{R}^n)$, $b = 0$ on F, $F \subset E$, with $m(E - F)$ small. It is easy to see that we can modify g and b in such a way so that g has compact support without changing any other of the stated properties. Thus g in addition belongs to $L_1^q(\mathbf{R}^n)$. Since T is a bounded operator on L^q, we have therefore that $T(g)$ belongs to $L_1^q(\mathbf{R}^n)$. In fact the question whether a given L^q function is in L_1^q is a matter completely determined by the L^q modulus of continuity of that function (see page 139); this is obviously preserved by T because T is bounded on L^q and is translation-invariant.

Now that we know that $T(g) \in L_1^q(\mathbf{R}^n)$ we obtain from Theorem 1 of this chapter that $T(g)$ has at almost every point of $\mathbf{R}^n$ a derivative in the L^q sense. It therefore suffices to consider $T(b)$ and show that it has an L^q derivative at almost every point of the set F. We shall show that it indeed happens at every x^0 where the following two hold:

$$(19) \qquad \frac{1}{r^n} \int_{|y| \le r} |b(x^0 + y)|^q \, dy = o(r^q), \quad \text{as} \quad r \to 0;$$

$$\int_{|y| \le 1} \frac{|b(x^0 + y)|}{|y|^{n+1}} \, dy < \infty.$$

We know by the discussion in §3.2 that (19) is valid for almost every x^0 in F. Observe that because $b \in L^q(\mathbf{R}^n)$ we get from the finiteness of the integral in (19) in addition that

$$(20) \qquad \int_{\mathbf{R}^n} \frac{|b(x^0 + y)|}{|y|^n} \, dy < \infty, \quad \text{and} \quad \int_{\mathbf{R}^n} \frac{|b(x^0 + y)|}{|y|^{n+1}} \, dy < \infty.$$

For simplicity of notation let us take $x^0 = 0$. Let us momentarily fix a positive number r. Then

$$(Tb)(x) = \int K(x - y)b(y) \, dy = \int_{|x-y| \ge 2r} K(x - y)b(y) \, dy$$

$$+ \lim_{\varepsilon \to 0} \int_{\varepsilon \le |x-y| \le 2r} K(x - y)b(y) \, dy = I_r^1(x) + I_r^2(x).$$

The integral giving $I_r^1(x)$ converges absolutely; I_r^2 converges in the L^q norm by assumption, and $I_r^2 = (T - T_{2r})(b)$. Let us look at I_r^2 more carefully. If we make the restriction that $|x| \leq r$ then the integrand in I_2^r involves only those y for which $|y| \leq 3r$. Thus we can modify $b(y)$ by replacing it with $b(y)\chi_{3r}(y)$, where χ_{3r} is the characteristic function of the ball $|y| \leq 3r$. Hence

$$\int_{|x| \leq r} |I_r^2(x)|^q \, dx \leq \int_{\mathbf{R}^n} |(T - T_{2r})(b\chi_{3r})|^q \, dx \leq A_q^q \int_{\mathbf{R}^n} |b(y)\chi_{3r}(y)|^q \, dy$$

$$= A_q^q \int_{|y| \leq 3r} |b(y)|^q \, dy = o(r^{n+q}).$$

Here we have used the uniform boundedness of the T_ε in $L^q(\mathbf{R}^n)$ norm, and the first property in (19) with $x^0 = 0$. We have therefore succeeded in showing that

$$(21) \qquad \int_{|x| \leq r} |I_r^2(x)|^q \, dx = o(r^{n+q}), \quad \text{as} \quad r \to 0.$$

To study I_r^1 we use Taylor's expansion on K in the form $K(x - y) = K(-y) + (x, \nabla K(-y)) + \varepsilon(r) |x|/|y|^{n+1}$. Here $|x| \leq r$, $|x - y| \geq r$, and $\varepsilon(r)$ tends to zero with r. We insert this in the integral defining I_r^1 and we have

$$(22) \quad I_r^1(x) = \int_{|x-y| \geq 2r} K(-y)b(y) \, dy + \int_{|x-y| \geq 2r} (x, \nabla K(-y))b(y) \, dy$$

$$+ \varepsilon'(r) |x| \int \frac{|b(y)|}{|y|^{n+1}} \, dy.$$

The first integral is equal to

$$\int_{\mathbf{R}^n} K(-y)b(y) \, dy + O\left(\int_{|y| \leq 3r} \frac{|b(y)|}{|y|^n} \, dy \right).$$

That $\int_{\mathbf{R}^n} K(-y)b(y) \, dy$ converges absolutely follows from the first inequality in (20) with $x^0 = 0$. We also get easily from (19) that

$$\int_{|y| \leq 3r} \frac{|b(y)|}{|y|^n} \, dy = o(r), \quad \text{as} \quad r \to 0.$$

The second integral in (22) is handled in the same way. It is equal to

$$\sum_{j=1}^{n} x_j \int_{\mathbf{R}^n} \frac{\partial K(-y)}{\partial y_j} b(y) \, dy + o(r),$$

as $r \to 0$ and these integrals converge by the second inequality in (20). For the same reason the third integral in (22) is also finite. Altogether then

$$(23) \qquad I_r^1(x) = A_0 + \sum_{j=1}^{n} x_j A_j + o(|x|), \quad \text{as} \quad x \to 0$$

with

$$A_0 = \int_{\mathbf{R}^n} K(-y)b(y)\, dy, \; A_j = \int_{\mathbf{R}^n} \frac{\partial K(-y)}{\partial y_j} b(y)\, dy, \quad j = 1, \dots, n.$$

Combined with (21), we see that (23) implies our desired result.

4. Desymmetrization principle

We shall consider the idea of desymmetrization first in a rather general but abstruse form. Afterwards we will make various comments and give several illustrations to help clarify its meaning.

4.1 A general theorem. We shall be concerned with a function $U(x, y)$, $(x, y) \in \mathbf{R}^n \times \mathbf{R}^1_+ = \mathbf{R}^{n+1}_+$; U is measurable in the $(n + 1)$ dimensional upper-half-space, and for simplicity (since we are concerned with the behavior near $y = 0$) we shall assume that U vanishes when $y \geq h$, for some fixed h, $h > 0$, and that U is square integrable on every bounded subset of $\mathbf{R}^{n+1}_+$ which is at a positive distance from the boundary $\mathbf{R}^n$.

THEOREM 5. *Suppose that we are given a set $E \subset \mathbf{R}^n$, and that for every $x^0 \in E$ the following two conditions hold:*

$$(24) \qquad \int_0^\infty y\, |U(x^0, y)|^2\, dy < \infty$$

$$(25) \qquad \iint_{|t| \leq y} y^{1-n} |U(x^0 + t, y) + U(x^0 - t, y)|^2\, dt\, dy < \infty$$

Then we can conclude that for almost every $x^0 \in E$ we have

$$(26) \qquad \iint_{|t| \leq y} y^{1-n} |U(x^0 + t, y)|^2\, dt\, dy < \infty.$$

By the usual arguments we may assume (upon reducing E to a possibly smaller subset F) that F itself is a compact set, and that the integrals (24) and (25) are uniformly bounded as x^0 ranges over F. It will then suffice to prove that (26) holds for almost every x^0 in F. That is we assume that

$$(24') \qquad \int_0^\infty y\, |U(x^0, y)|^2\, dy \leq M, \quad \text{if} \quad x^0 \in F$$

$$(25') \qquad \iint_{|t| \leq y} y^{1-n} |U(x^0 + t, y) + U(x^0 - t, y)|^2\, dt\, dy \leq M, \qquad x^0 \in F.$$

We integrate the inequality (25') over F and make the change of variables

$x^0 + t = u$ and $x^0 - t = v$. This gives

$$\iint\limits_{(u+v)/2\in F} du\, dv \int\limits_{|u-v|<2y} |U(u, y) + U(v, y)|^2\, y^{1-n}\, dy < \infty.$$

If we reduce the domains of integration by restricting the variable v to lie in F we certainly have

(26)
$$\iint\limits_{(u+v)/2\in F,\, v\in F} du\, dv \int\limits_{|u-v|<2y} |U(u, y) + U(v, y)|^2\, y^{1-n}\, dy < \infty.$$

Let us next integrate the inequality (24′) for x^0 over F, but making the change which is appropriate here, by setting v in place of x^0. This gives

(27)
$$\int_{v\in F}\int_0^\infty |U(v, y)|^2\, y\, dy\, dv < \infty.$$

Now let

(28)
$$\begin{cases} I = \iint\limits_{(u+v)/2\in F,\, v\in F} du\, dv \int\limits_{|u-v|<2y} |U(v, y)|^2\, y^{1-n}\, dy \\[2em] \mathscr{I} = \iint\limits_{(u+v)/2\in F,\, v\in F} du\, dv \int\limits_{|u-v|<2y} |U(u, y)|^2\, y^{1-n}\, dy. \end{cases}$$

We can majorize I by dropping the restriction that $(u + v)/2 \in F$. Hence

$$I \le \int_{\mathbf{R}^n} du \int_{v\in F} dv \int_{|u-v|<2y} |U(v, y)|^2\, y^{1-n}\, dy$$

$$= \int_{v\in F} \int_0^\infty |U(v, y)|^2 \left\{ \int_{|u-v|<2y} du \right\} dv\, y^{1-n}\, dy.$$

Since the integral in brackets equals cy^n we get that $I < \infty$ by (27). Therefore in view of (26) we have that $\mathscr{I} < \infty$.

Clearly $\mathscr{I}$ can be rewritten as a double integral,

(29)
$$\mathscr{I} = \int_{\mathbf{R}^n} \int_0^\infty |U(u, y)|^2\, \sigma(u, y) y^{1-n}\, du\, dy,$$

where

$$\sigma(u, y) = \int\limits_{\substack{(u+v)/2\in F,\, v\in F \\ |u-v|<2y}} dv.$$

That is for fixed $y > 0$, and $u \in \mathbf{R}^n$, $\sigma(u, y)$ denotes the measure of the set of points v in $\mathbf{R}^n$ lying in the ball of center u and radius $2y$, and further restricted by the conditions that $v \in F$ and $(u + v)/2 \in F$.

The turning point of the proof will be to show that if u^0 is a point of density of F, then $\sigma(u, y) \sim m(B(u, 2y)) = c_1 y^n$, as the variable point (u, y) tends to $(u^0, 0)$ non-tangentially.

Suppose for simplicity of notation that $u^0 = 0$ is a point of density of F. At this point u^0, the non-tangential approach will for us mean the restriction $|u| < y$. Let χ be the characteristic function of the set F and $\tilde{\chi} = 1 - \chi$ that of the complementary set. Then

$$\sigma(u, y) = \int_{|u-v|<2y} \chi(v)\chi((u+v)/2)\, dv = \int_{|u-v|<2y} dv - \int_{|u-v|<2y} \tilde{\chi}(v)\, dv$$

$$- \int_{|u-v|<2y} \tilde{\chi}((u+v)/2)\, dv + \int_{|u-v|<?y} \tilde{\chi}(v)\tilde{\chi}((u+v)/2)\, du\, dv.$$

The second integral on the right-side of course equals $m(B(u, 2y)) = c_1 y^n$; it is therefore enough to show that the last three integrals are each $o(y^n)$. Recall that $|u| \leq y$. Thus the second integral is dominated by

$$\int_{|v|\leq 3y} \tilde{\chi}(v)\, dv = m(^cF \cap B(0, 3y)).$$

which is $o(y^n)$ since 0 is a point of density of F. The third integral may be rewritten as $\int_{|u-x|<y} \tilde{\chi}(x)\, dx$, which is dominated by $\int_{|x|\leq 2y} \tilde{\chi}(x)\, dx$ and is likewise $o(y^n)$. The fourth integral is of course dominated by the second integral and so is $o(y^n)$. Thus $\sigma(u, y) = c_1 y^n + o(y^n)$, as $y \to 0$ with $|u| < y$ and the assertion that $\sigma(u, y) \sim c_1 y^n$ is proved.

It follows that there is a closed subset F_0 of F with $m(F - F_0)$ fixed but arbitrarily small, so that $\sigma(u, y) \geq c_2 y^n$, if $|u - u^0| < y$ and $0 < y < c_3$, for appropriate positive constants c_2 and c_3. Now let $\mathscr{R} = \bigcup_{u^0 \in F^0} \Gamma_1^{c_3}(u^0)$, where $\Gamma_1^{c_3}(u^0)$ is the truncated cone given by $\Gamma_1^{c_3}(u_3) = \{(u, y) : |u - u_0| < y, 0 < y < c_3\}$. The finiteness of (29) and what we have proved about σ then implies the finiteness of

$$\iint_{\mathscr{R}} |U(u, y)|^2\, y\, du\, dy.$$

However, by a very simple reasoning we have already used in Chapter VI, (see page 208) the finiteness of the last integral implies the finiteness for almost every $u^0 \in F_0$ of

$$\iint_{\substack{|u-u^0|<y \\ y<c_3}} |U(u, y)|^2\, y^{1-n}\, du\, dy = \iint_{\substack{|t|<y \\ y<c_3}} |U(u^0 + t, y)|^2\, y^{1-n}\, dt\, dy.$$

The full integral, $\iint_{|t|\leq y} |U(u^0 + t, y)|^2\, y^{1-n}\, dt\, dy$ then converges because of the local square-integrability of U that we assumed. Since the set F_0 differed from F by a subset of arbitrarily small measure the proof of the theorem is therefore complete.

4.2   Remarks.

4.2.1   The first set of comments are of a rather trifling nature. By the usual arguments (or by a closer examination of the proof) it can be seen that the assumption (25) can be replaced by the weaker assumption that

$$\iint_{|t|<\alpha y} y^{1-n} |U(x^0 + t, y) + U(x^0 - t, y)|^2 \, dt \, dy < \infty$$

where the constant α which determines the opening of the cone may vary with the point x^0. One can also obtain the conclusion (which is only stronger in appearance) that

$$\iint_{|t|<\beta y} y^{1-n} |U(x^0 + t, y)|^2 \, dt \, dy < \infty$$

for every β, as x^0 ranges over almost all points of E.

There are also simple (and immediate) variants of the theorem where the sum $U(x^0 + t, y) + U(x^0 - t, y)$ is replaced by other combinations such as $U(x^0 + t, y) - U(x^0 - t, y)$; also the square which appears may be replaced in (24), (25), and (26) with $y|U(x^0, y)|^p$, $y^{1-n} |U(x^0 + t, y) + U(x^0 - t, y)|^p$ and $y^{1-n} |U(x^0 + t, y)|^p$ respectively, with $p < \infty$. The variant for $p = \infty$ is stated separately below.

4.2.2   Suppose that for every x^0 in a set E

$$\sup_{y > 0} |U(x^0, y)| < \infty \quad \text{and} \quad \sup_{|t| \leq y} |U(x^0 + t, y) + U(x^0 - t, y)| < \infty.$$

Then for almost every x^0 in E, $\sup_{|t| \leq y} |U(x^0 + t, y)|$ is also finite. The proof of this statement is similar to that of Theorem 5, and is conceptually even simpler. The reader should have no difficulty in carrying out the details of the argument. The one point here which may involve non-measurable sets is disposed of as in §3.1.1.

4.2.3   There is a consequence of the statement in §4.2.2 which is worth treating separately. Suppose f is defined near x^0. We consider the significant condition

$$(30) \qquad f(x^0 + t) + f(x^0 - t) - 2f(x^0) = O(|t|) \quad \text{as} \quad t \to 0.$$

This condition is certainly satisfied if f has an ordinary derivative at x^0, but the converse is not true. It can be shown, moreover, that there are f which satisfy (30) uniformly for all x^0 without having an ordinary derivative at any x^0.

It turns out that while (30) by itself does not imply differentiability, it does play the role of a "Tauberian condition," allowing one to refine one form of differentiability to another form. We formulate one such result.

COROLLARY. *Suppose that f has a derivative in the harmonic sense at every point x^0 of a given set E. (This would hold in particular if f were differentiable in the L^q sense for each $x^0 \in E$). Assume also that (30) holds for each $x^0 \in E$. Then f has an ordinary derivative for almost every $x^0 \in E$.*

The proof is an easy application of what we have already done. By the splitting theorem in §2.2 we can write $f = g + b$, where g is differentiable almost everywhere in the ordinary sense, and where b vanishes on F, with $F \subset E$, and $m(E - F)$ small. It will suffice to show that b has an ordinary derivative almost everywhere on F. Since g is differentiable almost everywhere it satisfies condition (30) almost everywhere, and so b satisfies condition (30) at almost every point of F. The crucial fact is that b vanishes on F. Therefore the conditions becomes

$$b(x^0 + t) + b(x^0 - t) = O(|t|), \qquad t \to 0.$$

Now let $U(x, y) = \dfrac{b(x)}{y}$. Thus by the vanishing of b, we have

$$\sup_{y > 0} |U(x^0, y)| = 0$$

for almost every $x^0 \in F$, and

$$\sup_{|t| \le y} |U(x^0 + t, y) + U(x^0 - t, y)| < \infty,$$

for almost every $x^0 \in F$. By the statement in §4.2.2 we conclude that

$$\sup_{|t| = y} |U(x^0 + t, y)| < \infty,$$

which is

$$\sup_{t \ne 0} \frac{|b(x^0 + t)|}{|t|} < \infty,$$

i.e.

$$b(x^0 + t) = O(|t|), \quad \text{as} \quad t \to 0,$$

again for almost all $x^0 \in F$. Theorem 3 in §3.3 now allows us to conclude that b has an ordinary derivative almost everywhere in F.

5. Another characterization of differentiability

5.1   The theorem we intend to prove is as follows.

THEOREM 6.   *Suppose $f \in L^2(\mathbf{R}^n)$. Then for almost every $x^0 \in \mathbf{R}^n$ the following two conditions are equivalent:*
   (i) *f has a derivative in the L^2 sense at x^0.*

(ii) $$\int_{\mathbf{R}^n} \frac{|f(x^0 + t) + f(x^0 - t) - 2f(x^0)|^2 \, dt}{|t|^{n+2}} < \infty.$$

A consequence of this theorem is the following corollary.

COROLLARY.   *Suppose f is given in an open neighborhood of a set E. Then f has an ordinary derivative at almost every $x^0 \in E$, if and only if the following two conditions hold for almost every $x^0 \in E$.*

   (a) $f(x^0 + t) + f(x^0 - t) - 2f(x^0) = O(|t|), \quad \text{as} \quad t \to 0$

   (b) $\displaystyle\int_{|t| \leq \delta} \frac{|f(x^0 + t) + f(x^0 - t) - 2f(x^0)|^2 \, dt}{|t|^{n+2}} < \infty$

where δ is a sufficiently small positive number (depending on x^0).

5.2   We shall need to establish a fact which may be viewed as the "global" analogue of our theorem (the relevant notions of global and local are discussed in the introductory remarks of this chapter).

PROPOSITION.   *Suppose $f \in L_1^2(\mathbf{R}^n)$. Then*

(31) $$\int_{\mathbf{R}^n} \int_{\mathbf{R}^n} \frac{|f(x + t) + f(x - t) - 2f(x)|^2}{|t|^{n+2}} \, dx \, dt = a_n \int_{\mathbf{R}^n} \sum_{j=1}^{n} \left| \frac{\partial f}{\partial x_j} \right|^2 dx.$$

The proof of this is best carried out by the Fourier transform. Thus let

$$\hat{f}(y) = \int_{\mathbf{R}^n} f(x) e^{2\pi i x \cdot y} \, dx, \qquad f(x) = \int_{\mathbf{R}^n} \hat{f}(y) e^{-2\pi i x \cdot y} \, dy,$$

taken in the L^2 sense. By Plancherel's theorem, and the properties of the Sobolov spaces $L_1^2(\mathbf{R}^n)$ we then have

$$\int_{\mathbf{R}^n} \left| \frac{\partial f}{\partial x_j} \right|^2 dx = 4\pi^2 \int_{\mathbf{R}^n} |x_j|^2 |\hat{f}(x)|^2 \, dx.$$

Also

$$\int_{\mathbf{R}^n} |f(x + t) + f(x - t) - 2f(x)|^2 \, dx = 4 \int_{\mathbf{R}^n} |\hat{f}(x)|^2 |1 - \cos 2\pi x \cdot t|^2 dx.$$

Hence

$$\int_{|t|\geq\varepsilon}\int_{\mathbf{R}^n} |f(x+t)+f(x-t)-2f(x)|^2\,dx\,\frac{dt}{|t|^{n+2}} = \int_{\mathbf{R}^n}|\hat{f}(x)|^2\,\mathscr{I}_\varepsilon(x)\,dx,$$

with

$$\mathscr{I}_\varepsilon(x) = 4\int_{|t|\geq\varepsilon}\frac{|1-\cos 2\pi x\cdot t|^2}{|t|^{n+2}}\,dt.$$

Now it is clear that as $\varepsilon\to 0$, $\mathscr{I}_\varepsilon(x)$ increases monotonically to

$$\mathscr{I}(x) = 4\int_{\mathbf{R}^n}\frac{|1-\cos 2\pi x\cdot t|^2}{|t|^{n+2}}\,dt,$$

which is obviously radial in x and homogeneous of degree 2. Thus

$$\mathscr{I}(x) = b_n\,|x|^2,\quad\text{with}\quad b_n = 4\int_{\mathbf{R}^n}\frac{|1-\cos 2\pi t_1|^2}{|t|^{n+2}}\,dt.$$

Hence the left-side of (31) is $b_n\int_{\mathbf{R}^n}|x|^2\,|\hat{f}(x)|^2\,dx$, while the right side is $a_n 4\pi^2\int_{\mathbf{R}^n}|x|^2\,|\hat{f}(x)|^2\,dx$. The proposition is therefore proved with $a_n = b_n/4\pi^2$. The reader should compare this proposition with Proposition 5 in §3.5 of Chapter V.

5.3 A moment's reflection shows that it suffices to prove the theorem, and its corollary, under the assumption that f vanishes outside a bounded set. We shall therefore make this assumption about f. Assume that f has a derivative in the L^2 sense at each point of a set E. Then in view of the fact that this implies that f has a harmonic derivative at each such point we get by Theorem 2 in §2.2 that we can split f as $f = g + b$ with $g\in L_1^\infty(\mathbf{R}^n)$, and $b = 0$ on F, where $F\subset E$ and $m(E-F)$ is small. Nothing will be changed if we assume that g (and thus b) also vanishes outside a bounded set. Since g has bounded support then $g\in L_1^\infty(\mathbf{R}^n)$, and therefore by the above proposition

$$(32)\qquad \int_{\mathbf{R}^n}\frac{|g(x^0+t)+g(x^0-t)-2g(x^0)|^2}{|t|^{n+2}}\,dt < \infty$$

for almost every $x^0\in\mathbf{R}^n$.

Further, g also has an ordinary derivative at almost every point in $\mathbf{R}^n$ (see Theorem 1), and therefore b has a derivative in the L^2 sense for almost every point of E, and hence for almost every point of F. Since b vanishes on F we get by (17) in §3.2 that $\int_{|t|\leq 1}\frac{|b(x^0+t)|^2}{|t|^{n+2}}\,dt < \infty$, for almost every x^0 in F. Since in any case $b \in L^2(\mathbf{R}^n)$ we see that for these x^0,

$$(33) \qquad \int_{\mathbf{R}^n} \frac{|b(x^0 + t)|^2}{|t|^{n+2}} \, dt < \infty.$$

If we combine this with the fact that $b(x^0) = 0$ and (32) we obtain the conclusion (ii) of the theorem.

5.4   We now come to the converse part of Theorem 6. Let $u(x, y) = P_y * f$ be the Poisson integral of f, and consider $\dfrac{\partial^2 u}{\partial y^2}$. For this purpose we need a favorable estimate of $\dfrac{\partial^2 P_y}{\partial y^2}$. Since

$$\frac{\partial^2 P_y}{\partial y^2} = - \sum_{j=1}^{n} \frac{\partial^2}{\partial x_j^2} (P_y(x)),$$

it suffices to observe that

$$\left| \frac{\partial^2}{\partial x_j^2} P_y(x) \right| \le y^{-n-2} \psi(x/y) \quad \text{where} \quad \psi(x) \le A(1 + |x|)^{-n-3}.$$

We remark also that

$$\int_{\mathbf{R}^n} \frac{\partial^2}{\partial y^2} P_y(x) \, dx = 0,$$

and what is crucial, that $\dfrac{\partial^2 P_y}{\partial y^2}(x)$ is radial, and hence even in x. In view of these facts

$$(34) \qquad \frac{\partial^2 u}{\partial y^2} = \frac{1}{2} \int_{\mathbf{R}^n} \frac{\partial^2}{\partial y^2} P_y(t)[f(x + t) + f(x - t) - 2f(x)] \, dt$$

and hence

$$(35) \qquad \left| \frac{\partial^2 u}{\partial y^2} \right| \le A' y^{-n-2} \int_{|t| \le y} |\Delta_t| \, dt + A' y \int_{|t| \ge y} \frac{|\Delta_t|}{|t|^{n+3}} \, dt = I_1(y) + I_2(y).$$

Here we have set $\Delta_t = f(x + t) + f(x - t) - 2f(x)$. We estimate $\displaystyle\int_0^\infty y \left| \frac{\partial^2 u}{\partial y^2} \right|^2 dy$ in terms of similar integrals for I_1 and I_2. By Schwarz's inequality,

$$|I_1(y)|^2 \le (A')^2 y^{-2n-4} \left(\int_{|t| \le y} \frac{|\Delta_t|^2 \, dt}{|t|^{n+1}} \right) \left(\int_{|t| \le y} |t|^{n+1} \, dt \right)$$

$$= By^{-3} \int_{|t| \le y} \frac{|\Delta_t|^2}{|t|^{n+1}} \, dt.$$

Therefore

$$\int_0^\infty y\,|I_1(y)|^2\,dy \le B\int_0^\infty y^{-2}\int_{|t|\le y}\frac{|\Delta_t|^2}{|t|^{n+1}}\,dt\,dy = B\int_{R^n}\frac{|\Delta_t|^2}{|t|^{n+2}}\,dt.$$

A similar argument works for I_2 and altogether this gives

(36) $$\int_0^\infty y\left|\frac{\partial^2 u}{\partial y^2}\right|^2 dy \le B'\int_{R^n}\frac{|f(x+t)+f(x-t)-2f(x)|^2}{|t|^{n+2}}\,dt.$$

In view of the evenness of $\dfrac{\partial^2 P}{\partial y^2}$ we can also modify (34) and write it in a more extended form

(34') $$\frac{\partial^2 u}{\partial y^2}(x+\tau,y) + \frac{\partial^2 u}{\partial y^2}(x-\tau,y)$$
$$= \int_{R^n}\frac{\partial^2}{\partial y^2}P_y(t+\tau)[f(x+t)+f(x-t)-2f(x)]\,dt,$$

where τ is arbitrary. Now if $|\tau|\le y$ we can make similar estimates on $\dfrac{\partial^2}{\partial y^2}P_y(t+\tau)$ and get that

(35') $$\left|\frac{\partial^2 u}{\partial y^2}(x+\tau,y) + \frac{\partial^2 u}{\partial y^2}(x-\tau,y)\right| \le A\{I_1(y)+I_2(y)\}, \quad\text{if } |\tau|\le y.$$

Therefore by the same argument we obtain instead of (36) the conclusion that

(36') $$\iint_{|\tau|\le y}\left|\frac{\partial^2 u}{\partial y^2}(x+\tau,y) + \frac{\partial^2 u}{\partial y^2}(x-\tau,y)\right|^2 y^{1-n}\,d\tau\,dy$$
$$\le B\int_{R^n}\frac{|f(x+t)+f(x-t)-2f(x)|^2}{|t|^{n+2}}\,dt.$$

We can now invoke the desymmetrization theorem in §4.1 at each point where the condition (ii) of the statement of Theorem 6 is valid. Here we have set $U(x,y) = \dfrac{\partial^2 u}{\partial y^2}(x,y)$ with $U = 0$, if $y > 1$. The conclusion is that at almost all such points x^0

(37) $$\iint_{|\tau|\le y, y\le 2}\left|\frac{\partial^2 u}{\partial y^2}(x^0+\tau)\right|^2 y^{1-n}\,d\tau\,dy < \infty$$

and therefore by the theory of the previous chapter (see §2.5, in particular p. 213), $\dfrac{\partial u}{\partial y}$ has a non-tangential limit for almost all such x^0; and finally because of §2.5 in Chapter VII, u has a harmonic derivative at almost all points where condition (ii) holds.

We can now split f as $g + b$, where g is in $L_1^\infty(\mathbf{R}^n)$, and by the above we know that the integral condition holds almost everywhere for g instead of f. Thus it holds for almost all relevant points for b; but b also vanishes for these points. In summary we obtain that

$$(38) \quad \int_{\mathbf{R}^n} \frac{|b(x^0 + t) + b(x^0 - t)|^2}{|t|^{n+2}} \, dt < \infty, \qquad b(x^0) = 0, \quad x^0 \in F$$

where $F \subset E$, and $m(E - F)$ is small.

We now again involve the desymmetrization theorem, this time with $U(x, y) = y^{-2}b(x)$, if $y < 1$. Then we in fact have

$$\int_0^\infty y \, |U(x^0, y)|^2 \, dy = 0, \qquad x^0 \in F$$

and

$$\iint_{|t| \le y} y^{1-n} \, |U(x^0 + t, y) + U(x^0 - t, y)|^2 \, dt \, dy < \infty,$$

as easy consequence of (38). The result is then that

$$\int_{|t| \le 1} \frac{|b(x^0 + t)|^2}{|t|^{n+2}} \, dt < \infty,$$

for almost all x^0 in F, and at such points clearly

$$\int_{|t| \le r} |b(x^0 + t)|^2 \, dt = o(r^{n+2}), \quad \text{as} \quad r \to 0$$

which means in particular that b has a derivative in the L^2 sense. The converse is therefore also proved. The corollary is an immediate consequence of the theorem and the corollary in §4.2.3, page 261.

6. Further results

6.1 Most of the results in this chapter have analogues for derivatives of higher order. The appropriate definitions are as follows. Let k be integral, $k \ge 1$. We say that f has an *ordinary derivative of order* k at x^0 if there exists a polynomial of degree $\le k$ in y, $P_{x_0}(y)$, so that

$$f(x^0 + y) - P_{x_0}(y) = o(|y|^k), \quad \text{as} \quad y \to 0.$$

Similarly f will be said to have a *derivative of order* k *in the* L^q *sense* if

$$h^{-n} \int_{|y| \le h} |f(x^0 + y) - P_{x_0}(y)|^q \, dy = o(h^{kq}),$$

as $h \to 0$. Finally if we assume f to be integrable near x^0, and set equal to zero outside a neighborhood of x^0, we shall say that it has a *harmonic derivative of order k at x^0 if the* $\left\{ \dfrac{\partial^\alpha u}{\partial x^\alpha}(x, y) \right\}_{|\alpha| \le k}$, all have non-tangential limits at x^0.

The generalization of Theorem 1 is the statement: If $1 < p \le \infty$, and $f \in L_k^p(\mathbf{R}^n)$, then f has an ordinary derivative of order k for almost all points of $\mathbf{R}^n$, if $p > n/k$; however if $p < n/k$, then f has a derivative of order k in the L^q sense for almost all points, where $1/q = 1/p - k/n$. The generalization of Theorem 2 is the splitting $f = g + b$, where $g \in L_k^\infty(\mathbf{R}^n)$, b vanishes on F, where $F \subset E$, $m(E - F) < \varepsilon$, and f is assumed to have a derivative of order k in the harmonic sense for all points of E. In generalizing Theorem 3 it suffices to assume that for each $x_0 \in E$, there exists a polynomial of degree $\le k - 1$ in y, $\tilde{P}_{x_0}(y)$, so that

$$f(x + y) - \tilde{P}_{x_0}(y) = O(|y|^k), \quad \text{as} \quad y \to 0.$$

Then f has an ordinary derivative of order k for almost every $x^0 \in E$.

The condition (ii) for the kernel $K(x)$ which arises in Theorem 4 should be amplified to read

$$|\nabla^r K(x)| \le A/|x|^{n+r}, \quad 0 \le r \le k;$$

with this modification the singular integrals in question also preserve almost everywhere the differentiability of order k in the L^q sense.

Finally if $f \in L^2(\mathbf{R}^n)$ the condition characterizing the existence of derivatives of order k in the L^2 sense is given by the finiteness for almost all points in question of

$$\int_{\mathbf{R}^n} \frac{|\Delta_{x_0}^k(t)|^2}{|t|^{n+2k}} \, dt.$$

Here $f(x^0 + t) - P_{x_0}(t) = R_{x_0}(t)$, and $\Delta_{x_0}^k(t) = R_{x_0}(t) + (-1)^{k-1} R_{x_0}(-t)$. For the above see Calderón and Zygmund [7], Stein and Zygmund [1], and Stein [8].

6.2 The result of Theorem 1 holds also for $p = 1$, but requires a different argument. (Compare however the inequality in §2.5 of Chapter V, and the identity (18) in §2.3 of that chapter.) A further consequence of the above is the following theorem. If f is of bounded variation on $\mathbf{R}^n$ in the sense of Tonelli, then for almost every point f has a derivative in the L^q sense, where

$$q = n/(n - 1).$$

See Calderón and Zygmund [6].

6.3 The splitting theorem (of §2.2) can be given in somewhat sharper form if we assume differentiability in the sense of L^q. We formulate the result for derivatives of order 1. Suppose $f \in L^q(\mathbf{R}^n)$, $1 \le q$, and for each point $x^0 \in F$, with F compact,

$$h^{-n} \int_{|y| \le h} |f(x^0 + y) - f(x^0)|^q \, dy \le A h^q, \quad 0 < h < \infty$$

with A independent of x^0. Then $f = g + b$, where g is continuously differentiable and it and its first partial derivatives are bounded; $b = 0$ on F. For the general formulation for higher derivatives and further details see Calderón and Zygmund [8].

6.4 In view of the measurability difficulties discussed in §3.1.1 it may be of interest to state the following theorem. Suppose f is Lebesgue measurable on $\mathbf{R}^n$. Then the set of points where f has an ordinary derivative is Lebesgue measurable (see Haslam-Jones [1]).

The fact that the set $E = \left\{ x : \limsup_{h \to 0} \left| \dfrac{f(x + h) - f(x)}{h} \right| < \infty \right\}$ is Lebesgue measurable may be seen as follows: For each integer k let

$$E_k = \{ x : |f(x + h) - f(x)| \le k\,|h|, \quad \text{for all } |h| < 1/k \}.$$

Then $E = \bigcup E_k$; also because we have taken the open balls $|h| < 1/k$, it follows that $f\big|_{E_k}$ is continuous and E_k is closed.*

6.5 Suppose k is odd. We shall say that f has a *symmetric derivative of order* k, *in the L^q sense*, at x^0, if there exists a polynomial $P_{x_0}(y)$, in y of degree $\le k$, such that

$$h^{-n} \int_{|y| \le h} |f(x_0 + y) - f(x_0 - y) - P_{x_0}(y)|^q \, dy = o(h^{kq})$$

as $y \to 0$. If k is even we make the same definition except that $f(x_0 + y) - f(x_0 - y)$ must be replaced by $f(x_0 + y) + f(x_0 - y)$.

THEOREM: *Suppose that for each $x^0 \in E$, f has a symmetric derivative of order k in the L^q sense. Then for almost every x^0 in E, f has a derivative of order k in the L^q sense.*

For $n = 1, k = 1$, and for the ordinary notion of differentiability, this theorem goes back to Khintchine [1]. For the one-dimensional form see M. Weiss [1]. The general formulation can be deduced from the methods of the present chapter. Assume for example that $k = 1$. Let u be the Poisson integral of f and consider $\dfrac{\partial u}{\partial x_j} = \dfrac{\partial P_y}{\partial x_j} * f$. Since the kernel $\dfrac{\partial P}{\partial x_j}$ is odd we can write

$$\frac{\partial u}{\partial x_j}(x + \tau, y) + \frac{\partial u}{\partial x_j}(x - \tau, y) = \int_{\mathbf{R}^n} \frac{\partial}{\partial x_j} P_y(t + \tau)[f(x + t) - f(x - t)]\, dt$$

and our assumption shows that $\dfrac{\partial u}{\partial x_j}(x^0 + \tau, y) + \dfrac{\partial u}{\partial x_j}(x^0 - \tau, y)$ tends to a limit as $|\tau| \le y$ and $y \to 0$, for $x^0 \in E$. Thus by the desymmetrization theorem in §4.2.2 we get that $\dfrac{\partial u}{\partial x_j}$ is non-tangentially bounded at x^0, for almost all $x^0 \in E$. Therefore f has a harmonic derivative of order 1 almost everywhere in E and hence by the splitting lemma one can reduce matters to the special case where f vanishes on E.

* I am indebted to H. Federer for this argument.

More generally for k even we consider $\dfrac{\partial^k u}{\partial y^k}(x + \tau, y) + \dfrac{\partial^k}{\partial y^k} u(x - \tau, y)$;

while for odd k one takes

$$\frac{\partial^k}{\partial y^{k-1}\,\partial x_j} u(x + \tau, y) + \frac{\partial^k}{\partial y^{k-1}\,\partial x_j} u(x - \tau, y), \qquad j = 1, \ldots, n;$$

then one applies the results of Chapter VII to show that f has almost everywhere on E a harmonic derivative of order k, thus reducing matters to the special case where f vanishes on E. For this special case, which is much easier, the kind of arguments used in §4 apply; see for instance Stein and Zygmund [1, Lemma 14].

6.6 Suppose f has a derivative in the L^q sense at each $x^0 \in E$. Then the first partial derivatives in the L^q sense of f exist for almost every $x^0 \in E$. More particularly if

$$\frac{1}{h^n} \int_{|y| \leq h} |f(x^0 + y) - f(x^0) - \sum \alpha^j_{x_0} y_j|^q \, dy = o(h^q), \qquad h \to 0$$

for $x^0 \in E$, then if e_j is the unit vector $(0, \ldots, 1, 0, \ldots, 0)$

$$\frac{1}{h} \int_{|y_j| \leq h} |f(x^0 + e_j y_j) - f(x^0) - \alpha^j_{x_0} y_j|^q \, dy_j = o(h^q)$$

for almost all $x^0 \in E$. For this and related results see M. Weiss [2] and [3].

6.7 Suppose $2 \leq q < \infty$. Let $f \in L^q(\mathbf{R}^n)$. Then f has a derivative of order one in the L^q sense for almost all $x^0 \in E$ if and only if the following two conditions are satisfied for almost all $x^0 \in E$:

(1) $\displaystyle \int_{\mathbf{R}^n} \frac{|f(x^0 + t) + f(x^0 - t) - 2f(x^0)|^q}{|t|^{n+q}} \, dt < \infty$

(2) $\displaystyle \int_{\mathbf{R}^n} \frac{|f(x^0 + t) + f(x^0 - t) - 2f(x^0)|^2}{|t|^{n+2}} \, dt < \infty.$

See Stein and Zygmund [1]; also Wheeden [1], Neugebauer [1].

6.8 The following example shows that Theorem 4 in §3.3 does not extend to the notion of the ordinary derivative (i.e. the case $q = \infty$). We consider the case of $\mathbf{R}^1$, with T the Hilbert transform.

Consider first a function $F_0(x)$ defined on $\mathbf{R}^1$ with the properties: (1) $F_0(x) = (\log 1/|x|)^{1-\varepsilon}$ for $|x| \leq 1/2$, $(0 < \varepsilon < 1)$; $F_0(x)$ vanishes outside a compact set and is smooth away from the origin; also $F_0(x) \geq 0$, $x \in \mathbf{R}^1$. Let $\tilde{F}_0(x)$ denote the Hilbert transform of F_0. It is not difficult to see that $\tilde{F}_0$ (after a suitable modification on a set of measure zero) is absolutely continuous on $\mathbf{R}^1$ and $\dfrac{d\tilde{F}_0}{dx} \in L^1(\mathbf{R})$. Write $F(x) = \sum 2^{-k} F_0(x + r_k)$, where $r_1, r_2, r_k, \ldots$ is an enumeration of the rationals. Then $\tilde{F} = \sum 2^{-k} \tilde{F}_0(x + r_k)$ is the Hilbert transform of F

and is absolutely continuous, with $\dfrac{d\tilde{F}}{dx} \in L^1(\mathbf{R}^1)$. Thus $\tilde{F}$ has an ordinary derivative for almost every x, but since F is unbounded near every point, it has an ordinary derivative nowhere.

For the periodic analogue of F_0 we may take f_0 with

$$f_0(x) \sim \sum_{n>1} \frac{\cos nx}{n(\log n)^\varepsilon}, \qquad \tilde{f}_0(x) \sim \sum_{n>1} \frac{\sin nx}{n(\log n)^\varepsilon}.$$

For a treatment of these series see Zygmund [8, Chapter V].

Notes

Section 1. The notion of a function differentiable at a given point in the L^p sense was first studied systematically by Calderón and Zygmund [7]. Part (b) of Theorem 1 is due to them, but part (a) dealing with the ordinary derivative is older; see for example Cesari [1]. The reader is also referred to the work of Federer [1] for a variety of topics related to the material in this chapter.

Section 2. The idea of splitting of functions, in the context of ordinary differentiability of one variable appears first in Marcinkiewicz [1]. This basic technique was extended to n-dimensions in Calderón and Zygmund [7]; the presentation given here, which is based on the theory of harmonic functions, has several essential advantages over the previous methods. It is due to Zygmund and the author, and is sketched in the author's survey, Stein [8].

Section 3. Theorem 3 is a famous theorem of Denjoy, Rademacher and Stepanov; see Saks [2, Chapter IX].

The proof given here is, of course, not the standard one, relying as it does on the notion of the harmonic derivative. For §3.2, as well as variants of Theorem 4, see Calderón and Zygmund [7].

Section 5. The original proof of Theorem 6 and its corollary is in Stein and Zygmund [1]; see also Wheeden [1]. The argument given here is also due to Zygmund and the author, and is sketched in Stein [8].

Appendices

A. Some inequalities

We shall summarize here some well-known inequalities that are used systematically above. Further details may be found in Zygmund [8], Chapter I, and Hardy, Littlewood, and Polya [1].

A.1 Minkowski's inequality for integrals states in effect that the norm of an integral is not greater than the integral of the corresponding norms. In explicit form, for the case of L^p spaces, this can be restated as follows. Let $1 \leq p < \infty$, then

$$\left(\int_{\mathcal{Y}} \left(\int_{\mathcal{X}} |F(x, y)| \, dx \right)^p dy \right)^{1/p} \leq \int_{\mathcal{X}} \left(\int_{\mathcal{Y}} |F(x, y)|^p \, dy \right)^{1/p} dx.$$

Here $F(x, y)$ is a measurable function on the σ-finite product measure space $\mathcal{X} \times \mathcal{Y}$; dx and dy are the measures on $\mathcal{X}$ and $\mathcal{Y}$ respectively.

A.2 Young's inequality for convolutions is as follows: Let $h = f * g$, then

$$\|h\|_q \leq \|f\|_p \|g\|_r,$$

where $1 \leq p, q, r \leq \infty$ and $1/q = 1/p + 1/r - 1$.

Two noteworthy special cases arise; first $r = 1$, then $p = q$; also when r is the conjugate index to p (namely $1/p + 1/r = 1$), then $q = \infty$. In this case it can also be shown that h is continuous.

A.3 The following is a general integral inequality of wide application.

Let $(Tf)(x) = \int_0^\infty K(x, y) f(y) \, dy$. Here K is assumed to be homogeneous of degree -1; that is $K(\lambda x, \lambda y) = \lambda^{-1} K(x, y)$, for $\lambda > 0$. In addition we assume that $\int_0^\infty |K(1, y)| y^{-1/p} \, dy = A_K < \infty$. Then

$$\|Tf\|_p \leq A_K \|f\|_p.$$

Here the norms $\|\cdot\|_p$ are those of $L^p(0, \infty; dx)$, and $1 \leq p \leq \infty$.

To prove this we write $(Tf)(x) = \int_0^\infty K(1, y) f(yx) \, dy$ and use Minkowski's inequality for integrals.

An interesting special case, the *Hilbert integral*, arises if $K(x, y) = \dfrac{1}{x + y}$.

A.4   Another useful instance of §A.3 is the pair of inequalities due to Hardy

$$\left(\int_0^\infty \left(\int_0^x f(y)\, dy\right)^p x^{-r-1}\, dx\right)^{1/p} \le p/r \left(\int_0^\infty (yf(y))^p y^{-r-1}\, dy\right)^{1/p},$$

$$\left(\int_0^\infty \left(\int_x^\infty f(y)\, dy\right)^p x^{r-1}\, dx\right)^{1/p} \le p/r \left(\int_0^\infty (yf(y))^p y^{r-1}\, dy\right)^{1/p}.$$

Here $f \ge 0$, $p \ge 1$, and $r > 0$.

B. The Marcinkiewicz interpolation theorem

B.1   We extend here the theorem given in §4 of Chapter I. We assume that p_0, p_1, q_0, q_1 are given exponents, with $1 \le p_i \le q_i \le \infty$, $p_0 < p_1$, and $q_0 \ne q_1$. T is a sub-additive transformation which is defined on $L^{p_0}(\mathbf{R}^n) + L^{p_1}(\mathbf{R}^n)$. We recall the definition that T is of *weak type* (p_i, q_i). This means that there is a constant A_i, so that for every $f \in L^{p_i}(\mathbf{R}^n)$

$$m\{x: |Tf(x)| > \alpha\} \le \left(\frac{A_i \, \|f\|_{p_i}}{\alpha}\right)^{q_i}, \quad \text{all} \quad \alpha > 0.$$

If $q_i = \infty$, it means that $\|Tf\|_{q_i} \le A_i \|f\|_{p_i}$.

THEOREM.   *Suppose T is simultaneously of weak-types (p_0, q_0) and (p_1, q_1). If $0 < \theta < 1$, and $1/p = \dfrac{(1 - \theta)}{p_0} + \theta/p_1$, $1/q = \dfrac{(1 - \theta)}{q_0} + \theta/q_1$, then T is of type (p, q), namely*

$$\|Tf\|_q \le A \|f\|_p, \qquad f \in L^p(\mathbf{R}^n).$$

Here $A = A(A_i, p_i, q_i, \theta)$, but it does not otherwise depend on T or f.

From the proof given below it is easy to see that the theorem can be extended to the following situations: First the underlying measure space $\mathbf{R}^n$ of the $L^{p_i}(\mathbf{R}^n)$, can be replaced by a general measure space (and the measure space occurring in the domain of T need not be the same as the one entering in the range of T). Secondly the sub-additivity condition can be replaced by the more general condition $|T(f_1 + f_2)(x)| \le K\{|Tf_1(x)| + |Tf_2(x)|\}$. A less superficial generalization of the theorem can be given in terms of the notion of Lorentz spaces, which unify and generalize the usual L^p spaces and the weak-type spaces. For a discussion of this more general form of the Marcinkiewicz interpolation theorem see *Fourier Analysis*, Chapter V.

B.2   Suppose h is a given measurable function on $\mathbf{R}^n$. We have already used the notion of its distribution function $\lambda(\alpha)$, defined by $\lambda(\alpha) = m\{x: |h(x)| > \alpha\}$, with m Lebesgue measure on $\mathbf{R}^n$. For the proof of the above theorem and in other situations it is useful to consider the concept of the non-increasing re-arrangement of h. This is a function h^*, defined on $(0, \infty)$ which has the same distribution function as h, but which is non-increasing on $(0, \infty)$. h^* is defined as $h^*(t) = \inf \{\alpha, \lambda(\alpha) \le t\}$. Both h^* and λ are non-negative non-increasing functions, and are continuous on the right. h^* and h have the same distribution function, since the set where $h^*(t) > \alpha$ is the interval $0 \leq t < \lambda(\alpha)$, which of course has measure $\lambda(\alpha)$. Thus

$$\|h^*(t)\|_p = \left(\int_0^\infty |h^*(t)|^p \, dt \right)^{1/p} = \left(\int_{R^n} |h(x)|^p \, dx \right)^{1/p} = \|h\|_p.$$

B.3 We shall also make use of some integral inequalities for functions on $(0, \infty)$.

The first is the set of Hardy inequalities given in Appendix A.4. In addition we also need the observation that

$$\left(\int_0^\infty [t^{1/p} h(t)]^{q_2} \frac{dt}{t} \right)^{1/q_2} \leq A \left(\int_0^\infty [t^{1/p} h(t)]^{q_1} \frac{dt}{t} \right)^{1/q_1}$$

wherever h is a non-negative non-increasing function on $(0, \infty)$, $0 < p \leq \infty$, and $q_1 \leq q_2 \leq \infty$. $A = A(p, q_1, q_2)$ does not depend on h. To prove this inequality assume that the integral on the right side is normalized to be one. Integrating only between $t/2$ and t in this integral (and using the fact that h is at least $h(t)$ in that interval) we get $\sup_{0 < t} t^{1/p} h(t) \leq A_1$. This is the desired result for $q_2 = \infty$. The general result, where $q_1 \leq q_2 < \infty$, then follows by writing

$$\int_0^\infty [t^{1/p} h(t)]^{q_2} \frac{dt}{t} \leq \sup_t [t^{1/p} h(t)]^{q_2 - q_1} \int_0^\infty [t^{1/p} h(t)]^{q_1} \frac{dt}{t}.$$

B.4 We come now to the proof of the Marcinkiewicz theorem. We let σ denote the slope of the line segment in $\mathbf{R}^n$ joining the points $(1/p_0, 1/q_0)$ with $(1/p_1, 1/q_1)$. Since $(1/p, 1/q)$ lies on this segment, we have

$$\sigma = \frac{1/q_0 - 1/q}{1/p_0 - 1/p} = \frac{1/q - 1/q_1}{1/p - 1/p_1}.$$

For any $t > 0$, we split an arbitrary function $f \in L^p(\mathbf{R}^n)$ as follows:

$$f = f^t + f_t$$

where $f^t(x) = f(x)$, if $|f(x)| > f^*(t^\sigma)$, and $f^t(x) = 0$ otherwise; $f_t = f - f^t$.
It follows easily that

$$(f^t)^*(y) \leq f^*(y), \quad \text{if} \quad 0 \leq y \leq t^\sigma \quad \text{and}$$

$$(f^t)^*(y) = 0, \quad \text{if} \quad y > t^\sigma.$$

Also

$$(f_t)^*(y) \leq f^*(t^\sigma), \quad \text{if} \quad y \leq t^\sigma \quad \text{and}$$

$$(f_t)^*(y) \leq f^*(y), \quad \text{if} \quad y \geq t^\sigma.$$

Now if $f = f_1 + f_2$, then as is easily verified

$$(T(f))^*(t) \leq (Tf_1)^*(t/2) + (Tf_2)^*(t/2).$$

The weak-type (p_0, q_0) assumption on T implies that whenever $f \in L^{p_0}(\mathbf{R}^n)$, then $(fT)^*(t) \leq A_0 t^{1/q_0} \|f\|_{p_0}$. Similarly if $f \in L^{p_1}(\mathbf{R}^n)$, $(Tf)^*(t) \leq A_1 t^{1/q_1} \|f\|_{p_1}$. We now use the decomposition $f = f^t + f_t, f \in L^p(\mathbf{R}^n)$, and since $p_0 < p < p_1$, $f^t \in L^{p_0}(\mathbf{R}^n)$, and $f_t \in L^{p_1}(\mathbf{R}^n)$. Inserting this in the above gives

$$T(f)^*(t) \leq A_0 (2/t)^{1/q_0} \|f\|_{p_0} + A_1 (2/t)^{1/q_1} \|f_t\|_{p_1}.$$

However, $\|T(f)\|_q = \|Tf^*\|_q$, and by the inequality in §B.3 the latter is majorized by a constant multiple of

$$\left(\int_0^\infty (t^{1/q} (Tf)^*(t))^p \frac{dt}{t} \right)^{1/p},$$

since $p \leq q$, because $p_i \leq q_i$.

Applying the previous estimate for $(Tf)^*(t)$, reduces the majorization of $\|Tf\|_q$ to a constant multiple of

$$(1) \quad \left\{ \int_0^\infty [t^{1/q - 1/q_0} \|f^t\|_{p_0}]^p \frac{dt}{t} \right\}^{1/p} + \left\{ \int_0^\infty [t^{1/q - 1/q_1} \|f_t\|_{p_1}]^p \frac{dt}{t} \right\}^{1/p}.$$

In view of the estimate $(f^t)^*(y) \leq f^*(y)$, if $y \leq t^\sigma$ and $(f^t)^*(y) = 0$, if $y > t^\sigma$ made earlier, we have that

$$\|f^t\|_{p_0} = \left(\int_0^\infty (y^{1/p_0}(f^t)^*(y))^{p_0} \frac{dy}{y} \right)^{1/p_0} \leq \text{constant} \times \int_0^\infty y^{1/p_0}(f^t)^*(y) \frac{dy}{y}$$

$$\leq \text{constant} \times \int_0^{t^\sigma} y^{1/p_0} f^*(y) \frac{dy}{y}.$$

We insert this estimate for $\|f^t\|_{p_0}$ in the first bracket in (1) above. After the indicated change of variables ($t^\sigma \to t$), and the application of the first of the Hardy inequalities (in Appendix A.4), we see that the first term in (1) is majorized by a constant $\times \|f\|_p$. A similar argument for the second term concludes the proof of the theorem.

C. Some elementary properties of harmonic functions*

C.1 A useful form of the maximum principle can be stated as follows. *Suppose u is a real-valued function of class C^2 in a bounded region $\mathcal{R}$, and let u be continuous in $\overline{\mathcal{R}}$. We assume that $\Delta(u) \geq 0$ in $\mathcal{R}$. If $u \leq 0$ on the boundary of $\mathcal{R}$, then $u \leq 0$, throughout $\mathcal{R}$.*

In proving this assertion it is convenient to make the stronger assumption that $\Delta u > 0$ in $\mathcal{R}$. We can reduce to this special case by considering instead of u, $u + \varepsilon |x|^2 - \delta$, where $\varepsilon > 0$, $\delta > 0$ and both ε and δ are small. Let us assume then that $\Delta u > 0$ in $\mathcal{R}$. If it were not true that $u \leq 0$ in $\mathcal{R}$, then u would have to attain a positive maximum at some point $x^0 \in \mathcal{R}$. Since $(\Delta u)(x^0) > 0$, it must be

* For a discussion of the elementary properties of harmonic functions see also *Fourier Analysis*, Chapter II.

true that $\dfrac{\partial^2 u}{\partial x_j^2}(x^0) > 0$ for at least one j. By the maximum property $\dfrac{\partial u}{\partial x_j}(x^0) = 0$ and so by Taylor's theorem

$$u(x^0 + \xi e_j) - u(x^0) = \frac{\xi^2}{2}\frac{\partial^2 u}{\partial x_j^2}(x^0) + o(\xi^2),$$

where e_j is the unit vector along the x_j direction and ξ is small. This of course contradicts the fact that $u(x^0)$ is the maximum value of u in $\mathscr{R}$.

C.2 Suppose that u is harmonic in $\mathscr{R}$. Then

$$u(x^0) = \frac{1}{\omega_{n-1}}\int_{S^{n-1}} u(x^0 + y'r)\,d\sigma(y').$$

Here $d\sigma(y')$ is the element of volume on the unit sphere S^{n-1} in $\mathbf{R}^n$; ω_{n-1} is the total volume of that sphere; and r is sufficiently small so that the ball of radius r with center x^0 lies entirely in $\mathscr{R}$. This is the *mean-value property* of harmonic functions. This fact follows, as is well-known, from Green's identity

$$\int_{\mathscr{G}}(u\,\Delta v - v\,\Delta u)\,dx = \int_{\partial\mathscr{G}}\left(u\frac{\partial v}{\partial n} - v\frac{\partial u}{\partial n}\right)d\tau$$

where $\mathscr{G}$ is the region contained between the spheres of radius r and radius ε centered at the point x^0 (ε is small); $d\tau$ is the surface element on $\partial\mathscr{G}$, which is the union of these two spheres, and $\dfrac{\partial}{\partial n}$ is the differentiation in the outward normal direction. We take $v(x) = |x - x^0|^{-n+2} = r^{-n+2}$. So $\Delta v = 0$ in $\mathscr{G}$ and $v = 0$ on larger of the two boundary spheres. Letting $\varepsilon \to 0$ and collecting terms gives the desired result.

C.3 The mean-value property and the device of regularization lead immediately to a variety of useful estimates for harmonic functions. Let φ be a fixed C^∞ function in $\mathbf{R}^n$, which is radial, is supported in the unit ball, and is normalized, namely $\int_{\mathbf{R}^n}\varphi(x)\,dx = 1$. Then by integration we get as an immediate consequence of the mean-value property,

$$u(x^0) = \int_{\mathbf{R}^n} u(x^0 - y)\varphi_r(y)\,dy,$$

where $\varphi_r(y) = r^{-n}\varphi(y/r)$.

Here we assume that as before the distance of x^0 from the boundary of $\mathscr{R}$ exceeds r. The above can be rewritten as $u(x^0) = \int_{\mathbf{R}^n} u(y)\varphi_r(x^0 - y)\,dy$ and thus $\left(\dfrac{\partial}{\partial x}\right)^\alpha u(x^0) = \int_{\mathbf{R}^n} u(y)\left(\dfrac{\partial}{\partial x^0}\right)^\alpha \varphi_r(x^0 - y)\,dy$. From this and Schwarz's inequality we get the following inequality for harmonic functions in n variables,

$$\left|\left(\frac{\partial}{\partial x}\right)^\alpha u(x^0)\right| \le A_\alpha r^{-n/2-|\alpha|}\left(\int_{B_r}|u(y)|^2\,dy\right)^{1/2}.$$

B_r denotes the ball of radius r centered at x^0.

C.4 We prove here the assertion made in §3.1.5 of Chapter III, namely that if f has the spherical harmonic development

$$f(x) \sim \sum Y_k(x)$$

and $\int_{S^{n-1}} |Y_k(x)|^2 \, d\sigma(x) = O(k^{-N})$, for every N as $k \to \infty$, then f can be corrected on a set of measure zero so as to be a C^∞ function on S^{n-1}. To prove this it suffices to verify the inequality

$$(*) \qquad \sup_{|x|=1} \left| \frac{\partial^\alpha Y_k(x)}{\partial x^\alpha} \right| \le A'_\alpha k^{(n/2+|\alpha|)} \left(\int_{S^{n-1}} |Y_k(x)|^2 \, d\sigma(x) \right)^{1/2}.$$

Let us normalize Y_k by assuming that $\int_{S^{n-1}} |Y_k(x)|^2 \, d\sigma(x) = 1$. Y_k is of course homogeneous of degree 0. If we set $P_k(x) = |x|^k Y_k(x)$, then P_k is a solid harmonic of degree k. Now

$$\int_{|x| \le 1+\varepsilon} |P_k(x)|^2 \, dx = \left(\int_{S^{n-1}} |Y_k(x)|^2 \, d\sigma(x) \right) \int_0^{1+\varepsilon} r^{2k+n-1} \, dr \le (1+\varepsilon)^{2k+n}.$$

We invoke here the final inequality of §C.3, and take $u = P_k$ with x^0 any point on S^{n-1} and B_r a ball of radius ε centered at x^0. The result is

$$\left| \left(\frac{\partial}{\partial x} \right)^\alpha P_k(x^0) \right| \le A_\alpha \varepsilon^{-n/2-|\alpha|} (1+\varepsilon)^{k+n/2}.$$

ε is at our disposal and we choose it to be equal to $1/k$. Then $(1+\varepsilon)^{k+n/2} \le$ constant, and we have

$$(**) \qquad \sup_{|x|=1} \left| \left(\frac{\partial}{\partial x} \right)^\alpha P_k(x) \right| \le A''_\alpha k^{n/2+|\alpha|}.$$

Finally since $Y_k(x) = |x|^{-k} P_k(x)$, the Leibnitz identity shows that $(**)$ implies $(*)$.

D. Inequalities for Rademacher functions

It is our purpose here to provide a proof for the inequalities for Rademacher functions asserted in §5.2 of Chapter IV. See also Zygmund [8], chapter V.

D.1 Let $\mu, a_0, a_1, \ldots, a_N$, be real numbers. Then because the Rademacher functions are mutually independent variables we have

$$\int_0^1 \exp \mu \sum_{m=0}^N a_m r_m(t) \, dt = \prod_{m=0}^N \int_0^1 \exp \mu a_m r_m(t) \, dt.$$

However in view of their definition, $\int_0^1 e^{\mu a_m r_m(t)} \, dt = \cosh \mu a_m$. If we now make use of the simple inequality $\cosh x \le e^{x^2}$, we obtain

$$\int_0^1 e^{\mu F(t)} \, dt \le \exp \mu \sum a_m^2,$$

with $F(t) = \sum a_m r_m(t)$.

D.2 Let us make the normalizing assumption that $\sum_{n=0}^{N} a_m^2 = 1$. Then since $e^{\mu|F|} \leq e^{\mu F} + e^{-\mu F}$, we have

$$\int_0^1 e^{\mu|F(t)|} dt \leq 2e^{\mu^2}.$$

Let $\lambda(\alpha) = m\{t:|F(t)| > \alpha\}$, be the distribution function of $|F|$. If we take $\mu = \alpha/2$ in the above inequality we have $\lambda(\alpha) \leq 2e^{(\alpha/2)^2 - (\alpha/2)\alpha}$ and so $\lambda(\alpha) \leq 2e^{-\alpha^2/4}$. From this it follows immediately that

$$\left(\int_0^1 |F(t)|^p dt\right)^{1/p} = \|F\|_p \leq A_p \leq Ap^{1/2}, \quad \text{for} \quad p < \infty$$

and so in general

$$\|F(t)\|_p \leq A_p \left(\sum_{m=0}^{N} |a_m|^2\right)^{1/2}, \quad p < \infty.$$

D.3 We shall now extend the last inequality to several variables. The case of two variables is entirely typical of the inductive procedure used in the proof of the general case.

We can also limit ourselves to the situation when $p > 2$, since for the case $p \leq 2$ the desired inequality is a simple consequence of Hölder's inequality and the orthogonality of the Rademacher functions.

We have

$$F(t_1, t_2) = \sum_0^N \sum_0^N a_{m_1 m_2} r_{m_1}(t_1) r_{m_2}(t_2) = \sum_0^N F_{m_1}(t_2) r_{m_1}(t_1).$$

Now by what has just been proved

$$\int_0^1 |F(t_1, t_2)|^p dt_1 \leq A_p^p \left(\sum_{m_1} |F_{m_1}(t_2)|^2\right)^{p/2}.$$

Integrate this with respect to t_2, and use the fact that for any sequence of functions

$$\int_0^1 \left(\sum_{m_1} |F_{m_1}(t_2)|^2\right)^{p/2} dt_2 \leq \left(\sum_{m_1} \left(\int_0^1 |F_{m_1}(t_2)|^p dt_2\right)^{2/p}\right)^{p/2}.$$

This assertion is merely a restatement of a special case of Minkowski's inequality (in Appendix A.1) for the exponent $p/2$, and the function $|F_{m_1}(t_2)|^2$.

However $F_{m_1}(t_2) = \sum_{m_2} a_{m_1 m_2} r_{m_2}(t_2)$, and therefore the case already proved shows that

$$\left(\int_0^1 |F_{m_1}(t_2)|^p dt_2\right)^{2/p} \leq A_p^2 \sum_{m_2} a_{m_1 m_2}^2.$$

Inserting this in the above gives

$$\int_0^1 \int_0^1 |F(t_1, t_2)|^p dt_1 dt_2 \leq A_p^{2p} \sum_{m_2} \sum_{m_1} a_{m_1 m_2}^2$$

which leads to the desired inequality

$$\|F\|_p \leq A'_p \|F\|_2, \qquad p < \infty.$$

D.4　The converse inequality

$$\|F\|_2 \leq B_p \|F\|_p, \qquad p > 0$$

is a simple consequence of the direct inequality.

In fact for any $p > 0$, (here we may assume $p < 2$) define the exponent $r > 2$ by the rule that the exponent 2 should be midway (in the appropriate sense), between p and r. That is, take $1/2 = (1/2)(1/p + 1/r)$. By Hölder's inequality

$$\|F\|_2 \leq \|F\|_p^{1/2} \|F\|_r^{1/2}.$$

We already know that $\|F\|_r \leq A'_r \|F\|_2$, $r > 2$. We therefore get

$$\|F\|_2 \leq (A'_r)^2 \|F\|_p,$$

which is the required converse inequality.

Bibliography*

R. Adams, N. Aronszajn, and K. T. Smith
[1] "Theory of Bessel potentials," Part II, *Ann. Inst. Fourier 17* (1967), 1–135.

S. Agmon, A. Douglis, and L. Nirenberg
[1] "Estimates near the boundary for solutions of elliptic partial differential equations satisfying general boundary conditions. I," *Com. Pure Applied Math. 12* (1959), 623–727, (Chapter V).

N. Aronszajn, F. Mulla, and P. Szeptycki
[1] "On spaces of potentials connected with L^p spaces," *Ann. Inst. Fourier 13* (1963), 211–306.

N. Aronszajn and K. T. Smith
[1] "Theory of Bessel potentials I," *Ann. Inst. Fourier 11* (1961), 385–475.

F. Bagemhil and W. Seidel
[1] "Some boundary properties of analytic functions," *Math. Zeit. 61* (1954), 186–199.

A. Benedek, A. P. Calderón, and R. Panzone
[1] "Convolution operators on Banach space valued functions," *Proc. Nat. Acad. Sci. 48* (1962), 356–365.

A. Besicovitch
[1] "Sur la nature des fonctions à carré sommable mésurables," *Fund. Math. 4* (1923), 172–195.
[2] "On a general metric property of summable functions," *J. London Math. Soc. 1* (1926), 120–128.

O. V. Besov
[1] "On embedding and extension theorems for some function classes" (Russian), *Trudy Mat. Inst. Steklov 60* (1960), 42–81.

O. V. Besov, V. P. Il'in, and P. I. Lizorkin
[1] "The L^p estimates of a certain class of non-isotopic singular integrals," *Dok. Akad. Nauk SSSR, 69* (1966), 1250–1253.

S. Bochner
[1] *Vorlesungen über Fouriersche Integrale*, Leipzig, 1932.
[2] *Harmonic Analysis and the Theory of Probability*, Berkeley, 1955.

S. Bochner and K. Chandrasekharan
[1] *Fourier Transforms*, Princeton, 1949.

* The citation *Fourier Analysis* used throughout the text refers to Stein and Weiss [4].

279

H. Boerner

[1] *Representations of Groups*, Amsterdam, 1963.

H. Busemann and W. Feller

[1] "Zur Differentiation des Lebesguesche Integrale," *Fund. Math. 22* (1934), 226–256.

A. P. Calderón

[1] "On the behavior of harmonic functions near the boundary," *Trans. Amer. Math. Soc. 68* (1950), 47–54.

[2] "On a theorem of Marcinkiewicz and Zygmund," *Trans. Amer. Math. Soc. 68* (1950), 55–61.

[3] "Integrales singulares y sus aplicaciones a ecuciones diferenciales hiperbolicos," *Curos Mat.* no. 3, Univ. of Buenos Aires (1960).

[4] "Lebesgue spaces of differentiable functions and distributions," *Proc. Symp. in Pure Math. 5* (1961), 33–49.

[5] "Boundary value problems for elliptic equations," *Joint Soviet-American Symposium on partial differential equations*, Novosibirsk (1963).

[6] "Commutators of singular integral operators," *Proc. Nat. Acad. Sci. 53* (1965), 1092–1099.

[7] "Singular integrals," *Bull. Amer. Math. Soc. 72* (1966), 426–465.

A. P. Calderón, M. Weiss, and A. Zygmund

[1] "On the existence of singular integrals," *Proc. Symp. Pure Math. 10* (1967), 56–73.

A. P. Calderón and A. Zygmund

[1] "On the existence of certain singular integrals," *Acta Math. 88* (1952), 85–139.

[2] "Singular integrals and periodic functions," *Studia Math. 14* (1954), 249–271.

[3] "On singular integrals," *Amer. J. Math. 78* (1956), 289–309.

[4] "Algebras of certain singular integrals," *Amer. J. Math. 78* (1956), 310–320.

[5] "Singular integral operators and differential equations," *Amer. J. Math. 79* (1957), 801–821.

[6] "On the differentiability of functions which are of bounded variation in Tonelli's sense," *Revista Union Mat. Arg. 20* (1960), 102–121.

[7] "Local properties of solutions of elliptic partial differential equations," *Studia Math. 20* (1961), 171–225.

[8] "On higher gradients of harmonic functions," *Studia Math. 26* (1964), 211–226.

L. Carleson

[1] "On the existence of boundary values of harmonic functions of several variables," *Arkiv. Mat. 4* (1962), 339–393.

[2] "Interpolation of bounded analytic functions and the corona problem," *Ann. of Math. 76* (1962), 547–559.

[3] "On convergence and growth of partial sums of Fourier series," *Acta Math. 116* (1966), 135–157.

L. Cesari
[1] "Sulle funzioni assolutamente continue in due variabili," *Annali di Pisa 10* (1941), 91–101.

M. Cotlar
[1] "Some generalizations of the Hardy-Littlewood maximal theorem," *Rev. Mat. Cuyana 1* (1955), 85–104.
[2] "A unified theory of Hilbert transforms and ergodic theory," *Rev. Mat. Cuyana 1* (1955), 105–167.

K. deLeeuw
[1] "On L^p multipliers," *Ann. of Math. 81* (1965), 364–379.

R. E. Edwards
[1] *Fourier Series*, Vol. II, New York, 1967.

R. E. Edwards and E. Hewitt
[1] "Pointwise limits for sequences of convolution operators," *Acta Math. 113* (1965), 181–218.

E. B. Fabes and N. M. Rivière
[1] "Singular integrals with mixed homogeneity," *Studia Math. 27* (1966), 19–38.

H. Federer
[1] *Geometric Measure Theory*, Berlin, 1969.

C. L. Fefferman
[1] "Inequalities for strongly singular convolution operators," *Acta Math. 124* (1970), 9–36.
[2] "Estimates for double Hilbert transforms," to appear.

C. L. Fefferman and E. M. Stein
[1] "Some maximal inequalities," to appear in *Amer. J. Math.*

K. O. Friedrichs
[1] "A theorem of Lichtenstein," *Duke Math. J. 14* (1947), 67–82.

E. Gagliardo
[1] "Caratterizzazioni delle trace sulla frontiera relative ad alcune classi di funzioni in n variabili," *Rend. Sem. Mat. Padoa 27* (1957), 284–305.
[2] "Proprietà di alcune classi di funzioni in più variabili," *Richereche di Mat. Napoli 7* (1958), 102–137.

G. Gasper, Jr.
[1] "On the Littlewood-Paley and Lusin functions in higher dimensions," *Proc. Nat. Acad. Sci. 57* (1967), 25–28.

G. Glaeser
[1] "Étude de quelques algèbres Tayloriennes," *Jour. d'Analyse Math. 6* (1958), 1–125.

L. S. Hahn
[1] "On multipliers of p-integrable functions," *Trans. Amer. Math. Soc. 128* (1967), 321–335.

G. H. Hardy and J. E. Littlewood
[1] "A maximal theorem with function-theoretic applications," *Acta Math. 54* (1930), 81–116.
[2] "Some properties of fractional integrals I," *Math. Zeit. 27* (1927), 565–606; "II" (*ibid.*), *34* (1932), 403–439.
[3] "Theorems concerning mean values of analytic or harmonic functions," *Quart. J. of Math.* (Oxford) *12* (1942), 221–256.

G. H. Hardy, J. E. Littlewood, and G. Polya
[1] *Inequalities*, Cambridge, 1934.

U. S. Haslam-Jones
[1] "Derivative planes and tangent planes of a measureable function," *Quart. J. Math.* (Oxford) *3* (1932), 120–132.

E. Hecke
[1] *Mathematische Werke*, Göttingen, 1959.

C. S. Herz
[1] "On the mean inversion of Fourier and Hankel transforms," *Proc. Nat. Acad. Sci. 40* (1954), 996–999.

E. Hewitt and K. A. Ross
[1] *Abstract Harmonic Analysis I*, Berlin, 1963.

I. I. Hirschman, Jr.
[1] "Fractional integration," *Amer. J. of Math. 75* (1953), 531–546.
[2] "Multiplier transformations I," *Duke Math. J. 26* (1956), 222–242; "II" (*ibid*), *28* (1961), 45–56.

K. Hoffman
[1] *Banach Spaces of Analytic Functions*, Englewood Cliffs, N.J., 1962.

L. Hörmander
[1] "Estimates for translation invariant operators in L^p spaces," *Acta Math. 104* (1960), 93–139.
[2] "Pseudo-differential operators," *Comm. Pure Appl. Math. 18* (1965), 501–507.
[3] "Psuedo-differential operators and hypoelliptic equations," *Proc. Symp. in Pure Math. 10* (1967), 138–183.
[4] "L^p estimates for (pluri-) subharmonic functions," *Math. Scand. 20* (1967), 65–78.

J. Horváth
[1] "Sur les fonctions conjuguées à plusieurs variables," *Indag. Math. 15* (1953), 17–29.

R. Hunt
[1] "An extension of the Marcinkiewicz interpolation theorem to Lorentz spaces," *Bull. Amer. Math. Soc. 70* (1964), 803–807.

R. Hunt and R. L. Wheeden
[1] "On the boundary valus of harmonic functions," *Trans. Amer. Math. Soc. 132* (1968), 307–322.

F. John and L. Nirenberg
[1] "On functions of bounded mean oscillation," *Comm. Pure and Applied Math. 14* (1961), 415–426.

B. J. Jones, Jr.
[1] "A class of singular integrals," *Amer. J. of Math. 86* (1964), 441–462.

A. Khintchine
[1] "Recherches sur les structures des fonctions mesurables," *Fund. Math. 9* (1923), 212–279.

J. J. Kohn and L. Nirenberg
[1] "An algebra of pseudo-differential operators," *Comm. Pure and Applied Math. 18* (1965) 269–305.

P. Kree
[1] "Sur les multiplicateurs dans $\mathscr{F}L^p$," *Ann. Inst. Fourier 16* (1966), 31–89.

P. I. Lizorkin
[1] "(L_p, L_q) multipliers of Fourier integrals" (Russian), *Dok. Akad. Nauk SSSR 145* (1962), 527–530.
[2] "Characteristics of boundary values of functions of $L_p^{\ r}(E_n)$ on hyperplanes" (Russian), *Dok. Akad. Nauk SSSR 150* (1963) 986–989.

J. Marcinkiewicz
[1] "Sur les séries de Fourier," *Fund. Math. 27* (1936), 38–69.
[2] "Sur quelques intégrales du type de Dini," *Ann. Soc. Pol. Math. 17* (1938), 42–50.
[3] "Sur la sommabilité forte des séries de Fourier," *J. London Math. Soc. 14* (1939), 162–168.
[4] "Sur les multiplicateurs des séries de Fourier," *Studia Math. 8* (1939), 78–91.
[5] "Sur l'interpolation d'opérations," *C.R. Acad. Sci. Paris 208* (1939), 1272–1273.

J. Marcinkiewicz and A. Zygmund
[1] "Quelques inégalités pour les opérations linéaires," *Fund. Math. 32* (1939), 115–121.
[2] "On the summability of double Fourier series," *Fund. Math. 32* (1939), 122–132.

S. G. Mihlin
[1] *Singular Integrals*, Amer. Math. Soc. Translation no. 24, 1950.
[2] "On the multipliers of Fourier integrals" (Russian), *Dok. Akad. Nauk. 109* (1956), 701–703; also, "Fourier integrals and multiple singular integrals (Russian)," Vest. *Leningrad Univ. Ser. Mat. 12* (1957), 143–145.

B. Muckenhoupt
[1] "On certain singular integrals," *Pacific J. Math. 10* (1960), 239–261.

B. Muckenhoupt and E. M. Stein
[1] "Classical expansions and their relation to conjugate harmonic functions," *Trans. Amer. Math. Soc. 118* (1965), 17–92.

C. J. Neugebauer
[1] "Differentiability almost everywhere," *Proc. Amer. Math. Soc. 16* (1965), 1205–1210.

O. Nikodym
[1] "Sur les ensembles accessibles," *Fund. Math. 10* (1927), 116–168.

S. M. Nikolskïi
[1] "On the embedding, continuity, and approximation theorems for differentiable functions in several variables," (Russian), *Uspehi Mat. Nauk 16* (1961), 63–114.

L. Nirenberg
[1] "On elliptic partial differential equations, *Ann. di Pisa 13* (1959), 116–162.

R. O'Neil
[1] "Convolution operators and $L(p, q)$ spaces," *Duke Math. J. 30* (1963), 129–142.

J. Privalov
[1] "Sur les fonctions conjuguées," *Bull. Soc. Math. France 44* (1916), 100–103.

F. Riesz and B. Sz. Nagy
[1] *Functional Analysis*, New York 1955.

M. Riesz
[1] "Sur les fonctions conjuguées," *Mat. Zeit. 27* (1927), 218–244.
[2] L'intégrale de Riemann-Liouville et le problème de Cauchy," *Acta Math. 81* (1949), 1–223.

W. Rudin
[1] *Fourier Analysis on Groups*, New York, 1962.

S. Saks
[1] "Remark on the differentiability of the Lebesgue indefinite integral," *Fund. Math. 22* (1934), 257–261.
[2] *Theory of the Integral*, Warsaw, 1937.

J. Schwartz
[1] "A remark on inequalities of Calderón-Zygmund type for vector valued functions," *Comm. Pure and Applied Math. 14* (1961), 785–799.

R. T. Seeley
[1] "Singular integrals on compact manifolds," *Amer. J. of Math. 81* (1959), 658–690; also, "Refinement of the functional calculus of Calderón and Zygmund," *Indag. Math. 27* (1965), 167–204.
[2] "Elliptic singular integrals," *Proc. Symp. Pure Math. 10* (1967), 308–315.

C. Segovia
[1] "On the area function of Lusin," *Studia Math. 33* (1969), 312–343.

K. T. Smith
[1] "A generalization of an inequality of Hardy and Littlewood," *Canad. J. Math. 8* (1956), 157–170.

S. L. Sobolov
[1] "On a theorem in functional analysis" (Russian), *Mat. Sob. 46* (1938), 471–497.
[2] *Applications of Functional Analysis in Mathematical Physics, Amer. Math. Soc. Transl. of Monographs 7* (1963).

K. Sokol-Sokolowski
[1] "On trigonometric series conjugate to Fourier series in two variables," *Fund. Math. 33* (1945), 166–182.

E. M. Stein
[1] "Interpolation of linear operators," *Trans. Amer. Math. Soc. 83* (1956), 482–492.
[2] "Note on singular integrals," *Proc. Amer. Math. Soc. 8* (1957), 250–254.
[3] "On the functions of Littlewood-Paley, Lusin and Marcinkiewicz," *Trans. Amer. Math. Soc. 88* (1958), 430–466.
[4] "A maximal function with applications to Fourier series," *Ann. of Math. 68* (1958), 584–603.
[5] "On the theory of harmonic functions of several variables II," *Acta Math. 106* (1961), 137–174.
[6] "On some functions of Littlewood-Paley and Zygmund," *Bull. Amer. Math. Soc. 67* (1961), 99–101.
[7] "The characterization of functions arising as potentials I," *Bull. Amer. Math. Soc. 67* (1961), 102–104; "II" (*ibid*), *68* (1962), 577–582.
[8] "Singular integrals, harmonic functions, and differentiability properties of functions of several variables," *Proc. Symp. in Pure Math. 10* (1967), 316–335.
[9] "Classes H^p, multiplicateurs et fonctions de Littlewood-Paley," *C.R. Acad. Sci., Paris 263* (1966) 716–719; 780–71; also 264 (1967), 107–108.
[10] *Intégrales singulières et fonctions différentiables de plusieurs variables,* Lecture notes by Bachvan and A. Somen of a course given at Orsay, academic year 1966–67.
[11] "The analogues of Fatou's theorem and estimates for maximal functions," *Proceedings C.I.M.E.,* held at Urbino, July 5 to 13, 1967.
[12] "Note on the class LlogL," Studia Math. *31* (1969), 305–310.
[13] *Topics in Harmonic Analysis Related to the Littlewood-Paley Theory,* Annals of Math. Study no. 63, Princeton (1970).

E. M. Stein and G. Weiss
[1] "An extension of a theorem of Marcinkiewicz and some of its applications," *J. Math. Mech. 8* (1959), 263–284.
[2] "On the theory of harmonic functions of several variables, I The theory of H^p spaces," *Acta Math. 103* (1960), 25–62.
[3] "Generalization of the Cauchy-Riemann equations and representations of the rotation group," *Amer. J. Math. 90* (1968), 163–196.
[4] *Introduction to Fourier Analysis on Euclidean Spaces,* Princeton (1971). Referred to as *Fourier Analysis* in text.

E. M. Stein and A. Zygmund
[1] "On the differentiability of functions," *Studia Math. 23* (1964), 247–283.

[2] "Boundedness of translation invariant operators on Hölder and L^p spaces," *Ann. of Math. 85* (1967), 337–349.

R. S. Strichartz
[1] "Multipliers on fractional Sobolov spaces," *J. Math. Mech. 16* (1967), 1031–1060.

M. H. Taibleson
[1] "The preservation of Lipschitz spaces under singular integral operators," *Studia Math. 24* (1963), 105–111.
[2] "On the theory of Lipschitz spaces of distributions on Euclidean *n*-space, I," *J. Math. Mech. 13* (1964), 407–480; "II," (*ibid*) *14* (1965), 821–840; "III," (*ibid*) *15* (1966), 973–981.

E. C. Titchmarsh
[1] "On conjugate functions," *Proc. London Math. Soc. 29* (1929), 49–80.
[2] *Introduction to the Theory of Fourier Integrals*, Oxford, 1937.

J. Unterberger and J. Bokobza
[1] "Les opérateurs de Calderón-Zygmund précisés, *C.R. Acad. Sci. Paris 259* (1964), 1612–1614.

S. Wainger
[1] *Special Trigonometric Series in k Dimensions*, Mem. Amer. Math. Soc. no. 59 (1965).

A. Weil
[1] *L'intégration dans les groupes topologiques et les application*, Paris, 1951.

M. Weiss
[1] "On symmetric derivatives in L^p," *Studia Math. 24* (1964), 89–100.
[2] "Total and partial differentiability in L^p," *Studia Math. 25* (1964), 103–109.
[3] "Strong differentials in L^p," *Studia Math. 27* (1966), 49–72.

M. Weiss and A. Zygmund
[1] "A note on smooth functions," *Indag. Math. 62* (1959), 52–58.

H. Weyl
[1] "Bemerkungen zum Begriff der Differentialquotienten gebrochener Ordnung," *Vier. Natur. Gesellschaft Zürich 62* (1917), 296–302.
[2] *The Classical Groups*, Princeton (1939).

R. L. Wheeden
[1] "On the *n*-dimensional integral of Marcinkiewicz," *J. Math. Mech. 14* (1965), 61–70.
[2] "On hypersingular integrals and Lebesgue spaces of differentiable functions I," *Trans. Amer. Math. Soc. 134* (1968), 421–436.

H. Whitney
[1] "Analytic extensions of differentiable functions defined in closed sets," *Trans. Amer. Math. Soc. 36* (1934), 63–89.

N. Wiener
[1] "The ergodic theorem," *Duke Math. J. 5* (1939), 1–18.

A. Zygmund
[1] "On certain integrals," *Trans. Amer. Math. Soc. 55* (1944), 170–204.
[2] "Smooth functions," *Duke Math. J. 12* (1945), 47–76.
[3] "On the boundary values of functions of several complex variables," *Fund. Math. 36* (1949), 207–235.
[4] "On a theorem of Marcinkiewicz concerning interpolation of operators," *Jour. de Math. 35* (1956), 223–248.
[5] "On singular integrals," *Rend. di Mat. 16* (1957), 468–505.
[6] "On the preservation of classes of functions," *J. Math. Mech. 8* (1959), 889–895.
[7] *Trigonometrical Series*, Warsaw, 1935.
[8] *Trigonometric Series* (2nd edition), 2 vols., Cambridge, Eng. 1959.

INDEX